# IONIC LIQUIDS
# FURTHER UnCOILed

# IONIC LIQUIDS FURTHER UnCOILED

## Critical Expert Overviews

*Edited by*

**Natalia V. Plechkova**
*The Queen's University of Belfast*

**Kenneth R. Seddon**
*The Queen's University of Belfast*

Published by John Wiley & Sons, Inc., Hoboken, New Jersey
Published simultaneously in Canada

For general information on our other products and services or for technical support, please contact our Customer Care Department within the United States at (800) 762-2974, outside the United States at (317) 572-3993 or fax (317) 572-4002.

Wiley also publishes its books in a variety of electronic formats. Some content that appears in print may not be available in electronic formats. For more information about Wiley products, visit our web site at www.wiley.com.

*Library of Congress Cataloging-in-Publication Data:*

Ionic liquids further uncoiled : critical expert overviews / edited by Kenneth R. Seddon, Natalia V. Plechkova.
    1 online resource.
    Includes bibliographical references and index.
  Description based on print version record and CIP data provided by publisher; resource not viewed.
    ISBN 978-1-118-83961-4 (ePub) – ISBN 978-1-118-83971-3 (Adobe PDF) –
ISBN 978-1-118-43863-3 (hardback)  1. Ionic solutions.  I. Seddon, Kenneth R., 1950– editor of compilation.  II. Plechkova, Natalia V., editor of compilation.
    QD561
    541'.372–dc23

                                                            2013043854

Printed in the United States of America

# CONTENTS

COIL Conferences     vii
Preface     ix
Acknowledgements     xi
Contributors     xiii
Abbreviations     xv

1   **Ionic Liquid and Petrochemistry: A Patent Survey**     1
*Philippe Bonnet, Anne Pigamo, Didier Bernard, and*
*Hélène Olivier-Bourbigou*

2   **Supercritical Fluids in Ionic Liquids**     39
*Maaike C. Kroon and Cor J. Peters*

3   **The Phase Behaviour of 1-Alkyl-3-Methylimidazolium Ionic Liquids**     59
*Keiko Nishikawa*

4   **Ionic Liquid Membrane Technology**     87
*João G. Crespo and Richard D. Noble*

5   **Engineering Simulations**     117
*David Rooney and Norfaizah ab Manan*

6   **Molecular Simulation of Ionic Liquids: Where We Are and**
**the Path Forward**     149
*Jindal K. Shah and Edward J. Maginn*

7   **Biocatalytic Reactions in Ionic Liquids**     193
*Florian Stein and Udo Kragl*

8   **Ionicity in Ionic Liquids: Origin of Characteristic Properties**
**of Ionic Liquids**     217
*Masayoshi Watanabe and Hiroyuki Tokuda*

9   **Dielectric Properties of Ionic Liquids: Achievements**
**So Far and Challenges Remaining**     235
*Hermann Weingärtner*

10   **Ionic Liquid Radiation Chemistry**     259
*James F. Wishart*

11   **Physicochemical Properties of Ionic Liquids**     275
*Qing Zhou, Xingmei Lu, Suojiang Zhang, and Liangliang Guo*

Index     309

# COIL CONFERENCES

| | | | |
|---|---|---|---|
| COIL-1 | Salzburg | Austria | 2005 |
| COIL-2 | Yokohama | Japan | 2007 |
| COIL-3 | Cairns | Australia | 2009 |
| COIL-4 | Washington | USA | 2011 |
| COIL-5 | Algarve | Portugal | 2013 |
| COIL-6 | Jeju Island | Korea | 2015 |
| COIL-7 | Ottawa* | Canada | 2017 |
| COIL-8 | Belfast* | UK | 2019 |

* Precise location still to be confirmed.

# PREFACE

This is the second of three volumes of critical overviews of the key areas of ionic liquid chemistry. The first volume is entitled *Ionic Liquids UnCOILed* (Wiley 2013), the current volume is *Ionic Liquids Further UnCOILed*, and the final volume, called *Ionic Liquids Completely UnCOILed*, will be published later this year. The history and rationale behind this trilogy was explained in the preface to the first volume, and so will not be repeated here.

Instead, we will use this space to expand on the subtitle, constant for all three volumes: Critical Expert Overviews.

**critical**, adjective

1. Involving or exercising careful judgement or judicious evaluation
2. Of decisive importance in relation to an issue; decisive, crucial

*Critical* has two, rather different, meanings—both are implied in the subtitle of this book. These reviews are both decisively important *and* written by top world experts (hence the second adjective), exercising the judicious evaluation that they are uniquely qualified to do.

**overview**, noun

1. A general survey; a comprehensive review of facts or ideas; a concise statement or outline of a subject. Also: a broad or overall view of a subject.
2. A view from above.

This book includes eleven critical expert overviews of differing aspects of ionic liquids. We look forward to the response of our readers (we can be contacted at quill@qub.ac.uk). It is our view that, in the second decade of the 21st century, reviews that merely regurgitate a list of all papers on a topic, giving a few lines or a paragraph (often the abstract!) to each one, have had their day—five minutes with an online search engine will provide that information. Such reviews belong with the slide rule, the fax machine, and the printed journal—valuable in their day, but of little value now. The value of a review lies in the expertise and insight of the reviewer—and their willingness to share it with the reader. It takes moral courage to say "the work of [. . .] is irreproducible, or of poor quality, or that the conclusions are not valid," but in a field expanding at the prestigious rate of ionic liquids, it is essential to have this honest feedback. Otherwise, errors are propagated. Papers still appear using hexafluorophosphate or tetrafluoroborate ionic liquids for synthetic or catalytic chemistry, and calculations on "ion pairs" are still being used to rationalise liquid state properties! We trust this volume, containing eleven excellently perceptive reviews, will help guide and secure the future of ionic liquids.

NATALIA V. PLECHKOVA
KENNETH R. SEDDON

# ACKNOWLEDGEMENTS

This volume is a collaborative effort. We, the editors, have our names emblazoned on the cover, but the book would not exist in its present form without support from many people. Firstly, we thank our authors for producing such splendid, critical chapters, and for their open responses to the reviewers' comments and to editorial suggestions. We are also indebted to our team of expert reviewers, whose comments on the individual chapters were challenging and thought provoking, and to Ian Gibson for producing the central image on the front cover. The backing from the team at Wiley, led by Dr. Arza Seidel, has been fully appreciated—it is always a joy to work with such a professional group of people. Finally, this book would never have been published without the unfailing, enthusiastic support from Deborah Poland and Sinead McCullough, whose patience and endurance never cease to amaze us.

N.V.P.
K.R.S.

# CONTRIBUTORS

NORFAIZAH AB MANAN, QUILL Research Centre, School of Chemistry and Chemical Engineering, David Keir Building, Stranmillis Road, Belfast, BT9 5AG, UK

DIDIER BERNARD, IFP Energies nouvelles, Rond-point de L'échangeur de Solaize, 69360 Solaize, France

PHILIPPE BONNET, ARKEMA Centre de Recherche Rhône-Alpes, Rue Henri Moissan, BP63, 69493 Pierre-Bénite cedex, France

JOÃO G. CRESPO, REQUIMTE/CQFB, Department of Chemistry, Faculdade de Ciencias e Tecnologia, Universidade Nova de Lisboa, 2829-516 Caparica, Portugal

LIANGLIANG GUO, Beijing Key Laboratory of Ionic Liquids Clean Process, Key Laboratory of Green Process and Engineering, State Key Laboratory of Multiphase Complex System, Institute of Process Engineering, Chinese Academy of Sciences, Beijing 100190, People's Republic of China

UDO KRAGL, Department of Chemistry, University of Rostock, 18051, Rostock, Germany

MAAIKE C. KROON, Department of Chemical Engineering and Chemistry, Eindhoven University of Technology, Eindhoven, The Netherlands

XINGMEI LU, Beijing Key Laboratory of Ionic Liquids Clean Process, Key Laboratory of Green Process and Engineering, State Key Laboratory of Multiphase Complex System, Institute of Process Engineering, Chinese Academy of Sciences, Beijing 100190, People's Republic of China

EDWARD J. MAGINN, Department of Chemical and Biomolecular Engineering, University of Notre Dame, 182 Fitzpatrick Hall, Notre Dame, IN 46556-5637, USA

KEIKO NISHIKAWA, Division of Nanoscience, Graduate School of Advanced Integration Science, Chiba University, Chiba, 263-8522, Japan

RICHARD D. NOBLE, Alfred T. & Betty E. Look Professor, University of Colorado, Chemical Engineering Department, UCB 424, Boulder, CO 80309, USA

**HÉLÈNE OLIVIER-BOURBIGOU**, IFP Energies nouvelles, Rond-point de L'échangeur de Solaize, 69360 Solaize, France

**COR J. PETERS**, Chemical Engineering Program, The Petroleum Institute, P.O. Box 2533, Abu Dhabi, United Arab Emirates, and Department of Chemical Engineering and Chemistry, Eindhoven University of Technology, Eindhoven, The Netherlands

**ANNE PIGAMO**, ARKEMA Centre de Recherche Rhône-Alpes, Rue Henri Moissan, BP63, 69493 Pierre-Bénite cedex, France

**DAVID ROONEY**, QUILL Research Centre, School of Chemistry and Chemical Engineering, David Keir Building, Stranmillis Road, Belfast, BT9 5AG, UK

**JINDAL K. SHAH**, The Center for Research Computing, University of Notre Dame, Notre Dame, Indiana, USA

**FLORIAN STEIN**, Department of Chemistry, University of Rostock, 18051, Rostock, Germany

**HIROYUKI TOKUDA**, Department of Chemistry and Biotechnology, Yokohama National University, 79-5 Tokiwadai Hodogaya-ku, Yokohama 240-8501, Japan

**MASAYOSHI WATANABE**, Department of Chemistry and Biotechnology, Yokohama National University, 79-5 Tokiwadai Hodogaya-ku, Yokohama 240-8501, Japan

**HERMANN WEINGÄRTNER**, Physical Chemistry II, Faculty of Chemistry and Biochemistry, Ruhr-University Bochum, D-44780 Bochum, Germany

**JAMES F. WISHART**, Chemistry Department, Brookhaven National Laboratory, Upton, New York, 11973-5000, USA

**SUOJIANG ZHANG**, Beijing Key Laboratory of Ionic Liquids Clean Process, Key Laboratory of Green Process and Engineering, State Key Laboratory of Multiphase Complex System, Institute of Process Engineering, Chinese Academy of Sciences, Beijing 100190, People's Republic of China

**QING ZHOU**, Beijing Key Laboratory of Ionic Liquids Clean Process, Key Laboratory of Green Process and Engineering, State Key Laboratory of Multiphase Complex System, Institute of Process Engineering, Chinese Academy of Sciences, Beijing 100190, People's Republic of China

# ABBREVIATIONS

## IONIC LIQUIDS

| | |
|---|---|
| GNCS | guanidinium thiocyanate |
| GRTIL | gemini room temperature ionic liquid |
| [HI-AA] | hydrophobic derivatised amino acid |
| IL | ionic liquid |
| poly(GRTIL) | polymerised gemini room temperature ionic liquid |
| poly(RTIL) | polymerised room temperature ionic liquid |
| [PSpy]$_3$[PW] | [1-(3-sulfonic acid)propylpyridinium]$_3$[PW$_{12}$O$_{40}$]·2H$_2$O |
| RTIL | room temperature ionic liquid |

## CATIONS

| | |
|---|---|
| [(allyl)mim]$^+$ | 1-allyl-3-methylimidazolium |
| [1-C$_m$-3-C$_n$im]$^+$ | 1,3-dialkylimidazolium |
| [C$_2$im]$^+$ | 1-ethylimidazolium |
| [C$_1$mim]$^+$ | 1,3-dimethylimidazolium |
| [C$_2$mim]$^+$ | 1-ethyl-3-methylimidazolium |
| [C$_3$mim]$^+$ | 1-propyl-3-methylimidazolium |
| [$^i$C$_3$mim]$^+$ | 1-isopropyl-3-methylimidazolium |
| [C$_4$mim]$^+$ | 1-butyl-3-methylimidazolium |
| [$i$-C$_4$mim]$^+$ | 1-isobutyl-3-methylimidazolium |
| [$s$-C$_4$mim]$^+$ | 1-secbutyl-3-methylimidazolium |
| [$^t$C$_4$mim]$^+$ | 1-tertbutyl-3-methylimidazolium |
| [C$_5$mim]$^+$ | 1-pentyl-3-methylimidazolium |
| [C$_6$mim]$^+$ | 1-hexyl-3-methylimidazolium |
| [C$_7$mim]$^+$ | 1-heptyl-3-methylimidazolium |
| [C$_8$mim]$^+$ | 1-octyl-3-methylimidazolium |
| [C$_9$mim]$^+$ | 1-nonyl-3-methylimidazolium |
| [C$_{10}$mim]$^+$ | 1-decyl-3-methylimidazolium |
| [C$_{11}$mim]$^+$ | 1-undecyl-3-methylimidazolium |
| [C$_{12}$mim]$^+$ | 1-dodecyl-3-methylimidazolium |
| [C$_{13}$mim]$^+$ | 1-tridecyl-3-methylimidazolium |
| [C$_{14}$mim]$^+$ | 1-tetradecyl-3-methylimidazolium |

| | |
|---|---|
| $[C_{15}mim]^+$ | 1-pentadecyl-3-methylimidazolium |
| $[C_{16}mim]^+$ | 1-hexadecyl-3-methylimidazolium |
| $[C_{17}mim]^+$ | 1-heptadecyl-3-methylimidazolium |
| $[C_{18}mim]^+$ | 1-octadecyl-3-methylimidazolium |
| $[C_n mim]^+$ | 1-alkyl-3-methylimidazolium |
| $[C_1 C_1 mim]^+$ | 1,2,3-trimethylimidazolium |
| $[C_2 C_1 mim]^+$ | 1-ethyl-2,3-dimethylimidazolium |
| $[C_3 C_1 mim]^+$ | 1-propyl-2,3-dimethylimidazolium |
| $[C_8 C_3 im]^+$ | 1-octyl-3-propylimidazolium |
| $[C_{12} C_{12} im]^+$ | 1,3-bis(dodecyl)imidazolium |
| $[C_1 OC_2 mim]^+$ | 1-(2-methoxyethyl)-3-methyl-3H-imidazolium |
| $[C_4 dmim]^+$ | 1-butyl-2,3-dimethylimidazolium |
| $[C_4 C_1 mim]^+$ | 1-butyl-2,3-dimethylimidazolium |
| $[C_6 C_{7O1} im]^+$ | 1-hexyl-3-(heptyloxymethyl)imidazolium |
| $[C_2 F_3 mim]^+$ | 1-trifluoroethyl-3-methylimidazolium |
| $[C_4 vim]^+$ | 3-butyl-1-vinylimidazolium |
| $[D_{mvim}]^+$ | 1,2-dimethyl-3-(4-vinylbenzyl)imidazolium |
| $[C_2 mmor]^+$ | 1-ethyl-1-methylmorpholinium |
| $[C_4 py]^+$ | 1-butylpyridinium |
| $[C_4 m_3 py]^+$ | 1-butyl-3-methylpyridinium |
| $[C_4 m_\gamma py]^+$ | 1-butyl-4-methylpyridinium |
| $[C_4 mpyr]^+$ | 1-butyl-1-methylpyrrolidinium |
| $[C_6 (dma)_\gamma py]^+$ | 1-hexyl-4-dimethylaminopyridinium |
| $[C_1 C_3 pip]^+$ | 1-methyl-1-propylpiperidinium |
| $[C_2 C_6 pip]^+$ | 1-ethyl-1-hexylpiperidinium |
| $[C_8 quin]^+$ | 1-octylquinolinium |
| $[DMPhim]^+$ | 1,3-dimethyl-2-phenylimidazolium |
| $[EtNH_3]^+$ | ethylammonium |
| $[Hmim]^+$ | 1-methylimidazolium |
| $[H_2 NC_2 H_4 py]^+$ | 1-(1-aminoethyl)-pyridinium |
| $[H_2 NC_3 H_6 mim]^+$ | 1-(3-aminopropyl)-3-methylimidazolium |
| $[Hnmp]^+$ | 1-methyl-2-pyrrolidonium |
| $[HN_{2\,2\,2}]^+$ | triethylammonium |
| $[N_{1\,1\,1\,2OH}]^+$ | cholinium |
| $[N_{1\,1\,2\,2OH}]^+$ | ethyl(2-hydroxyethyl)dimethylammonium |
| $[N_{1\,1\,1\,4}]^+$ | trimethylbutylammonium |
| $[N_{1\,4\,4\,4}]^+$ | methyltributylammonium |
| $[N_{1\,8\,8\,8}]^+$ | methyltrioctylammonium |
| $[N_{4\,4\,4\,4}]^+$ | tetrabutylammonium |
| $[N_{6\,6\,6\,14}]^+$ | trihexyl(tetradecyl)ammonium |
| $[NR_3 H]^+$ | trialkylammonium |
| $[P_{2\,2\,2(1O1)}]^+$ | triethyl(methoxymethyl)phosphonium |
| $[P_{4\,4\,4\,3a}]^+$ | (3-aminopropyl)tributylphosphonium |
| $[P_{6\,6\,6\,14}]^+$ | trihexyl(tetradecyl)phosphonium |
| $[P_{8\,8\,8\,14}]^+$ | tetradecyl(trioctyl)phosphonium |

| | |
|---|---|
| $[P_n mim]^+$ | polymerisable 1-methylimidazolium |
| $[PhCH_2 eim]^+$ | 1-benzyl-2-ethylimidazolium |
| $[pyH]^+$ | pyridinium |
| $[S_{222}]^+$ | triethylsulfonium |

## ANIONS

| | |
|---|---|
| $[Ala]^-$ | alaninate |
| $[\beta Ala]^-$ | $\beta$-alaninate |
| $[Al(hfip)_4]^-$ | tetra(hexafluoroisopropoxy)aluminate(III) |
| $[Arg]^-$ | arginate |
| $[Asn]^-$ | asparaginate |
| $[Asp]^-$ | asparatinate |
| $[BBB]^-$ | bis[1,2-benzenediolato(2-)-$O,O$']borate |
| $[C_1CO_2]^-$ | ethanoate |
| $[C_1SO_4]^-$, $[O_3SOC_1]^-$ | methyl sulfate |
| $[C_8SO_4]^-$, $[O_3SOC_8]^-$ | octyl sulfate |
| $[C_nSO_4]^-$ | alkyl sulfate |
| $[(C_n)(C_m)SO_4]^-$ | asymmetrical dialkyl sulfate |
| $[(C_n)_2SO_4]^-$ | symmetrical dialkyl sulfate |
| $[CTf_3]^-$ | tris{(trifluoromethyl)sulfonyl}methanide |
| $[Cys]^-$ | cysteinate |
| $[FAP]^-$ | tris(perfluoroalkyl)trifluorophosphate |
| $[Gln]^-$ | glutaminate |
| $[Glu]^-$ | glutamate |
| $[Gly]^-$ | glycinate anion |
| $[His]^-$ | histidinate |
| $[Ile]^-$ | isoleucinate |
| $[lac]^-$ | lactate |
| $[Leu]^-$ | leucinate |
| $[Lys]^-$ | lysinate |
| $[Met]^-$ | methionate |
| $[Nle]^-$ | norleucinate |
| $[NPf_2]^-$, $[BETI]^-$ | bis{(pentafluoroethyl)sulfonyl}amide |
| $[NTf_2]^-$, $[TFSI]^-$ | bis{(trifluoromethyl)sulfonyl}amide |
| $[O_2CC_1]^-$ | ethanoate |
| $[O_3SOC_2]^-$, $[O_3SOC_2]^-$ | ethylsulfate |
| $[OMs]^-$ | methanesulfonate (mesylate) |
| $[ONf]^-$ | perfluorobutylsulfonate |
| $[OTf]^-$ | trifluoromethanesulfonate |
| $[OTs]^-$ | 4-toluenesulfonate, $[4\text{-}CH_3C_6H_4SO_3]^-$ (tosylate) |
| $[Phe]^-$ | phenylalaninate |
| $[Pro]^-$ | prolinate |
| $[Ser]^-$ | serinate |

| [Suc]$^-$ | succinate |
|---|---|
| [tfpb]$^-$ | tetrakis(3,5-bis(trifluoromethyl)phenyl)borate |
| [Thr]$^-$ | threoninate |
| [Tos]$^-$ | tosylate |
| [Trp]$^-$ | tryphtophanate |
| [Tyr]$^-$ | tyrosinate |
| [Val]$^-$ | valinate |

## TECHNIQUES

| | |
|---|---|
| AES | Auger electron spectroscopy |
| AFM | atomic force microscopy |
| AMBER | assisted model building with energy refinement |
| ANN | associative neural network |
| ARXPS | angle resolved X-ray photoelectron spectroscopy |
| ASM | Associated-Solution Model |
| ATR-IR | attenuated total reflectance infrared spectroscopy |
| BPNN | back-propagation neural network |
| CADM | computer-aided design modelling |
| CC | Cole–Cole model |
| CCC | counter-current chromatography |
| CD | Cole–Davidson model |
| CE | capillary electrophoresis |
| CEC | capillary electrochromatography |
| CHARMM | Chemistry at HARvard Molecular Mechanics |
| COSMO-RS | **CO**nductor-like**S**creening**MO**del for Real Solvents |
| COSY | **CO**rrelation **S**pectroscop**Y** |
| CPCM | conductor-like polarisable continuum model |
| CPMD | Car–Parrinello molecular dynamics |
| DFT | density functional theory |
| DMH | dimethylhexene |
| DRS | dielectric relaxation spectroscopy |
| DSC | differential scanning calorimetry |
| ECSEM | electrochemical scanning electron microscopy |
| EC-XPS | electrochemical X-ray photoelectron spectroscopy |
| EFM | effective fragment potential method |
| EI | electron ionisation |
| EMD | equilibrium molecular dynamics |
| EOF | electro-osmotic flow |
| EPSR | empirical potential structure refinement |
| ES | electrospray mass spectrometry |
| ESI–MS | electrospray ionisation mass spectrometry |
| EXAFS | extended X-ray absorption fine structure |
| FAB | fast atom bombardment |
| FIR | far-infrared spectroscopy |

| | |
|---|---|
| FMO | fragment molecular orbital method |
| FTIR | Fourier transform infrared spectroscopy |
| GAMESS | general atomic and molecular electronic structure system |
| GC | gas chromatography |
| GGA | generalized gradient approximations |
| GLC | gas–liquid chromatography |
| GSC | gas–solid chromatography |
| HM | heuristic method |
| HPLC | high-performance liquid chromatography |
| HREELS | high-resolution electron energy loss spectroscopy |
| IGC | inverse gas chromatography |
| IR | infrared spectroscopy |
| IRAS | infrared reflection absorption spectroscopy |
| IR-VIS SFG | infrared visible sum frequency generation |
| ISS | ion scattering spectroscopy |
| L-SIMS | liquid secondary ion mass spectrometry |
| MAES | metastable atom electron spectroscopy |
| MALDI | matrix-assisted laser desorption |
| MBSS | molecular beam surface scattering |
| MC | Monte Carlo |
| MD | molecular dynamics |
| MIES | metastable impact electron spectroscopy |
| MLR | multi-linear regression |
| MM | molecular mechanics |
| MS | mass spectrometry |
| NEMD | non-equilibrium molecular dynamics |
| NMR | nuclear magnetic resonance |
| NR | neutron reflectivity |
| NRTL | non-random two liquid |
| OPLS | optimized potentials for liquid simulations |
| PCM | polarisable continuum model |
| PDA | photodiode array detection |
| PES | photoelectron spectroscopy |
| PGSE-NMR | pulsed-gradient spin-echo |
| PPR | projection pursuit regression |
| QM | quantum mechanics |
| QSAR | quantitative structure–activity relationship |
| QSPR | quantitative structure–property relationship |
| RAIRS | reflection absorption infrared spectroscopy |
| RI | refractive index |
| RNEMD | reverse non-equilibrium molecular dynamics |
| RNN | recursive neural network |
| RP-HPLC | reverse phase high-performance liquid chromatography |
| RST | regular solution theory |

| | |
|---|---|
| SANS | small-angle neutron scattering |
| SEM | scanning electron microscopy |
| SFA | surfaces forces apparatus |
| SFC | supercritical fluid chromatography |
| SFG | sum frequency generation |
| SFM | systematic fragmentation method |
| SIMS | secondary ion mass spectrometry |
| soft-SAFT | soft statistical associating fluid theory |
| STM | scanning tunnelling microscopy |
| SVN | support vector network |
| TEM | tunnelling electron microscopy |
| TGA | thermogravimetric analysis |
| THz-TDS | terahertz time-domain spectroscopy |
| TLC | thin layer chromatography |
| tPC-PSAFT | truncated perturbed chain polar statistical associating fluid theory |
| TPD | temperature programmed desorption |
| UHV | ultra-high vacuum |
| UNIFAC | UNIversal Functional Activity Coefficient |
| UNIQUAC | UNIversal QUAsiChemical |
| UPLC | ultra-pressure liquid chromatography |
| UPS | ultraviolet photoelectron spectroscopy |
| UV | ultraviolet |
| UV-Vis | ultraviolet-visible |
| XPS | X-ray photoelectron spectroscopy |
| XRD | X-ray powder diffraction |
| XRR | X-ray reflectivity |

## MISCELLANEOUS

| | |
|---|---|
| Å | 1 Ångstrom $= 10^{-10}$ m |
| ACS | American Chemical Society |
| ATMS | acetyltrimethylsilane |
| ATPS | aqueous two-phase system |
| BASF™ | Badische Anilin- und Soda-Fabrik |
| BASIL | Biphasic Acid Scavenging utilizing Ionic Liquids |
| BE | binding energy |
| BILM | bulk ionic liquid membrane |
| BNL | Brookhaven National Laboratory |
| b.pt. | boiling point |
| BSA | bovine serum albumin |
| BT | benzothiophene |
| calc. | calculated |
| CB | Cibacron Blue 3GA |

| | |
|---|---|
| CCD | charge coupled device |
| CE | crown ether |
| CEES | 2-chloroethyl ethyl sulphide |
| CFC MC | "continuous fractional component" Monte Carlo |
| CLM | charge lever momentum |
| CMC | critical micelle concentration |
| CMPO | octyl(phenyl)-$N,N$-diisobutylcarbamoylmethylphosphine oxide |
| [$C_n$MeSO$_4$] | alkyl methyl sulfate |
| CNTs | carbon nanotubes |
| COIL | Congress on Ionic Liquids |
| CPU | central processing unit |
| CWAs | chemical warfare agents |
| d | doublet (NMR) |
| $D°_{298}$ | bond energy at 298 K |
| 2D | two-dimensional |
| 3D | three-dimensional |
| DBT | dibenzothiophene |
| DC | direct current |
| DC18C6 | dicyclohexyl-18-crown-6 |
| DF | Debye and Falkenhagen |
| DH | Debye–Hückel |
| DIIPA | diisopropylamine |
| 4,6-DMDBT | 4,6-dimethyldibenzothiophene |
| DMF | dimethylmethanamide (dimethylformamide) |
| DNA | deoxyribonucleic acid |
| 2DOM | two-dimensional ordered macroporous |
| 3DOM | three-dimensional ordered macroporous |
| DOS | density of states |
| DPC | diphenylcarbonate |
| DRA | drag-reducing agent |
| DSSC | dye-sensitised solar cell |
| $E$ | enrichment |
| EDC | extractive distillation column |
| EE | expanded ensemble approach |
| EOR | enhanced oil recovery |
| EoS | equation of state |
| EPA | Environmental Protection Agency |
| EPSR | empirical potential structure refinement |
| eq. | equivalent |
| FCC | fluid catalytic cracking |
| FFT | fast Fourier transform |
| FIB | focussed ion beam |
| FSE | full-scale error |
| ft | foot |

| | |
|---|---|
| GDDI | generalised distributed data interface |
| GEMC | Gibbs ensemble Monte Carlo |
| HDS | hydrodesulfurisation |
| HEMA | 2-(hydroxyethyl) methacrylate |
| HOMO | highest occupied molecular orbital |
| HOPG | highly oriented pyrolytic graphite |
| HV | high vacuum |
| IgG | Immunoglobulin G |
| IPBE | ion-pair binding energy |
| IPE | Institute of Process Engineering, Chinese Academy of Sciences, Beijing |
| ITO | indium–tin oxide |
| IUPAC | International Union of Pure and Applied Chemistry |
| $J$ | coupling constant (NMR) |
| KWW | Kohlrausch–Williams–Watts |
| LCEP | lower critical end point |
| LCST | lower critical separation temperature |
| LEAF | Laser-Electron Accelerator Facility |
| LF-EoS | lattice-fluid model equation of state |
| LLE | liquid–liquid equilibria |
| LMOG | low molecular weight gelator |
| LUMO | lowest unoccupied molecular orbital |
| m | multiplet (NMR) |
| M | molar concentration |
| MBI | 1-methylbenzimidazole |
| MCH | methylcyclohexane |
| MDEA | methyl diethanolamine; bis(2-hydroxyethyl) methylamine |
| MEA | monoethanolamine; 2-aminoethanol |
| MFC | minimal fungicidal concentrations |
| MIC | minimal inhibitory concentrations |
| MMM | mixed matrix membrane |
| MNDO | modified neglect of differential overlap |
| m.pt. | melting point |
| MSD | mean square displacement |
| 3-MT | 3-methylthiophene |
| MW | molecular weight |
| MWCNTs | multi-walled carbon nanotubes |
| $m/z$ | mass-to-charge ratio |
| NBB | 1-butylbenzimidazole |
| NCA | $N$-carboxyamino acid anhydride |
| NE equation | Nernst–Einstein equation |
| NES | New Entrepreneur Scholarship |
| NFM | $N$-formylmorpholine |

| | |
|---|---|
| NIP | neutral ion pair |
| NIT | neutral ion triplet |
| NMP | $N$-methylpyrrolidone |
| NOE | nuclear Overhauser effect |
| NRTL | non-random two liquid |
| NRTL-SAC | non-random two liquid segmented activity coefficients |
| OKE | optical Kerr effect |
| $p$ | pressure |
| PAO | polyalphaolefin |
| PDMS | polydimethoxysilane |
| PEDOT | poly(3,4-ethylenedioxythiophene) |
| PEG | poly(ethyleneglycol) |
| PEM | polymer–electrolyte membrane |
| PEN | poly(ethylene-2,6-naphthalene decarboxylate) |
| PES | polyethersulfone |
| pH | $-\log_{10}([H^+])$; a measure of the acidity of a solution |
| PID | proportional integral derivative |
| $pK_b$ | $-\log_{10}(K_b)$ |
| PPDD | polypyridylpendant poly(amidoamine) dendritic derivative |
| (PR)-EoS | Peng-Robinson equation of state |
| PS | polystyrene |
| PSE | process systems engineering |
| psi | 1 pound per square inch = 6894.75729 Pa |
| PTC | phase transfer catalyst |
| PTFE | poly(tetrafluoroethylene) |
| PTx | pressure–temperature composition |
| $r$ | bond length |
| RDC | rotating disc contactor |
| REACH | Registration, Evaluation, Authorisation and restriction of CHemical substances |
| (RK) EoS | Redlich–Kwong equation of state |
| RMSD | root mean square deviation |
| RT | room temperature |
| s | singlet (NMR) |
| $S$ | entropy |
| scCO$_2$ | supercritical carbon dioxide |
| SDS | sodium dodecyl sulphate |
| SED | Stokes–Einstein–Debye equation |
| S/F | solvent-to-feed ratio |
| SILM | supported ionic liquid membrane |
| SILP | supported ionic liquid phase |
| SLE | solid liquid equilibrium |
| SLM | supported liquid membrane |

| | |
|---|---|
| t | triplet (NMR) |
| TBP | 4-(*t*-butyl)pyridine |
| TCEP | 1,2,3-tris(2-cyanoethoxy)propane |
| TEA | triethylamine |
| TEGDA | tetra(ethyleneglycol) diacrylate |
| THF | tetrahydrofuran |
| TIC | toxic industrial chemical |
| TMB | trimethylborate |
| TMP | trimethylpentene |
| TOF | time-of-flight |
| UCEP | upper critical end point |
| UCST | upper critical solution temperature |
| UHV | ultra-high vacuum |
| VFT | Vogel–Fulcher–Tammann equations |
| VLE | vapour–liquid equilibria |
| VLLE | vapour–liquid–liquid equilibria |
| VOCs | volatile organic compounds |
| v/v | volume for volume |
| w/w | weight for weight |
| wt% | weight pe rcent |
| $X$ | molar fraction |
| $\gamma$ | surface tension |
| $\delta$ | chemical shift in NMR |

# 1 Ionic Liquid and Petrochemistry: A Patent Survey

PHILIPPE BONNET and ANNE PIGAMO

ARKEMA Centre de Recherche Rhône-Alpes, Rue Henri Moissan, Pierre-Bénitecedex, France

DIDIER BERNARD and HÉLÈNE OLIVIER-BOURBIGOU

IFP Energies nouvelles, Rond-point de L'échangeur de Solaize, Solaize, France

## ABSTRACT

Industrial applications of ionic liquids in petrochemistry have been reviewed through the US and EP granted patents published from 1990 to 2010. A *Chemical Abstracts* search on the STN host retrieved about 300 patents, about 130 of them found relevant and are fully analysed in this chapter. This survey has been divided into six thematic sections: new formulations and methods of fabrication for an improved use of ionic liquids; separation processes using ionic liquids; use of ionic liquids as additives with specific properties; use of ionic liquids as both acidic catalysts and solvents; applications of ionic liquids as solvents of catalytic systems; and ionic liquids and biopolymers. Our study has been complemented by a short description of the emerging areas concerning ionic liquids using the patent applications published during the past five years.

## 1.1 INTRODUCTION

Interest in ionic liquids has been growing rapidly worldwide, as demonstrated by the increasing number of publications and patents these last years. The applications and the prospects for ionic liquids are vast. In the chemical and

*Ionic Liquids Further UnCOILed: Critical Expert Overviews*, First Edition.
Edited by Natalia V. Plechkova and Kenneth R. Seddon.
© 2014 John Wiley & Sons, Inc. Published 2014 by John Wiley & Sons, Inc.

petrochemical industries, numerous applications and benefits of using ionic liquids have been described. However, it is difficult to know which applications have been translated into viable industrial and commercialised processes.

As news releases and scientific publications are a part of company strategic communication, relevant information is difficult to assess. We assumed that granted patents could be one of the most relevant sources of information. From our perspective, companies generally only devote human resources, and pay all the necessary fees to have their patents granted, if they expect an actual industrial development of the claimed invention.

A bibliographic search was performed on the *Chemical Abstracts* database using the STN host. It retrieved about 4000 patent families dealing with "ionic liquids." Among these patent families, about 500 contain a US or EP granted patent during the period from 1990 to 2010. After a keyword restrictive search to the petrochemicals and oil area, we selected about 300 documents. We then fully analysed the most relevant documents, and these are reported in this chapter.

## 1.2   NEW FORMULATIONS AND METHODS OF FABRICATION FOR AN IMPROVED USE OF IONIC LIQUIDS

In recent patents, improved ionic liquid formulations and new mode of preparations have been disclosed. Some ionic liquids have been claimed as new products. The aim of these inventions is generally to provide either new cations or new anions or both for ionic liquids with higher purity, such as halogen-free ionic liquids. These formulations are claimed to be advantageous when ionic liquids are used as solvents in catalytic reactions. The most cited reactions are hydroformylation, hydrogenation, and oligomerisation or isomerisation. It appeared to be of interest to review here these new ionic liquids and their preparation processes.

### 1.2.1   Alkyl Sulfate Ionic Liquids

Several patents devoted to halogen-free ionic liquid synthesis, mainly based on sulfate anions, have been filed by Merck GmbH or Solvent Innovation. In these patents [1], the use of onium alkyl sulfate ($[C_nSO_4]^-$; $n = 3 - 36$) salts is claimed in various processes, including their use as solvents for catalytic reactions, such as hydroformylation, hydrogenation, oligomerisation, and isomerisation. Sulfate ionic liquids are described as being more friendly than halide ionic liquids, which often lead to corrosion and/or disposal issues. In particular, long chain alkyl sulfate ionic liquids are preferably claimed thanks to their improved stability to hydrolysis compared with the methyl sulfate analogues. Examples give a comparative hydrolysis stability study of 1-butyl-3-methylimidazolium methyl sulfate, $[C_4mim][C_1SO_4]$, and 1-butyl-3-methylimidazolium octyl sulfate, $[C_4mim][C_8SO_4]$. At $80\,°C$, the octyl sulfate is stable for more than 2 hours, whereas methyl sulfate exhibits rapid degradation. These long chain 1,3-dialkylimidazolium alkyl sulfates are prepared through ion exchange process between 1,3-dialkylimidazolium chloride and sodium alkyl sulfate salts.

New imidazolium and pyridinium ionic liquids bearing anions of general formula $[Me(OCH_2CH_2)_nOSO_3]^-$ or $[Me(OCH_2CH_2)_nSO_3]^-$ are also reported [2]. These sulfates and sulfonates are claimed to be more stable to hydrolysis than their methyl sulfate analogues, and to have higher thermal stability. Examples show a comparative hydrolysis stability study of 1-butyl-3-methylimidazoliummethyl sulfate and $[C_8mim][Me(OCH_2CH_2)_2OSO_3]$. As previously described, these imidazolium sulfates and sulfonates are prepared through ion exchange processes between 1,3-dialkylimidazolium chloride and $[Me(OCH_2CH_2)_nOSO_3]^-$ or $[Me(OCH_2CH_2)_nSO_3]^-$ salts, respectively. The application of $[C_4mim][Me(OCH_2CH_2)_2OSO_3]$ to hydroformylation of 1-octene with $[Rh(acac)(CO)_2]$ (Hacac = pentane-2,4-dione) pre-catalyst was illustrated.

A new scalable process to prepare high-purity imidazolium or pyridinium alkyl sulfates containing less than 3 ppm of halide contaminant has been granted [3]. This process includes the step of treating a compound of formula [cation][(RO)SO$_3$] with an alcohol R′OH to give [cation][(R′O)SO$_3$]. Compounds of formula [cation][(RO)SO$_3$] can be prepared by alkylating a tertiary or aromatic amine with a dialkyl sulfate. As described in the examples, dimethyl sulfate can be used to prepare [Rmim][(MeO)SO$_3$] ionic liquids, which are then treated with R′OH to give [Rmim][(R′O)SO$_3$] ionic liquids. A wide variety of R′OH alcohols may be used, such as long alkyl chain alcohols or alkyl chains containing heteroatoms:

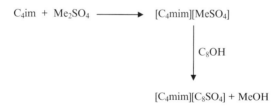

Merck GmbH describes an alternative route to onium alkyl sulfates [4, 5] by the reaction of an onium halide with a symmetrical dialkyl sulfate $([(C_n)_2SO_4]; n = 1 - 14)$ or an asymmetrical dialkyl sulfate $([(C_n)(C_m)SO_4];$ $n = 1$ or $2$, $m = 4 - 20$). Halogen can be removed as a volatile haloalkane, leading to low levels of halogen contaminant in the corresponding ionic liquids.

$$[C_4mim]Cl + [C_nMeSO_4] \longrightarrow [C_4mim][C_nSO_4] + MeCl\uparrow$$

This method has been extended to a large number of reactants: fluorinated alkyl sulfates, alkyl trialkylsilyl sulfates, alkyl acyl sulfates, alkyl sulfonyl sulfates, aryl or alkyl carboxylic acids, and anhydrides [4, 5].

Onium alkyl sulfates have also been used as starting material to prepare other onium ionic liquids. A patent by Wasserscheid et al., granted in 2004 [6], claims the preparation of various onium salts by anion exchange of an onium alkyl sulfate with metal salts. The alkyl sulfates are prepared by alkylation of the corresponding amines or phosphines with dialkyl sulfates.

Patent examples describe the reaction of 1,3-dialkylimidazolium alkyl sulfates with various alkaline salts. The illustrated anions are $[BF_4]^-$, $[PF_6]^-$, $[CF_3CO_2]^-$, $[CF_3SO_3]^-$, $[C_4F_9SO_3]^-$, and $[N(CF_3SO_2)_2]^-$. The preparation of a pyridinium hexafluorophosphate is also given.

BASF describes the reaction of dialkyl sulfates with 2.2 moles of alkylimidazoles in water or methanol at 180 °C for 6 hours under pressure, to prepare the corresponding 1,3-dialkylimidazolium sulfates in good yields (80–90%). This halogen-free process has been claimed, and broadened to pyridine derivatives [7]. These onium sulfates may react with various metal salts to give a wide range of onium ionic liquids such as ethanoate, tetraphenylborate, dihydrogenphosphate, and ordihydrogenborate.

Phosphonium alkyl sulfates are claimed as new products by Cytec [8]. The preparation process involves the alkylation of trialkylphosphines with a symmetrical dialkyl sulfate without solvent at 140–190 °C for several hours. This preparation procedure has been broadened to the reaction between trialkylphosphates and trialkylphosphines or alkylimidazoles. The obtained onium phosphates are also claimed as new products.

### 1.2.2 Other Ionic Liquids

#### 1.2.2.1 Ionic Liquids with Phosphorus-Containing Anions.
In two patents [9, 10] devoted to perfluoroalkyl phosphorus derivatives, Merck GmbH claims onium bis(perfluoroalkyl) phosphinates and perfluoroalkyl phosphonates as new ionic liquids. These compounds are prepared by anion exchange between an onium halide and the phosphorus-containing acid or its salts. Examples describe the preparation of both phosphonium and imidazolium ionic liquids using such a process.

#### 1.2.2.2 Alkylpyridinium Dicyanamide.
Lonza claims alkylpyridinium dicyanamides as new products [11].These compounds are prepared through ion exchange between an alkylpyridinium halide and an alkali dicyanamide. Among the claimed applications of these new ionic liquids is their use as reaction solvents, particularly as solvents for Suzuki reactions.

#### 1.2.2.3 Ionic Liquids with Cyanoborate Anions.
Oniumtetracyanoborates are described as being more stable than the corresponding tetrafluoroborate salts and thus they may be used as ionic liquids. Merck GmbH claims an effective and economical process for preparing these tetracyanoborates, $[B(CN)_4]^-$ [12]. In the first step, an alkali metal tetrafluoroborate is reacted with an alkali metal cyanide in the solid state at 100–500 °C, optionally in the presence of a

lithium halide. The obtained alkali metal tetracyanoborate is then reacted with an onium halide to give the expected onium tetracyanoborate. Examples include ammonium, phosphonium, imidazolium, and pyridinium tetrafluoroborate preparations. This patent furthermore relates to a process for oniumfluorocyanoborate preparation.

### 1.2.2.4 Guanidinium Ionic Liquids.

Merck GmbH claims a simple and inexpensive process for the preparation of guanidinium salts ionic liquids, according to Scheme 1.1 [13]:

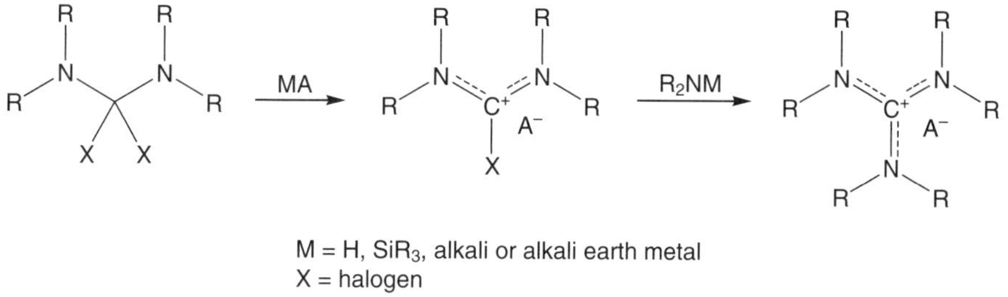

M = H, SiR$_3$, alkali or alkali earth metal
X = halogen

**Scheme 1.1**   Preparation of guanidinium ionic liquids.

This process gives high-purity salts without using a guanidine as starting material, the synthesis of which is frequently complicated and difficult. In the patent examples, only alkyl R groups are described, with a variety of the following anions: $[ClO_4]^-$, $[HSO_4]^-$, $[NO_3]^-$, $[CF_3COO]^-$, $[CF_3SO_3]^-$, and $[N(CF_3SO_2)_2]^-$.

### 1.2.2.5 Ionic Liquids with Lactam Cations.

Deng et al. claim protonated NH-lactams as new room temperature ionic liquids [14]. The preparation process of these compounds involves protonation of the corresponding NH-lactam with a Brønsted acid. These new ionic liquids appear advantageous because starting NH-lactams may exhibit a lower toxicity than pyridines or alkylimidazole bases. Furthermore, some of the lactams, such as caprolactam, are industrially available on a very large scale at reasonable costs.

## 1.2.3  High-Purity Ionic Liquids

Usual ionic liquid preparation methods can lead to products containing acids or halide contaminants, which may be detrimental for some applications. In order to obtain high-purity ionic liquids, several processes have been developed.

ExxonMobil proposes a rather simple process in order to remove acidic contaminants [15]. Ionic liquids are prepared by ion exchange with an onium halide in the presence of inert liquids. Depending on the exchange reaction conditions, a decantation or a filtration is sufficient to give a high-purity ionic liquid phase. Examples describe dialkylimidazolium chloride exchange with various alkali salts such as tetrafluoroborate, hexafluorophosphate, and bis{(trifluoromethyl)sulfonyl}amide.

BASF claims a process for purifying an ionic liquid which comprises partial crystallisation of the ionic liquid, followed by separation of the crystals from the residual melt [16]. This process can be carried out continuously or discontinuously. Examples describe the application of this method to the recycling of used ionic liquids.

Merck GmbH claims a process for the preparation of onium salts having a low chloride content by reaction of an onium chloride with an aqueous solution of Brønsted acid. The hydrochloric acid formed is bound to 1,4-dioxane and can be separated by azeotropic distillation [17]. Examples describe the application of this process to the reaction of alkylpyridinium or dialkylimidazolium chloride with various acids such as $HBF_4$, $CF_3SO_3H$, and $H_2SiF_6$.

### 1.2.4   Production of Ionic Liquids under Ultrasonication

Roche claims a process for the manufacture of onium ionic liquids by an anion exchange reaction between an onium halide and an acid or a salt, where the reaction is carried out under ultrasound conditions [18]. Preferred ionic liquids prepared according to this process are pyridinium or imidazolium tetrafluoroborates, hexafluorophosphates, or trifluoromethanesulfonates. According to the examples, ultrasonication allows a significant reduction of reaction time.

### 1.2.5   Immobilised Ionic Liquids

ExxonMobil Research and Engineering Company claims [19] a new range of immobilised ionic liquids. These ionic liquids are based on the reaction between a polymeric supports having hydroxy groups on their surface with ionic liquids. The typical reactions involved anchoring the ionic liquid by condensation reactions or by ring opening reactions or more preferably by hydrosilylation reaction through a terminal alkoxysilane group, such as $Si(OEt)_3$, grafted on the cationic part of the ionic liquid. Another alternative pathway can also be the reaction between a double bond present on the polymeric support and a silane Si-H group. A large range of materials is described, but 1,3-dialkylimidazolium is preferred, with one of the alkyl groups bearing an alkoxysilane function at one end. Such materials are applied for catalytic purposes.

## 1.3   SEPARATION PROCESSES USING IONIC LIQUIDS

### 1.3.1   Introduction

More and more attention is given to the use of ionic liquids for separation processes. Their particular property of having no measurable vapour pressure makes them ideal for separation by extraction. They do not form azeotropes, as usually observed for more common solvents. Polar and ionic compounds are very soluble in ionic liquids, but alkanes and non-polar organic compounds are generally very poorly miscible. This provides an opportunity for applications involving purification processes. Tuning the cations and the anions can modify the solubility properties. Because of the wide range of possible combinations, ionic liquids offer a large area of investigation.

In this section, we will discuss applications for ionic liquids in separation processes based on granted patents. Reported applications are related to several areas such as:

- separation of olefins from hydrocarbon mixtures
- close-boiling mixture treatment
- acid removal from organic mixtures
- sulfur compound elimination from hydrocarbon streams
- natural gas purification
- oxygen- or nitrogen-containing compound separation

### 1.3.2   Separation of Olefins

There are many streams in petroleum chemistry that include mixtures of various alkenes and alkanes. Separation of these products is difficult because of their close-boiling points. Chevron USA Inc. [20] describes a method for selectively separating mono-olefins from di-olefins. The method implies the use of metal salts, typically silver or copper salts, in ionic liquids. Metal salts, such as silver(I) salts, are known to form complexes with di-olefins. Ionic liquids are used in this case to dissolve, suspend, or disperse the olefin-complexing metal salts. The salt $Ag[BF_4]$ dissolved in 1-butyl-3-methylimidazolium tetrafluoroborate exemplifies these separations. The non-complexed olefin can be separated by decantation or distillation. The alkenes are regenerated by separation from the metal complex by temperature, pressure change, or application of a stripping gas such as an inert gas.

Others patents claim that the ionic liquid itself can be an extracting agent. A simplified process that does not require the presence of a metal complex is described in two patents [21, 22] by Oxeno Olefinchemie and Nova Chemicals Society, respectively. The process is preferably used when the mixtures comprise organic compounds with the same number of carbon atoms, for example, for the extraction of butanes from mixtures of butenes and butanes, or for the separation of propene and propane. Liquid–liquid or gas–liquid extraction can be useful processes. Shell discloses an alternative process type for separating olefins from paraffinic hydrocarbons using a supported ionic liquid membrane [23]. This process can be operated under ambient conditions, thus being highly energy efficient and producing very few waste by-products. The supported ionic liquid membrane is prepared by the immersion of a suitable membrane in an ionic liquid composition. The ionic liquid is drawn into and held within the pores of the membrane by capillary forces. The ionic liquid composition can comprise a metal salt dissolved, dispersed, or suspended. This metal salt is supposed to form complexes with olefinic hydrocarbons. Examples mainly refer to a polycarbonate membrane and an ionic liquid composition comprising a silver(I) nitrate solution with 1-butyl-3-methylimidazolium nitrate.

### 1.3.3    Close-Boiling Mixtures Treatment

BASF discloses broad patents for using ionic liquids as entraining agents in an extraction process to separate azeotropic or close-boiling mixtures or to perform reactive distillation [24–26]. Ionic liquids appear to be more selective than conventional extracting agents. In principle, it allows the possibility of reducing mass flow, or the number of theoretical plates in the extractive column. Examples concern an azeotropic mixture of trimethylborate and methanol. A 1-ethyl-3-methylimidazolium-based salt is used to break the azeotrope and separate the constituents of the mixture. Several mixtures were investigated, all of them dealing with an alcohol-containing mixture (methanol, ethanol, cyclohexanol, or isopropanol). Recovery of the ionic liquid from the extractive distillation and recycling is also described [27]. An integrated process includes the extraction with the use of the ionic liquid, and the evaporation stage for the regeneration of the ionic liquid. This step is performed at low pressure. An alternative possibility is the use of a stripper that is operated through use of inert gas or steam.

### 1.3.4    Acid Removal from Organic Mixtures

*1.3.4.1    Recovery of Hydrochloric Acid.*    Ionic liquids are not miscible with certain organics. BASF takes advantage of this property and claims a simplified process to remove acids from reaction mixtures [28]. Examples of reactions in which the process of the invention may be applied include alkylations, silylations, phosphorylations, and sulfurisations, or whatever process produces an acid as a by-product. Usually, an auxiliary base is added and forms a salt by reaction with the acid. Salts are generally insoluble and precipitate, so that they form suspensions that are more difficult to handle and separate. BASF has commercialised a process to remove HCl by adding a specific compound such as an alkylimidazole. The reaction between the alkylimidazole and HCl leads to the formation an alkylimidazolium chloride salt ionic liquid. This salt is liquid above 100 °C and is immiscible with the organic phase. A liquid–liquid phase separation is simple from a process-engineering point of view. BASF commercialises this technology, called BASIL™, offering the supply of the 1-alkylimidazole for many reactions where acid scavenging is needed, and offering to recover the used alkylimidazolium salt for recycling. Higher yields are claimed when using this separation technology.

Alternatively, the ionic liquid itself can be a separating agent due to its affinity with the hydrogen chloride. It has been investigated more specifically for the separation of HCl from phosgene. For example, BASF claims a process [29] in which at least part of the hydrogen chloride is dissolved by contact with an ionic liquid and then separated off. A process scheme illustrates the separation of HCl from phosgene, $COCl_2$, in a tray column and the regeneration of the ionic liquid before returning to the column. The recycling is based on an evaporator stage, where hydrogen chloride is driven off from the solvent. One or more evaporation stages can be used in which different pressure levels can prevail. This case utilises the miscibility property of the 1-ethyl-3-methylimidazolium chloride with HCl over phosgene, the HCl presumably forming the stable $[HCl_2]^-$ anion in this process.

Solvay applies a similar approach for the preparation of $COF_2$, organic acid fluorides, or phosphorus(V) fluoride. All these products are useful as raw materials [30]. Hydrochloric acid, produced as a by-product, is usually removed with acid scavengers like amines. The invention is based on the fact that HCl(or HF) is retained in the ionic liquid while $COF_2$, $PF_5$, or acid fluorides are not. Any suitable device able to favour the contact area can treat the gas mixture. A mixture of ionic liquids is preferred. Cations are chosen from imidazolium derivatives, and anions are chosen between triflate and tosylate anions. Higher selectivity is evidenced using ionic liquids over ethanenitrile solvent. The constituents retained in the ionic liquid can be recovered in a reconditioning treatment, by application of a vacuum, heating, or passing inert gases through the ionic liquid. The desorption temperature is preferably not higher than 100 °C. In the case of a vacuum, 1 mbar is a preferred limit. Depressurisation and heating can be combined to emphasise the desorption efficiency.

*1.3.4.2   Recovery of Sulfuric Acid.*    HaldorTopsoe A/S claims [31] a process for recovering a sulfuric acid stream. An ionic liquid is used as an adsorbent with the adsorption process occurring by counter current between the ionic liquid and the sulfuric acid and/or $SO_3$-containing stream. Then the sulfuric acid is separated in a further separation step from the ionic liquid, the latter being recycled to the absorption step. Claimed ionic liquids are preferably based on the hydrogensulfate, $[HSO_4]^-$, anion, and $[NH_4]^+$, or alkali cation as a countercation. This process is useful for treating exhaust gases form metallurgical operations and from the combustion of hydrogen-containing fuels (petrochemical alkylation, petroleum coke, $H_2S$, etc.)

### 1.3.5   Sulfur Compound Elimination from Hydrocarbon Streams

Ionic liquids can be used to reduce the sulfur content in fuels for reducing $SO_2$ emissions. The deep desulfurisation is a major problem in the production of fuel. Merck has disclosed the extraction of the sulfur-containing impurities from fuels by means of ionic liquids, in order to comply with the regulatory requirements [32]. By several extraction steps the residual sulfur content can be reduced. Efficiency of the method is illustrated for the removal of a model stimulant, such as dibenzothiophene dissolved in dodecane, and for real fuels.

Ionic liquid recycling is claimed. Examples of possible processes can be steam distillation, sublimation, absorption, and purification using active charcoal or zeolite. Similarly, Extractica discloses a process for extracting sulfur-containing compounds from gasoline and fuels with an ionic liquid [33]. These compounds can be partially oxidised to sulfoxides and/or sulfones to increase their solubility in ionic liquids. A desulfurisation technology based on liquid–liquid extraction sounds attractive over conventional hydrotreating technology, since generally mild conditions would result in lower capital and operating costs. The pathway and its operating conditions for recycling the ionic liquid are not disclosed, although the step is claimed.

IFP [34] obtained a granted patent dedicated to the elimination of sulfur compounds by alkylation reaction, the alkylation agent being dissolved in an ionic liquid. It appears from the literature that the final desulfurisation of gas oils makes this finishing process necessary, to transform the sulfur-containing molecules that are particularly refractory to standard hydrodesulfurisation precursors. Instead of oxidation, alkylation is described in the IFP patent. Here, the charged sulfur-containing derivatives, such as the sulfonium derivatives formed through the alkylation process, have an increased solubility in ionic liquids. For example, extraction of butanethiol is achieved with trimethyloxonium tetrafluoroborate in the ionic liquid $[C_4mim][NTf_2]$.

More recently, a way to improve the hydrodesulfurisation process is described by UOP LL [35]. A treatment of a diesel fuel with an ionic liquid such as $[C_4mim][HSO_4]$ is efficient for denitrogenation and allows a reduction of the amount of the catalyst used in the subsequent hydrodesulfurisation process. Similarly, the life time of this catalyst can be increased by up to about 50%, to about 100%, when compared to desulfurisation without performing denitrogenation. An integrated process with two steps of denitrogenation (by extraction), two regenerators for the liquid ionic recovery by steam stripping, and the subsequent desulfurisation zone is proposed.

### 1.3.6 Natural Gas Purification

Carbon dioxide and hydrogen sulfide from natural gas are usually removed thanks to the use of amines. Amines have a natural affinity for both $CO_2$ and $H_2S$, allowing this to be an efficient process. However, amine treatments present several main issues and challenges, such as their intensive energy requirement, corrosivity of the amine, and its possible degradation. Use of a physical solvent (such as $N$-formylmorpholine) is an alternative option, which requires less energy for regeneration but tends to have lower $CO_2$ capacities and co-solubilises the hydrocarbon in some extent.

Chevron Texaco Company claims a new process for removing $CO_2$ from hydrocarbon containing streams using an ionic liquid absorbent [36]. The new method provides several advantages, such as high $CO_2$ capacity and low hydrocarbon solubility, and requires low energy for regeneration of the ionic liquid absorbent. $CO_2$ loading curves in solvents, evidence that ionic liquid behaviour can be adjusted to a chemical solvent (such as methyldiethanolamine) or a physical solvent (such as water) by changing the anion. Physical solvents dissolve $CO_2$ into the liquid without any chemical forces and have low volumetric $CO_2$ loading, while chemical solvents reversibly bind $CO_2$ as hydrogen carbonate species and have a high volumetric $CO_2$ loading. 1-Butyl-3-methylimidazolium ethanoate ionic liquid shows loadings that are intermediate between the aqueous amines and physical solvents. Thus, the patent claims the use of an ionic liquid comprising a cation and an anion having a carboxylate moiety.

IFP discloses three complementary patents [37–39] for processing natural gas with a solvent that removes the acid compounds such as hydrogen sulfide and carbon dioxide. The ionic liquid is brought into contact with the purified gases or/and with the acid gases (obtained after regeneration of the charged solvent) in order to trap the residual content of solvent from these streams. The solvent can be a mixture of water, methanol and diethanolamine. The contact of gases (either purified stream or acidic stream) with 1-butyl-3-methylimidazolium bis{(trifluoromethyl)sulfonyl}amide enables methanol and water contained in gases to be recovered. Methanol contained in the solvent acts as an antihydrate compound. Compounds selected from alcohols, glycols and glycol ethers compounds can be useful to prevent the formation of hydrocarbon hydrates during transportation and storage.

### 1.3.7 Oxygen- or Nitrogen-Containing Polar Compound Separation

Polar compounds in organic mixtures can be successfully extracted with ionic liquids by taking advantage of the immiscibility between the ionic liquid and the organic phase. The traditional liquid-liquid separation would not be possible for the separation of a cycloalkanol and a cycloalkanone from a cycloalkane because all of them would dissolve in conventional solvents. The oxidation of cyclohexane, and its subsequent separation, is a key-step in the manufacture of adipic acid and caprolactam. Invista Technology [40] claims a separation process by contacting an ionic liquid with a mixture comprising a non-polar solvent and at least one of an alcohol and a ketone. Similarly, BASF describes a general method using ionic liquids for extracting impurities selected from a wide range of polar compounds such as phenols, alcohols, amines, and acids. These impurities can be removed with good efficiency from hydrocarbons. Removal of water is described using 1-methylimidazolium sulfate. A drying process using this technology could be envisaged [41].

### 1.3.8 Other Applications

In addition to the above examples, we can find the use of ionic liquids to improve purification processes. Arkema [42] discloses a hydrogenation process to convert chlorolactams to the corresponding lactams in presence of a metal catalyst, such as palladium deposited on carbon. Chlorolactams are impurities that impact the properties of the polymers prepared from lactams (especially the colour), and must be removed. The reaction is performed in an ionic liquid that avoids the deactivation of the catalyst and improves the yield of the reaction.

Similarly, ionic liquids are used as additive solvents for the selective hydrogenation of block copolymers (typically triblock copolymers formed from polystyrene, polybutadiene and polymethylmethacrylate). The presence of the block comprising olefinic double bonds renders them sensitive to light, oxidising agents and to heat. Arkema [43] describes a process for the selective hydrogenation of the olefinic double bonds of block copolymers using a catalyst based on a metal from Group 8–10 in a medium comprising an organic solvent and a water-immiscible ionic liquid. In the process according to the invention, the catalyst is dissolved in the ionic liquid and the copolymer to be hydrogenated in an organic solvent. After the reaction, the hydrogenated copolymer can be isolated by precipitation, by introducing the reaction medium into a large amount of a non-solvent for the hydrogenated copolymer or, when there are two phases, by decantation and subsequent isolation of the copolymer. The yield is enhanced by the dissolution of the catalyst in the ionic liquid, such as $[C_4mim][PF_6]$.

## 1.4   USE OF IONIC LIQUIDS AS ADDITIVES WITH SPECIFIC PROPERTIES

Beyond the typical reactions where ionic liquids are expected to be applied, some companies seem to have developed some very specific knowledge where ionic liquids are involved in physical processes rather than pure chemical interactions.

### 1.4.1   Ionic Liquids as Lubricants

UT-Battelle LLC and University of Tennessee Research Foundation claim [44] the use of an ionic liquid made of tertiary ammonium with long alkyl linear chains (preferred octyl) and bis{(perfluoroalkyl)sulfonyl}amide (preferred-with $CF_3$, viz. $[NTf_2]^-$) anions that displays excellent lubrication performances, alone or in combination, between two metal surfaces (such as aluminium), thanks to the excellent thermal stability of the ionic liquid and its low affinity with water. Thirty per cent of extra performance can be therefore achieved compared with conventional oil in applications such as car engines.

### 1.4.2   Ionic Liquids as Antistatic Agents in Polymers

Evonik Goldschmidt claims [45] the use of ionic liquids as a solvent for alkali metal salts, the latter providing antistatic polymers properties. This addresses the issue of poor solubility of antistatic additives in the polymer matrix. A special focus deals with polyurethane. Although a large number of cations and anions are claimed, preferred compositions involve 1,3-dialkyimidazolium cations with alkyl sulfate anions in combination with sodium or potassium dicyanamide or thiocyanate salts. An alternative combination involves a third solvent based on adiol such as glycol.

### 1.4.3  Ionic Liquids as Additives for Oil Drilling/Oil Wells

Schlumberger Technology Corporation claims [46] new fluids for stimulation of hydrocarbons wells, more specifically for a technique called "matrix acidising," which consists of injecting acids that dissolve a small portion of the formation and create alternate paths for the oil. An acidic ionic liquid, such as one based on a chloroaluminate as an anion, is used as a retarding agent to produce HCl after reaction with water while using the exothermic reaction resulting from the ionic liquid formation to improve solubility and therefore lower melting points of the material. The kinetics of dissolution can be therefore much better adjusted, with the possibility of drilling much deeper. Also, this reduction in kinetics has a significant advantage regarding corrosion, and in overcoming most of the issues encountered during matrix treatment, such as organic deposits and avoidance of costly additives.

Schlumberger Technology Corporation also claims [47] a drilling fluid that combines a non-aqueous feature of an ionic liquid with an enhancement in electrical conductivity that is very important for telemetry purposes. The fluid contains hydrocarbons and a portion of water-stable ionic liquid.

### 1.4.4  Carbon Nanotubes

In the nanotechnology area, Fuji Xerox Co. [48] claims the production a film of carbon nanotube by dropping a composition of ionic liquid and carbon nanotube onto a liquid surface. Preferred ionic liquids for this application are hydrophobic ones, such as 1,3-dialkylimidazolium as a cation and hexafluorophosphate as an anion.

### 1.4.5  Fine Particles Recovery

Japan Science and Technology Agency [49] claims an original method for recovering fine particles below 300 nm of different natures: polyacetylene, polydiacetylene, oxides (titanium oxide), and metals (silver, gold). The originality of the patent lies in a very specific (and narrow) ratio between the concentration of the particles and the quantity of ionic liquid to be added for a complete recovery. Ionic liquids made of typical 1-butyl-3-methylimidazolium cation and hydrophobic perfluorinated anions, such as hexafluorophosphate, seem to be preferred.

### 1.4.6  Anionic Surfactants

Procter & Gamble Company claims the use of compositions of ionic liquids for surface treatment, including soft surfaces such as textiles and hard surfaces such as dishware, floors, and glassware. The ionic liquids are made of an amine oxide cation such as $[R_1R_2(OR_3)NOH]^+$ and an anion made of an alkyl aryl sulfonates [50], or of mid-chain length, branched alkyl sulfates or of mid-chain length polyoxyalkylene sulfates [51].

### 1.4.7   Pressure-Sensitive Compositions

In the field of surface-protecting films obtained in the form of sheets or tape, Nitto Denko Corporation claims a pressure-sensitive adhesive composition with good antistatic properties that can prevent electrification of the surface to be protected upon peeling. Different patents [52–55] cover compositions that include an ionic liquid, and different combinations displaying a polymer containing as a monomer, a methacrylate ester, and an ethylene oxide group. These patents claim the use of ionic liquids with nitrogen-, sulfur- or phosphorus-containing onium salts. Examples include the use of ionic liquids made of 1-butyl-3-methylimidazolium chloride and lithium bis{(trifluoromethyl) sulfonyl}amide.

## 1.5   USE OF IONIC LIQUIDS AS BOTH ACIDIC CATALYSTS AND SOLVENTS

### 1.5.1   Introduction

The most catalytically interesting ionic liquid acid catalysts are those derived from ammonium halides and Lewis acids. Among the Lewis acids reported, aluminium(III) chloride is the most commonly used. Acidic chloroaluminates are the most claimed ionic liquids acid catalysts to be used in petrochemical processes.

Mixtures of alkylpyridinium chloride [56] or alkylimidazolium halides [57, 58] with aluminium(III) chloride form fused chloroaluminates salts that were described in earlier patents for their applications as electrolytes. More recently, their use as solvents in catalytic reactions was disclosed [59].

Liquid clathrates composed of a mixture of aluminium(III) chloride, a quaternary ammonium or phosphonium salt, and at least one aromatic compound were also described as useful reusable aluminium(III) chloride catalysts for Friedel–Crafts reactions [60].

Mixing of a metal halide solid (such as aluminium(III) chloride) and an alkyl (most commonly methyl or ethyl) or aromatic ammonium halide solid salt can form ionic liquid compositions at low temperatures:

$$[Al_2Cl_6] + [NR_3H]Cl \longrightarrow [NR_3H][Al_2Cl_7]$$
$$(R = Me \text{ or } Et)$$

$$[Al_2Cl_6] + [pyH]Cl \longrightarrow [pyH][Al_2Cl_7]$$

These can be suitable for use as acidic catalysts in some transformations.

It seems that imidazolium halides, while largely described, are not so well used or popular for industrial applications in the petrochemical field because their price is too high. An economical method for the production of commercial amounts of these ionic liquids is needed.

Due to their ease of preparation, the commercial availability of their components, and their low costs, Lewis acidic ionic liquid have been advantageously used for carrying out many acid-catalysed processes such as:

- olefin oligomerisation for lube base stock manufacture
- olefin–paraffin alkylation for alkylate oil production for fuel
- aromatic alkylation
- paraffin carbonylation
- liquid phase fluorination process

These reactions will be detailed below. Chevron Chemical has been very active in this field. Chloroaluminate(III) ionic liquids are used to replace aluminium(III) chloride. One main advantage of using ionic liquid catalyst is that it forms a separate phase from the organic phase, which contains the reaction products and the residual olefin feed. The reaction products can then be easily separated by conventional means, such as decantation or distillation, and the ionic liquid that remains after recovery of the products may be recycled. Methods to regenerate the used ionic liquids are also largely disclosed.

### 1.5.2 Economical Preparation of Chloroaluminate(III) Ionic Liquids

Considering the usefulness of these low temperature ionic liquids, an economical method for manufacturing them has been especially disclosed [61]. This method is performed in two steps. In a first step an alkylamine, such as the triethylamine, is contacted with a hydrogen halide, such as HCl, in the gaseous state in the presence of a hydrocarbon diluent. Then, in a second step, the metal halide is added, as a solid or as a slurry. This two-step process avoids the difficult handling of the alkylammonium halide salts. It is interesting to note that the composition of the ionic liquid (molar ratio of the metal halide to the alkylammonium halide salt ranging from 1.5 to 1.9) can be determined simply and quickly from a correlation to the ionic liquid specific gravity, which is determined with a hydrometer tube.

### 1.5.3 Applications of Acidic Ionic Liquids: Catalysts and Solvents

*1.5.3.1 Olefin Oligomerisation.* In years before 2000, ionic liquid acid catalysts were described to catalyse the oligomerisation of olefins, such as 1-decene, to produce polyalphaolefins (PAOs) having a viscosity index of at least 120 and a poor point of $-45\,°C$ or less [62]. It has been found now that it is also possible to make PAOs, used as lube base stock, with higher viscosity using the ionic liquid catalyst in the absence of solvent [63]. In these processes, olefin feeds can originate from dehydrogenated Fischer–Tropsch paraffinic feedstock [64] or from Fischer–Tropsch dehydrated alcohols [65]. Because ionic liquids can be quite costly for this application, there was a need for a method to increase their efficiency for improving the economics of the process. This was achieved by the introduction of a control amount of dioxygen in the catalytic reaction headspace zone while controlling the water content of the olefinic feedstock [66]. The examples illustrate the impact of the concentration of dioxygen added: an increase in dioxygen concentration leads to an increase in olefin conversion and product viscosity.

***1.5.3.2 Alkylation of Paraffins.*** In the field of petrochemical catalysis, alkylation reactions of alkanes and olefins, such as isobutane and butenes, are important industrial processes to produce alkylate oil, mainly composed of branched octenes. These processes are catalysed by strong, highly corrosive acids, such as concentrated sulfuric or hydrofluoric acids. In recent years, most of the studies have focussed on the search of new solid acid catalysts and technologies in order to solve the problem of equipment corrosion and pollution. However, the problem with the new solids was still their quick deactivation, and none of these solid acids have been put into practice at commercial refineries. Another challenge of these processes is to improve the C8 content in the alkylate products and the trimethylpentene (TMP) to dimethylhexene (DMH) molar ratio in the C8 fraction. The use of acidic chloroaluminates as alternatives to strong liquid acids was disclosed several years ago for this application [67–70]. The more recent patents disclose improved processes. In particular, the ionic liquid is a composite consisting of a cation coming from alkylammonium or pyridinium chlorides, and an anion coming from two or more metal compounds [71]. The preferred metal compounds are aluminium(III) chloride, and a copper and/or a nickel salt. Continuous manufacturing of alkylate oil in a static mixer reaction apparatus was also described, with a reaction temperature of 30 °C and a ratio of isobutane to butene in the reactor of 10:1. The olefin conversion can be up to 98 wt%, with a C8 content in the alkylate oil of 76 wt% and a TMT/DMH ratio of up to 6.7.

Another interest in using chloroaluminates ionic liquids versus conventional acids is their ability to catalyse the alkylation of isoparaffins (such as isopentane) with ethylene, which is not possible in the existing processes. In this case, it was shown that pyridinium- or imidazolium-based chloroaluminates were more effective than aliphatic ammonium chloroaluminates in terms of ethylene conversion. The preferred ionic liquid was composed of 1-butylpyridinium chloride and aluminium(III) chloride (1:2 molar ratio). To maintain high ethylene conversion, HCl needs to be co-fed into the reactor. In that specific case, in a continuous run, ethylene conversion can be around 95 wt%, with a selectivity for the C7 fraction of up to 79 wt% [72]. Haloalkanes, such as chloroethane, have also been disclosed to be good promoters of chloroaluminates and can be used instead of HCl [73, 74]. In that case, the alkylate produced can contain a high amount of organic chloride. To dechlorinate the product, a conventional hydrotreatment can be applied [75]. The ethylene/isopentane alkylation process has been integrated in a refinery for the production of high-quality gasoline blending components from low value components, such as fluid catalytic cracking (FCC) off-gas and isopentane [76].

A novel way to reduce the concentration of double bonds in the olefinic hydrocarbon effluent, and at the same time enhance the quality of the fuel or lubricant, was provided by performing the alkylation reaction of a stream comprising at least one C8+ olefin with an isoparaffin (e.g., isobutane) with an acidic chloroaluminate in the presence of a Brønsted acid such as HCl [77], or by subjecting the alkylation effluent to a hydrogenation step.

### 1.5.3.3 Alkylation of Aromatic Ring.

Aromatic hydrocarbon alkylation with an olefin catalysed by acidic chloroaluminates was disclosed in earlier patents [78]. More recently, Chevron disclosed the alkylation of non-hydroxyl-containing aromatic compounds with olefin oligomers coming from light olefin oligomerisation in the presence of an ionic liquid. The olefinic oligomers are typically propylene oligomers (C12+). The acidic ionic liquid used for olefin oligomerisation and aromatic alkylation with the olefin may be the same, based on a trialkylammonium cation and a chloroaluminate(III) anion [79].

Aromatic alkylation of the ring of anilines or alkylated aniline derivatives is also of interest, in order to improve the solubility of anilines in hydrocarbons. Prior art in this field related an improvement of the ring alkylation selectivity by using acidic zeolites. Chemtura Corporation disclosed the use of ionic liquids as both solvents and acid catalysts, with the advantage of permitting a convenient separation of the alkylated anilines from the reaction mixture. The ionic liquids preferably used are acid chloroaluminates based on quaternary ammonium or 1,3-dialkylimidazolium cations, and the olefin can be either an alpha olefin or branched olefins. The inventors claim that the ionic liquid may be recycled by simple phase decantation and that the ionic liquid can be reused by itself or in combination of additional fresh one [80].

### 1.5.3.4 Carbonylation of Aromatic Hydrocarbons and Paraffins.

The carbonylation of aromatic hydrocarbons (the Gatterman–Koch reaction) to form alkylaromatic aldehydes has been described in the presence of acidic ionic liquids (Hammett acidity value less than −10) based on Lewis acids such as aluminium(III) chloride or gallium(III) chloride [81–83]. After the reaction, the aromatic aldehyde is normally present as a complex with the acid present in the system. The use of ionic liquids can provide for more convenient separation of the aldehyde products. This process can be applied on a mixture of *ortho-*, *meta-*, and *para-* xylenes. *m*-Xylene is the most reactive, and produces 2,4-dimethylbenzaldehyde with selectivity depending on the conversion. The difference in conversion rate of the different xylenes may be used to separate *p*-xylene (the least reactive) from *m*- and *o*-xylenes. After separation of the products, the alkyl aromatic aldehydes can be subjected to an oxidation reaction to form the corresponding acids. This claimed method could be applied to the reaction of toluene to form terephthalic acid, but is not exemplified.

The carbonylation of saturated hydrocarbons to give oxygenated saturated hydrocarbons is known to be more difficult to perform, and is generally catalysed by liquid super acids such as HF-BF$_3$. UOP LLC [84] disclosed the use of ionic liquids for the replacement of these liquid superacids, with the aim of being more environmentally friendly. In this reaction, the ionic liquid serves both as the catalyst and the solvent. 1-Butylpyridinium chloride–aluminium(III) chloride ionic liquid was exemplified to catalyse the carbonylation of isobutane to methyl isopropylketone with very high selectivity but very low conversion (<10%).

### 1.5.3.5  Isomerisation.

*Exo*-tetrahydrodicyclopentadiene is a high energy fuel that is usually obtained from synthetic reactions through an isomerisation of *endo*-tetrahydrodicyclopentadiene. This isomerisation is catalysed with strong acids, such as aluminium(III) chloride, but side reactions can occur such as ring-opening degradation and polymerisation, with the formation of many undesired by-products. Chinese Petroleum Corp. [85] discloses a novel method for producing *exo*-tetrahydrodicyclopentadiene with the use of acidic chloroaluminate(III) ionic liquids. Different cations are related, based on 1-alkylpyridinium, tetraalkylammonium, or 1,3-dialkylimidazolium. High conversions and selectivities can be reached (>99%). The ionic liquid may be recyclable, but no example is given.

In a similar way, CPC Corporation relates the production of adamantane by isomerising *exo*-tetrahydrodicyclopentadiene [86].

### 1.5.3.6  Fluorination of Halogenated Compounds.

Arkema discloses a patent [87] using a particular ionic liquid for liquid phase fluorination catalysis. The ionic liquids results from the reaction of a halogenated Lewis acid, based on titanium, niobium, tantalum, tin, or antimony. These liquids can be used in liquid phase fluorination, using hydrogen fluoride (HF), of saturated or unsaturated compounds containing C–Cl groups. Examples evidence the use of 1-butyl-3-methylimidazolium chloride associated with an antimony salt for the liquid phase fluorination of dichloromethane or trichloroethylene.

### 1.5.3.7  Polyamide Hydrolysis.

Invista North America S.A.R.L. claims [88] a process of hydrolysis with water of polyamide made of a diamine and a dicarboxylic acid such Nylon 6,6. The process is carried out at relatively low temperature (100 °C) with an ionic liquid based on hydrogensulfate [HSO$_4$]$^-$ or hydrogenphosphate [HPO$_4$]$^-$ and preferably a phosphonium cation. A complete process is claimed including the separation of the aqueous phase comprising the products of hydrolysis, such the diamine and the dicarboxylic acid, and a process of regeneration of the ionic liquid with either H$_2$SO$_4$ or H$_3$PO$_4$, depending on the nature of the anion used.

### 1.5.4   Regeneration of Chloroaluminate(III) Ionic Liquids

One of the unsolved problems impeding the commercial use of chloroaluminate(III) ionic liquid catalysts has been the inability to regenerate and recycle them. Chevron has provided different processes for regenerating the used and deactivated acidic chloroaluminates.

For example, in alkylation reactions, one of the major catalyst deactivation mechanisms is the formation of by-products known as "conjunct polymers," which are unsaturated cyclic or acyclic molecules formed by acidic side reactions, such as polymerisation, cyclisation, or hydride transfer. These conjunct polymers proved to be a cause of deactivation of acidic chloroaluminates by weakening their acidity.

Hydrogenation of these conjunct polymers to remove them from the chloroaluminate(III) ionic liquids has been disclosed. The hydrogenation can be operated over a metal, such as aluminium [89] in the presence of a Brønsted acid (such as HCl), or over a supported transition metal (such as nickel) [90], a metal alloy [91], or a homogeneous catalyst [92], under dihydrogen. The saturated polymeric compounds can then be separated in a second upper phase, leaving a denser ionic liquid phase that can be recovered and reused after filtration.

## 1.6   APPLICATIONS OF IONIC LIQUIDS AS SOLVENTS FOR CATALYTIC SYSTEMS

### 1.6.1   Introduction

The patents in this field can be classified according to the catalytic applications claimed. With our strategy of research, the main related applications can be divided into the following:

- olefin oligomerisation, co-dimerisation, and dimerisation
- olefin hydroformylation and hydrogenation
- alcohol carbonylation
- olefin metathesis
- hydrosilylation

In these varied applications, the use of ionic liquids may increase reaction rates and yields. It could permit the recovery, and sometimes the reuse, of catalytic systems. The use of ionic liquids as solvents is also usually claimed to reduce environmental impacts and to lead to more energy-efficient separation. Ionic liquids may appear as novel solutions to chemical industry. However, it is still very difficult to deduce from these patents if the ionic liquids have been really translated into viable industrial processes. This information is generally not made public.

### 1.6.2   Transition-Metal Catalysed Olefin Oligomerisation

At the beginning of the 1990s, IFP found that the mixture of an alkyl aluminium chloride and a quaternary ammonium or phosphonium chloride, such
as a 1,3-dialkylimidazolium chloride, formed a liquid at low temperature
(below 80 °C), and that these mixtures can be used as novel solvents for
transition-metal catalysis [93]. Nickel-catalysed olefin oligomerisation is one
of the main applications described in chloroaluminate ionic liquids. The reaction is performed in a two-phase medium consisting of (1) the ionic liquid,
where the nickel catalyst is dissolved, and (2) a hydrocarbon upper phase,
where the reaction products are separated. IFP described a process for carrying out the nickel-catalysed olefin oligomerisation in two catalytic steps of
different types: in a first step, it undergoes catalytic oligomerisation without
an ionic liquid, then the effluent of this first step is sent to a second reaction
zone in which oligomerisation is carried out in a two-phase medium containing
the ionic liquid [94].

Different nickel-catalysed compositions have been claimed to be active in
chloroaluminate ionic liquids. One can mention the use of a nickel compound
that contains at least one heterocyclic mono- or bi-carbene ligand [95]. The
nickel compound can be activated not only by alkyl aluminium derivatives but
also by methylaluminoxane [96].

A novel nickel catalyst composition that consists of a mixture of a nickel(0)
compound, a Brønsted acid, a nitrogen ligand, and an ionic liquid has been
discovered. In this case the ionic liquid is preferably based on a weak coordinating anion such as $[SbF_6]^-$ or $[NTf_2]^-$, and the ligand is preferably a diimine
[97].

### 1.6.3   Hydroformylation

The hydroformylation reaction of olefinic compounds is a reaction of great
industrial importance. Industrial processes use homogeneous catalysts based
either on cobalt or rhodium complex. One of the main challenges is the separation of the reaction products, and the recovery and reuse of the catalyst.
Many solutions have been claimed in patents. The use of ionic liquids in
this reaction was first described by IFP and then by Celanese. In order to
retain the metal in the ionic phase, ligands carrying ionic function must be
added to the catalytic composition. Nitrogen [98] or phosphorus ligands are
claimed [99].

It has been discovered and claimed by IFP that, when using cobalt as the
metal in presence of an ionic liquid, recycling of the metal is improved when
a Lewis base is present, and when a depressurisation step is realised before
decantation and separation of the products. The Lewis bases are preferably
chosen from the family of the pyridines [100]. In order to improve the reaction
rate and the cobalt recovery, a new implementation of the system has been
described. It consists of a hydroformylation stage carried out in the presence

of the cobalt, the ionic liquid, and a Lewis base as the ligand. This stage is followed by depressurisation and decantation stages, and then a recycling stage wherein the ionic liquid phase, which contains the cobalt, is sent back to hydroformylation first stage. An improvement can be achieved by adding some ligand in the post-reaction stage [101].

Celanese [102] disclosed that ionic liquids can act as solvents, but also as ligands, forming complexes with rhodium. Surprisingly, the reaction can be performed with a large excess of ligand, which can stabilise the active species and leads to a reduction of noble metal losses. One example of the "ligand liquid" is a sulfonated or carboxylated phosphorus triester. The use of ionic liquids based on organic sulfonates or sulfate anions, associated with sulfonated arylphosphines and a rhodium compound, has also been disclosed by Celanese [103]. These ionic liquids appear to allow a better retention of the rhodium in the ionic phase.

### 1.6.4    Carbonylation of Alcohols

The carbonylation of alcohols, more specifically methanol for the production of ethanoic acid, is extensively operated industrially. All the commercial processes are performed in liquid phase and are catalysed with a homogeneous catalyst system comprising a Group 8–10 metal, such as rhodium, and an iodine-containing compound. The separation of the volatile reaction products and starting materials from catalyst components is still a main issue for these processes. There is a need for a process that provides simple product separation while maintaining a stable catalyst environment, and giving high reaction rates with efficient heat removal from the reaction zone.

Eastman Chemical Company [104] describes a new continuous process in which vapour phase reactants, methanol, iodomethane, and water, are contacted with a non-volatile rhodium or iridium catalyst solution that comprises an ionic liquid. The ionic liquid is preferably based on iodide anions, and an organic cation such as 1-butyl-3-methylimidazolium.

At the same time, IFP [105] disclosed a process for carrying out the carbonylation of alcohol in the liquid phase, in which the homogeneous catalyst system, based on rhodium or iridium, is stabilised in an ionic liquid. At high methanol conversion ($>99$ wt%), the selectivity for ethanoic acid can be high ($>95$ wt%). After distillation of the products, the ionic liquid containing the catalyst can be recycled. WackerChemie [106] described a process for the continuous carbonylation of methanol with carbon monoxide in the gaseous phase in the presence of a rhodium catalyst supported in an ionic phase confined on a support such as silica gel. The reaction is preferably performed at $180\,°C$ and 20 bar using 1-butyl-3-methylimidazolium iodide as the ionic liquid and a dicarbonyldiiodorhodate(I)anion, $[Rh(CO)_2I_2]^-$, as the catalyst. The observed products are ethanoic acid, methyl ethanoate (major product), and dimethyl ether. This catalyst technology is sometimes referred to as supported ionic liquid-phase (SILP) technology.

### 1.6.5 Metathesis

The first patent that described the use of ionic liquids for metathesis made use of a tungsten catalyst dissolved in chloroaluminate ionic liquids [107] and was applied to mono-olefin cross-metathesis. Bayer [108] later disclosed a process for preparing cyclic and/or polymeric compounds by ring closing metathesis of the starting alkenes or alkynes that contain at least two functional groups. The metathesis is performed in ionic liquids, preferably chloroaluminates, with a metal–carbene transition metal catalyst. The preferred catalyst precursors are ruthenium or molybdenum imido carbene complexes.

In 2000, Chevron [109] filed a patent that related to a process for manufacturing fatty acid nitrites and fatty amines by cross-metathesis, followed by hydrogenation of normal alpha olefins and acrylonitrile in the presence of a transition metal catalyst, preferably a Schrock molybdenum type catalyst. In a preferred embodiment of the invention, the reaction is performed in an ionic liquid such as $[C_4mim][PF_6]$. The ionic liquid containing the catalyst can be reused after separation of the reaction products, but nothing is disclosed about the leaching of the metal into the organic phase.

The combination of ionic liquids and compressed carbon dioxide has been described in a patent filed by Boehringer Ingelheim International, in 2006, for continuous olefin ring closing metathesis of both liquid and solid reactants. The carbon dioxide acts as solvent for the reactants and for the product obtained, while the ionic liquid acts as the metal catalyst solvent and support. The preferred catalyst is a ruthenium carbene complex, and a continuous process is exemplified. A particular application of homo-metathesis in the presence of at least one ionic liquid and a metal catalyst has been used for the co-production of olefins and diesters from unsaturated fats [110]. Ruthenium catalysts are preferred. Recycling experiments are described with bis{(trifluoromethyl)sulfonyl}amide ionic liquids.

### 1.6.6 Hydrosilylation

Among processes for producing and for modifying organopolysiloxanes, hydrosilylation in the presence of transition metal catalysts is of particular importance since it permits a variety of Si–C linkages (Scheme 1.2).

In order to make it possible for the hydrosilylation catalyst to be recycled in a simple fashion, Goldschmidt-Degussa (now Evonik) proposes the use of ionic liquids [111].

This patent claims a process for preparing organomodified polysiloxanes which comprises reacting a SiH-containing polysiloxane with a compound that contains carbon–carbon multiple bonds in the presence of transition metal catalysts, wherein (1) the reaction is carried out in the presence of an ionic liquid, and (2) after the reaction is completed, the ionic liquid containing the dissolved catalyst is separated from the reaction mixture.

Preferred ionic liquids are imidazolium or pyridinium salts and more specifically 1,2,3-trimethylimidazolium methylsulfate as well as pyridinium

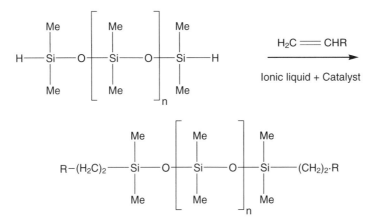

**Scheme 1.2**    Hydrosilylation for modification of organopolysiloxanes.

tetrafluoroborates. Typical examples describe the reaction of $\alpha,\omega$-(Si-H)-polydimethylsiloxane (20 to 50 silicon atom chain length) with a 1.3 molar equivalent of a multiple bond containing compound (such as unsaturated polyether 400–500 g $M^{-1}$ molecular weight) in the presence of a platinum catalyst dissolved in the ionic liquid at 80–100 °C for a period of 5 hours. After cooling to room temperature, the polyethersiloxane (free of metal, according to ICP determination) is separated from the ionic liquid catalyst phase by decantation or filtration. Preferred platinum catalysts used are hexachloroplatinic acid or $\mu$-chlorodichlorobis(cyclohexene)diplatinum(II).

### 1.6.7    Synthesis of Polymers

Polytrimethylene ether glycol and polytetramethylene ether glycol are used in various applications including fibres, films, or moulded products. A series of patents concerning preparation of such polymers in the presence of an ionic liquid has been granted to E.I. duPont de Nemours.

Use of ionic liquids for this process provides several advantages: the produced polyol polymer can be recovered in a separate phase from the ionic liquid and catalyst, and the polyol polymer molecular weight may be tuned thanks to appropriate choice of the ionic liquid.

Polytrimethylene ether glycols are prepared through polymerisation of 1,3-propanediol (or its oligomers) with an acid catalyst in the presence of an ionic liquid [112–114].

Preferred acid catalysts are polyfluoroalkylsulfonic acids. The patent claims contain a large number of compounds, but preferred ionic liquids are typical onium ionic liquids bearing a perfluoroalkylsulfonate anion, or specifically designed 1-(2-N,N,N-trialkylaminoalkyl-5-methyl)pyrrolidinium salts with a sulfonic acid. Separation of the ionic liquid phase containing the acid catalyst from the polymer phase through decantation and its recycling are also claimed. A typical example describes reaction of 1,3-propanediol with 1,1,2,2-tetrafl uoroethanesulfonic acid in the presence of 1-butyl-3-methylimidazolium1,1,2,2-tetrafluoroethanesulfonate at 160 °C for 9–10 hours. After cooling to 75 °C, the acid-containing ionic liquid phase is separated from the obtained polymer through decantation.

Polytetramethylene ether glycols are prepared through polymerisation of THF with an acid catalyst in the presence of an accelerator and an ionic liquid [115,116]. The preferred acid catalyst is polyfluoroalkylsulfonic acids and preferred accelerators are carboxylic anhydrides, acid chlorides, or carboxylic acids. The patent claims contain a large number of compounds, but preferred ionic liquids are classical onium ionic liquids containing a perfluoroalkyl-sulfonate anion, or a specifically designed 1-(2-*N,N,N*-trialkylaminoalkyl-5-methyl)pyrrolidinium salt with a sulfonic acid. A typical example describes the reaction of tetrahydrofuran with 1,1,2,2-tetrafluoroethanesulfonic acid and ethanoic anhydride in the presence of 1-butyl-3-methylimidazolium 1,1,2,2-tetrafluoroethanesulfonate at room temperature for 1 hour. The acid-containing ionic liquid phase is separated from the obtained polymer through decantation.

### 1.6.8 Microwave-Assisted Chemical Transformations with Ionic Liquids

A Swedish company (Personal Chemistry i Uppsala) claimed a method for performing a microwave-assisted chemical transformation, wherein an ionic liquid is used as solvent [117]. Ionic liquid preparation under microwave irra-diation is also claimed. Although the patent claims do not limit the scope of the chemical transformations involved, examples only illustrate the cases of alcohol or amine alkylation with an organic halide.

The main advantage of microwave irradiation is a dramatic reaction time reduction when compared to the same reaction carried out under more conventional conditions (several minutes vs. several hours).

### 1.6.9 Methane to Methanol

California Institute of Technology [118] claims a process for the conversion from methane to methanol in presence of $H_2SO_4$ and a catalyst made of plati-num dissolved in an ionic liquid. Rates are rather modest at high pressure and at 220 °C. 1,3-Dialkylimidazolium ionic liquids are preferred, but not with a butyl substituent, as this substituent can be oxidised during the reaction.

### 1.6.10 Fluorination

RhodiaChimie [119] claims a new fluorination method with KF as a fluorinat-ing agent and a tetraalkylphosphonium halide (preferably bromide), possibly in combination with a polar solvent, as reaction medium to undertake nucleophilic substitution reactions (F/Cl exchange). Different organic substrates are claimed, but the focus seems to be primarily on 1,3,5-trichlorobenzene and to a lesser extent phenylchloroform, $PhCCl_3$, as raw materials.

### 1.6.11  Aldol Condensation

Aldol condensation is claimed in a patent from ExxonMobil Research and Engineering Company [120]. The use of an ionic liquid as a reaction medium allows a much better design of the selectivity in terms of the number of additions. Neutral ionic liquids, such as those based on hexafluorophosphate or tetrafluoroborate anions, are claimed and used in combination with a catalyst, NaOH or KOH.

### 1.6.12  Acylation

Institut Universitari de Ciencia I Tecnologia claims [121] the use of an ionic liquid made up of the trifluoromethylsulfonate anion and, preferably,with 1,3-dialkylimidazolium as a cation to perform Friedel–Crafts acylation with, for example, anisole as a Friedel–Crafts substrate and carboxylic anhydride acids as an acylating agent. The reaction is performed without any other solvent, the ionic liquid playing both roles of solvent and catalyst. Reaction is carried out around $100\,^{\circ}C$ and the *ortho-/para*-selectivity is 100%.

### 1.6.13  Hydrocarbon Stream Drying

HaldorTopsoe A/S claims [122] a process for drying a hydrocarbon stream.The process is carried out by putting a wet hydrocarbon stream in contact with an ionic liquid, such as trialkylammonium triflate, and removing the water from the ionic liquid in a separate phase before recycling.

### 1.6.14  Photolysis

Fuchigami and Nisshinbo Industries claim [123] a process for photolysing organic matter by placing the waste water in contact with an ionic liquid in order to solubilise the organic matter of the waste stream into the ionic liquid phase, then treating this phase with a photocatalyst such as $TiO_2$ under conditions of light irradiation. The ionic liquid claimed is made up of a bis{(trifluoromethyl)sulfonyl}amide as anion and a quaternary ammonium cation.

### 1.6.15  Preparation of Alkoxyamines

Atofina claims [124] a biphasic process to prepare $\alpha,\beta,\beta$-trisubstituted hydroxylamines from nitroxide that can be used as radical polymerisation initiators. The biphasic medium comprises an ionic liquid containing a metal salt as a catalyst and an organic solvent that is immiscible with the ionic liquid. Preferably, ionic liquids are made of 1,3-dialkylimidazolium chloride or bromide, the catalyst is copper based, and the organic solvent is an aromatic hydrocarbon. The alkoxyamine can be then recovered from the organic phase and the ionic liquid phase can be recycled.

## 1.7  IONIC LIQUIDS AND BIOPOLYMERS

Biopolymers are polymers produced from biomass. They can be used as plastics, replacing the need for polystyrene or polyethylene-based plastics. The feedstocks for polymers derived from petrochemicals will eventually run out. In contrast, biopolymers are renewable, sustainable, and can be carbon neutral. Biopolymers have the potential to cut carbon emissions and reduce $CO_2$ quantities in the atmosphere: this is because when the $CO_2$ released when they degrade can be reabsorbed by crops grown to replace them.

Ionic liquid have now found a place in this field as a solvent for these biopolymers, helping dissolution, and helping processing and regeneration of used biopolymers.

### 1.7.1  Dissolution and Processing of Cellulose

Cellulose is the most abundant biorenewable material, and cellulose-derived products have been used in all cultures from the most primitive to highly developed modern technological society. Apart from the use of unmodified cellulose-containing materials, modern cellulose technology requires extraction and processing of cellulose. Cellulose and its derivatives can be substituted as a source for a number of chemicals. For example, oil feedstocks can be substituted with cellulose to prepare polymers for applications in paints, plastics, and other formulation materials. Cellulose is a major fraction of plant biomass, which is the feedstock for future "biorefineries" with the potential to replace the conventional petrochemical refineries.

Cellulose can be used unchanged, or after physical or chemical treatment. In the latter two cases, it is advantageous for cellulose to be preferably dissolved in a solvent. However, cellulose is insoluble in most solvents.

Ionic liquids have been known to be cellulose solvents for a very long time. Pioneering work [125] by Graenacher described cellulose dissolution in benzyl- or alkyl-pyridinium salts in the 1930s.

It was found at the University of Alabama [126] that cellulose can be dissolved in solvents described as ionic liquids that are substantially free of water. Cellulose displays high solubility in these ionic liquids. Viscous, liquid crystalline solutions are obtained at high concentrations, *ca.* 10–25 wt%. For example, fibrous cellulose was dissolved at 25 wt% in 1-butyl-3-methylimidazolium chloride by microwave heating to provide an optically clear, viscous solution.

The solubility of cellulose in ionic liquids can be controlled by changes in the anion and cation. The requirement for a small anion is indicated by the high solubility of cellulose in chloride-containing ionic liquids, with reduced solubility in the bromide systems and no solubility in tetrafluoroborate and hexafluorophosphate systems.

The solution of cellulose in ionic liquids can be processed simply, and cellulose can be regenerated from the ionic liquid solution by adding water. The recovered cellulose exhibits different physical properties, especially an amorphous structure evidenced from X-ray powder diffraction (XRD) measurements. This may indicate a slow breakdown of the polymer chains with time. The high crystallinity of primary cellulose renders it recalcitrant to hydrolysis aimed at producing glucose (the feedstock for producing fuels and chemicals). A pretreatment including the total or partial dissolution of the cellulose and the subsequent recovery of an amorphous state of cellulose allow a high enhancement of the saccharification [127], according to the University of Toledo. To accomplish the hydrolysis, enzymes such as cellulase are preferred; hydrolysing enzymes and water must penetrate the crystalline fibrils. The resulting amorphous regenerated cellulose is more accessible for the reaction and the yield is thus highly enhanced.

BASF discloses two related patents [128,129] in this area. The dissolution of cellulose in a very large list of possible ionic liquids is described, and the presence of 6–30 wt% of a nitrogen-containing base is claimed.

The ability of ionic liquids to dissolve cellulose has been described then for several applications:

1. Fabrication of nonwovens for textile applications [130] claimed by Fraunhofer-Gesellschaft. The particular properties of the nonwovens result in numerous possibilities for use in medicine for hygiene products, bed sheets, gauze, and so on.
2. Description of an absorbent paper sheet for tissue or towel [131] by Georgia Pacific Consumer Products LP, comprising 1–30 wt% of regenerated cellulose microfibres prepared from an ionic liquid mixture.
3. Procter & Gamble Company discloses a patent [132] for the sulfatation or sulfonation of cellulose and cellulose ethers in an ionic liquid. The resulting products are suitable in detergent compositions for fabric care, surface care, and air care applications.

### 1.7.2  Starch and Ionic Liquids

Similar to cellulose, starch is a natural biopolymer. A large number of organic starch esters have been prepared and described in the literature, but few are manufactured and used commercially. Practically, these are starch ethanoates and, to a smaller extent, starch succinates. Depending on their degree of substitution (the measure of average number of hydroxyl groups), they are used in applications such as the food area, paper industry, or packaging.

Kemira OYJ is a company interested in this field, and discloses the two following patents. According to the company, ionic liquids can dissolve starch to allow subsequent transformations such as esterification [133]. Both steps can be assisted by applying microwave irradiation and/or pressure. It has also been discovered that the contact between the ionic liquid and starch leads to its depolymerisation [134]. The depolymerisation of starch is achieved after dissolution in an ionic liquid, and treatment by agitating at a temperature and for a period of time, to obtain the desired depolymerisation products.

### 1.7.3 Other Biopolymers

Apart from cellulose and starch examples, Procter & Gamble Company discloses the use of ionic liquids for the dissolution of other biopolymers [135]. Certain biopolymers that are insoluble in water or conventional organic solvents can be extracted from their biological sources with ionic liquids. The concerned biopolymers are chitin, chitosan, elastin, collagen, keratin or polyhydroxyalkanoate. The process comprises the subsequent step of adding an effective amount of nonsolvent to the substantially anhydrous composition to reduce the solubility of the polymer and recover it. The biopolymer can easily be separated from the recoverable composition by known separation methods, such as centrifugation, sedimentation, crystallisation, decantation or filtration.

## 1.8 CONCLUSIONS AND PERSPECTIVES

Granted patents between 1990 and 2010 describe numerous applications of ionic liquids in separation and purification technologies, various catalytic processes, polymer and biopolymer synthesis, and transformation, and also as additives and fluids with very specific properties. The recovery and recyclability of the ionic liquids is not often disclosed in all these new technologies, even if it appears to be a key issue in obtaining economically viable ionic liquid processes.

Granted patents can be relevant indicators of the state of the art of a current industrial technological area. However, because of the time needed to secure a patent, granted patents are not always representative of the last new developments. Patent applications are published 18 months after the invention has been filed, and may be more relevant documents in order to access the most promising future developments. To complete our study, we thus decided to focus on the emerging areas described in the recent open literature on ionic liquids, and we look for the patent applications published during the last five years in these areas. Here we list the topics in which patent applications are the most numerous.

Concerning **enzymes and biomass valorisation**, about 60 patent applications using both ionic liquids and enzymes have been published since 2005 with a significant yearly increase from three publications in 2005 to 13 publications in 2010. These documents can be divided into two categories. About half of them deal with applications in which ionic liquids are used as solvents for the enzymes to produce fine chemicals with asymmetric synthesis being involved in most of the chemical transformations. The second half concerns processes in which ionic liquids and enzymes are used in two separate consecutive steps. The large majority of these documents deal with ionic liquids lignocellulosic biomass pretreatment, followed by enzymatic hydrolysis to produce glucose solutions as intermediates for biofuels.

Concerning biofuels and chemicals production from lignocellulosic biomass, about 40 other patent applications can be found in which ionic liquids are used to transform biomass into synthons. A large number of companies are found among these patent application assignees, but two main actors, Bioecon and Furanix, can be identified concerning the use of ionic liquids for biomass treatment or transformation.

The need to reduce the inventory of ionic liquids used in catalytic processes, and to go towards heterogenisation of the catalysts, can be detected by the numerous applications concerning **SILP** (supported ionic liquid phase) technology. About 50 patent applications concerning SILP have been published since 2005, with a significant yearly increase from two publications in 2005 to 15 publications in 2010. Various compounds are used as supports for ionic liquids, including minerals (activated carbon, silica, alumina, zeolite, etc.) and polymers. The obtained solids are described to be usable in various applications including the refining and petrochemical area. About half of these documents are Chinese applications. The applications have been filed by a large diversity of companies, including SudChemie, ConocoPhillips, Total, Johnson Matthey, BASF, and CPC.

**Nanomaterials** also emerge as a new area in the field of ionic liquids. About 30 patent applications concerning the use of ionic liquids as solvents to prepare nanomaterials have been published since 2005, with a significant yearly increase from two publications in 2005 to eight publications in 2010. Various compounds can be thus prepared, such as zeolites, minerals, polymers, metals, and hybrid materials, and yet very few data concerning their applications are found. About 60% of these documents are Chinese applications and their assignees are nearly all from academia. Thus, the actual industrial developments in this area could be doubtful.

**Desulfurisation** of oil remains a dominant application area for ionic liquids, with about 30 patent applications published since 2005, with a significant yearly increase from two publications in 2005 to eight publications in 2009, and six in 2010. About 60% of these documents are Chinese applications, and among the other assignees most of the oil industry majors can be found with one or two patent applications each.

About ninety patents applications concerning the use of ionic liquids as **lubricants** have been published since 2005, with a significant yearly increase from seven publications in 2005 to twenty-four publications in 2010. Ionic liquids are used as lubricant for engines, but they have found application also in the electronic industry. About half of these documents are Japanese applications, and the two main actors may be identified, NSK and the Chinese Lanzhou University.

Surprisingly, especially if we compare to the open literature, only 10 applications are found concerning the use of ionic liquids for carbon dioxide capture, with a rather constant number each year since 2005. Thus, the industrial interest in this application appears doubtful.

## REFERENCES

1 Wasserscheid, P., Boesmann, A., and Van Hal, R., *Ionic liquids*, US Pat.US7252791 (2007).

2 Wasserscheid, P., Boesmann, A., and Van Hal, R., *Halogen-free ionic liquids*, US.Pat, US7863458 (2011).

3 Wasserscheid, P., Van Hal, R., and Hilgers, C., *Process for the preparation of ionic liquids with alkyl sulfate and functionalized alkyl sulfate anions*, US. Pat., US7655803 (2010).

4 Ignatyev, N., Welz-Biermann, U.,Kucheryna, A., and Willner, H., *VerfahrenzurHerstellung von OniumAlkylsulfatenmitgeringemHalogenid-gehalt*, European Pat., EP1828142B1 (2010).

5 Ignatyev, N., Welz-Biermann, U., Kucheryna, A., and Willner, H., *VerfahrenzurHerstellung von Onium-saltzenmit Alkyl- oderArylsulfonat-anionenoder Alkyl- oderarylcarboxylat-anionenmitgeringenHalogenid-gehalt*, European Pat. EP1824827B1 (2009).

6 Wasserscheid, P., Hilgers, C., and Boesmann, A., *HalogenidfreieHerstellungionischerFlüssigkeiten*, European Pat., EP1182196B1 (2004).

7 Szarvas, L., Maase, M., and Massonne, K., *VerfahrenzurHerstellung von ionischenVerbindungen, derenKationeinquarternäres, sp²-hybridisiertes Stickstoffatomenhält*, European Pat., EP1723118B1 (2009).

8 Zhou, Y., Robertson, A.J., Hillhouse, J.H., and Baumann, D., *Phosphonium and imidazolium salts and methods of their preparation*,US Pat., US7638636 (2009).

9 Ignatyev, N., Weiden, M., Welz-Biermann, U., Heider, U., Sartori, P., Kucheryna, A., and Willner, H., *Method for the production of monohydro-perfluoroalkanes, bis(perfluoroalkyl)phosphinates and perfluoroalkylphosphonates*, US Pat., US7145004 (2006).

10 Welz-Biermann, U., Ignatyev, N., Weiden, M., Heider, U., Kucheryna, A., Willner, H., and Sartori, P., *Process for the preparation of bis(perfluoroalkyl)phosphinic acids and salts thereof*, US Pat., US7202379 (2007).

11 Taeschler, C., *AlkylpyridiniumcyanamidealspolareLösungmittel*, European Pat., EP1584617B1 (2010).

12 Welz-Biermann, U., Ignatyev, N., Bernhardt, E., Finze, M., and Willner, H., *Salts comprising cyanoborate anions*,US Pat. US7645434 (2010).

13 Ignatyev, N., Welz-Biermann, U., Bissky, G., and Willner, H., *Process for the preparation of guanidinium salts*, US Pat., US7439395 (2008).

14  Deng, Y., Du, Z., Guo, S., Li, Z., and Zhu, L., *Brønsted acidic room temperature ionic liquids each having a N-protonated lactam cation and method for preparing the same*, US Pat., US 7220869 (2007).

15  Mehnert, C.P., Dispenziere, N.C., and Cook, R.A., *Method for preparing high-purity ionic liquids*, US Pat., US6852229 (2005).

16  Fiene, M., Rust, H., Massonne, K., Stegmann, V., Huttenloch, O., and Hailek, J., *Reinigung von ionischenFlüssigkeiten*, European Pat., EP1824575B1 (2008).

17  Ignatyev, N., Welz-Biermann, U., Barthen, P., and Willner, H., Process for the preparation of onium salts having a low chloride content, US Pat., US7692007 (2010).

18  Bonrath, W., Leveque, J.M., Luche, J.L., and Petrier, C., *Production of ionic liquids*, European Pat., EP1453838B1 (2006).

19  Mehnert, C.P., and Cook, R.A., *Ionic liquid compositions*, US Pat., US6673737 (2004).

20  Boudreau, L.C., Driver, M.S., Munson, C.L., and Schinski, W.L., *Separation of dienes from olefins using ionic liquids*, US Pat., US6849774 (2005).

21  Roettger, D., Nierlich, F., Krissman, J., Wasserscheid, P., and Keim, W., *Method for separation of substances by extraction or by washing them with ionic liquids*, US Pat., US7304200 (2007).

22  Smith, R.S., Herrera, P.S., and Reynolds, S., *Use of ionic liquids to separate olefins, diolefins and aromatics*, US Pat., US7019188 (2006).

23  De Jong, F., and De With, J., *Process for the separation of olefins and paraffins*, US Pat., US7619129 (2009).

24  Beste, Y.A., and Schoenmakers, H., *Distillative method for separating narrow boiling or azeotropic mixtures using ionic liquid*, European Pat., EP1654046 (2009).

25  Arlt, W., Seiler, M., Jork, C., and Schneider, T., *Ionic liquids as selective additives for separation of close-boiling or azeotropic mixtures*, US Pat., US7435318 (2008).

26  Beste, Y.A., Eggersmann, M., and Schoenmakers, H., *Method for chemically reacting and separating a mixture in a column*, European Pat., EP1817090 (2008).

27  Beste, Y.A., Schoenmakers, H., Arlt, W., and Seiler, C.J., *Recycling of ionic liquids produces in extractive distillation*, US Pat., US7485208 (2009).

28  Maase, M., Massonne, K., Halbritter, K., Noe, R., Bartsch, M., Siegel, W., Stegmann, V., Flores, M., Huttenloch, O., and Becker, M., *Method for the separation of acids from chemical reaction mixtures by means of ionic fluids*, US Pat., US7767852 (2010).

29  Wolfert, A., Knosche, C., Pallasch, H.J., Sesing, M., Stroefer, E., Polka, H.M., and Heilig, M., *Method for separating hydrogen chloride and phosgene*, US Pat., US7659430 (2010).

30  Olschimke, J., Braukmüller, S., and Brosch, C., *Method for separating gas*, European Pat., EP1833758 (2009).

31  Hommeltoft, S.I., and Thellefsen, M., *Process for the recovery of sulphuric acid*, US Pat., US7595035 (2009).

32  Wasserscheid, P., Bosmann, A., Jess, A., Datsevich, L., Schmitz, C., and Wendt, A., *Process for removing polar impurities from hydrocarbons and mixtures of hydrocarbons*, US Pat., US7553406 (2009).

33   Schoonover, R.E., *Method for extraction of organosulfur compounds from hydrocarbons using ionic liquids*, US Pat., US7001504 (2006).

34   Olivier-Bourbigou, H., Uzio, D., and Magna, L., *Processing for eliminating sulfur-containing compounds and nitrogen-containing compounds from hydrocarbon*, US Pat., US7198712 (2007).

35   Serban, M., and Kocal, J.A., *Method of denitrogenating diesel fuel*, US Pat., US7749377 (2010).

36   Chinn, D., Vu, D., Driver, M.S., and Boudreau, L.C., $CO_2$ *removal from gas using ionic liquid absorbents*, US Pat., US7527775 (2009).

37   Cadours, R., Lecomte, F., Magna, L., and Barrere-Tricca, C., *Method for processing a natural gas with extraction of the solvent contained in the purified natural gas*, US Pat., US7470829 (2008).

38   Cadours, R., Lecomte, F., Magna, L., and Barrere-Tricca, C., *Method for processing a natural gas with extraction of the solvent contained in the acid gases*, US Pat., US7459011 (2008).

39   Cadours, R., Lecomte, F., Magna, L., and Barrere-Tricca, C., *Method for extracting an antihydrate contained in condensed hydrocarbons*, US Pat., US7470359 (2008).

40   Whiston, K., *Extraction process*, US Pat., US7442841 (2008).

41   Maase, M., Budich, M., Grossmann, G., and Szarvas, L., *Method for extracting impurities using ionic liquids*, US Pat., US7605297 (2009).

42   Hub, S., Lacroix, E., Bonnet, P., and Devic, M., *Method for catalytic hydrogenation purification of lactam containing chlorolactam impurities*, European Pat., EP1773765 (2008).

43   Bousand, B., Bonnet, P., Court, F., Devic, M., Hidalgo, M., and Navarro, C., *Hydrogenation method for unsaturated block copolymers and hydrogenated unsaturated block copolymers*, US Pat., US7202308 (2007).

44   Qu, J., Truhan, Jr., J.J., Dai, S., Luo, H., and Blau, P.J., *Lubricants or lubricant additives composed of ionic liquids containing ammonium cations*, US Pat., US7754664 (2010).

45   Hell, K., Hubel, R., and Weyershausen, B., *Use of solutions of metal salts in ionic liquids as anti-static agents for plastics*, European Pat., EP2038337 (2010).

46   Fu, D., and Card, R.J., *Fluids and techniques for matrix acidizing*, US Pat., US6350721 (2002).

47   Palmer, B.J., Fu, D., Card, R., and Volpert, E., *Wellbore fluids and their application*, US Pat., US6608005 (2003).

48   Watanabe, M., Manabe, C., Shigematsu, T., Hirakata, M., Okada, S., and Ooma, S., *Method for forming carbon nanotube thin film*, US Pat., US7592050 (2009).

49   Yokoyama, C., Kasai, H., Sarashina, E., Inomata, H., and Nakanishi, H., *Method of concentrating fine particle dispersion and method of recovering fine particle*, US Pat., US7732494 (2010).

50   Hecht, S.E., Cron, S.L., Scheibel, J.J., Miracle, G.S., Seddon, K.R., Earle, M., and Gunaratne, H.Q.N., *Ionic liquids derived from functionalized anionic surfactants*, US Pat., US7737102 (2010).

51   Hecht, S.E., Cron, S.L., Scheibel, J.J., Miracle, G.S., Seddon, K.R., Earle, M., and Gunaratne, H.Q.N., *Ionic liquids derived from functionalized anionic surfactants*, US Pat., US7786064 (2010).

52    Amano, T., Kobayashi, N., Ando, M., and Okumura, K., *Pressure-sensitive adhesive composition, pressure-sensitive adhesive sheets and surface protecting film*, US Pat., US7491758 (2009).

53    Ukei, N., Amano, T., and Ando, M., *Adhesive composition, adhesive sheet, and surface protective film*, US Pat., US7799853 (2010).

54    Amano, T., Kobayashi, N., and Ando, M., *Pressure-sensitive adhesive composition, pressure-sensitive adhesive sheets, and surface protecting film*, US Pat., US7842742 (2010).

55    Amano, T., Kobayashi, N., and Ando, M., *Pressure-sensitive adhesive composition, pressure-sensitive adhesive sheets, and surface protecting film*, US Pat., US7846999 (2010).

56    Nardi, J.C., Hussey, C.L., and King, L.A., *AlCl₃/1-alkyl pyridinium chloride room temperature electrolytes*, US Pat., US4122245 (1978).

57    Gifford, P.R., Shacklette, L.W., Toth, J.E., and Wolf, J.F., *Secondary batteries using room-temperature molten non-aqueous electrolytes containing 1,2,3-trialkylimidazolium halides or 1,3-dialkylimidazolium halide*, US Pat., US4463071 (1984).

58    Gifford, P.R., Palmisano, R.P., Shacklette, L.W., Chance, R.R., and Toth, J.E., *Secondary batteries containing room-temperature molten 1,2,3-trialkylimidazolium halide non-aqueous electrolyte*, US Pat., US4463072 (1984).

59    Chauvin, Y., Commereuc, D., Guibard, I., Hirschauer, A., Olivier, H., and Saussine, L., *Non-aqueous liquid composition with an ionic character and its use as a solvent*, US Pat., US5104840 (1992).

60    Park, W.S., *Liquid clathrate compositions*, US Pat., US6096680 (2000).

61    Hope, K.D., Stern, D.A., and Twomey, D.W., *Method for manufacturing ionic liquid catalysts*, US Pat., US6984605 (2006).

62    Atkins, M.P., Smith, M.R., Brian, E., *Lubricating oils*, EP Pat., EP0791643A1 (1997).

63    Hope, K.D., Driver, M.S., and Harris, T.V., *High viscosity polyalphaolefins prepared with ionic liquid catalyst*, US Pat., US6395948 (2002).

64    O'Rear, D.J., and Harris, T.V., *Process for making a lube base stock from a lower molecular weight feedstock*, US Pat., US6398946 (2002).

65    Johnson, D.R., Simmons, C.A., Mohr, D.H., Miller, S.J., Lee, S.K., Schinski, W.L., and Driver, M.S., *Process for increasing the yield of lubricating base oil from Fischer-Tropsch plant*, US Pat., US6605206 (2003).

66    Hope, K.D., Stern, D.A., and Benham, E.A., *Method and system to contact an ionic liquid catalyst with oxygen to improve a chemical reaction*, US Pat., US7309805 (2007).

67    Chauvin, Y., Hirschauer, A., and Olivier, H., *Catalytic composition and process for the alkylation of aliphatic hydrocarbons*, US Pat., US5750455 (1998).

68    Hirschauer, A., and Olivier-Bourbigou, H., *Catalytic composition and aliphatic hydrocarbon alkylation process*, US Pat., US6028024 (2000).

69    Hirschauer, A., and Olivier-Bourbigou, H., *Aliphatic hydrocarbon alkylation process*, US Pat., US6235959 (2001).

70    Chauvin, Y., Hirschauer, A., and Olivier, H., "Alkylation of isobutane with 2-butene using 1-butyl-3-methylimidazolium chloride-aluminum chloride molten salts as catalysts", *J.Mol.Catal.*, **92**, 155–165 (1994).

71    Liu, Z., Xu, C., and Huang, C., *Method for manufacturing alkylate oil with composite ionic liquid used as catalyst*, US Pat., US7285698 (2007).

72    Elomari, S., Trumbull, S., Timken, H-K.C., and Cleverdon, R., *Alkylation process using chloroaluminate ionic liquid catalysts*, US Pat., US7432409 (2008).

73    Harris, T.V., Driver, M., Elomari, S., and Timken, H-K.C., *Alkylation process using an alkyl halide promoted ionic liquid catalysts*, US Pat., US7431707 (2009).

74    Elomari, S., *Alkylation process using an alkyl halide promoted ionic liquid catalysts*, US Pat., US7495144 (2009).

75    Driver, M., and Dieckmann, G., *Reduction of organic halide in alkylate gasoline*, US Pat., US7538256 (2009).

76    Elomari, S., Trumbull, S., Timken, H-K.C., and Cleverdon, R., *Integrated alkylation process using ionic liquid catalysts*, US Pat., US7432408 (2008).

77    Elomari, S., *Alkylation of olefins with isoparaffins in ionic liquid to make lubricant or fuel blendstock*, US Pat., US7569740 (2009).

78    Abdul-Sada, A.K., Atkins, M.P., Ellis, B., Hodgson, P.K.G., Morgan, M.L.M., and Seddon, K.R., *Alkylation process*, US Pat., US5994602 (1999).

79    Driver, M., Campbell, C.B., and Harris, T.V., *Method of making an alkylated aromatic using acidic ionic liquid catalyst*, US Pat., US7732651 (2010).

80    Hobbs, S.J., Madabusi, V.K., Wang, J-Y., and Steiber, J.F., *Ring alkylation of aniline or an aniline derivative using ionic liquid catalysts*, US Pat., US7378554 (2008).

81    Saleh, R.Y., *Process for making aromatic aldehydes using ionic liquids*, US Pat., US6320083 (2001).

82    Saleh, R.Y., *Process for making aromatic aldehydes using ionic liquids*, World Pat., WO2000015594 (2000).

83    Knifton, J.F., *Process for producing p-tolualdehyde from toluene using an aluminum halide alkyl pyridinium halide "melt"; catalyst*, US Pat, US4554383 (1985).

84    Nemeth, L.T., Bricker, J.C., Holmgren, J.S., and Monson, L.E., *Direct carbonylation of paraffins using an ionic liquid catalyst*, US Pat., US6288281 (2001).

85    Huang, M-Y., Chang, J-C., Lin, J-C.and Wu, J-C., *Method for producing exotetrahydrodicyclopentadiene using ionic liquid catalyst*, US Pat., US7488860 (2009).

86    Huang, M-Y., Chang, J-C., Lin, J-C.and Wu, J-C., *Method for producing adamantane*, US Pat., US7488859 (2009).

87    Bonnet, P., Lacroix, E., and Schirmann, J-P., *Ion liquids derived from Lewis acid based on titanium, niobium, tantalum, tin or antimony, and uses thereof*, US Pat., US6881698 (2005).

88    Whiston, K., Forsyth, S., and Seddon, K.R., *Ionic liquid solvents and a process for the depolymerization of polyamides*, US Pat., US7772293 (2010).

89    Elomari, S., and Harris, T.V., *Regeneration of ionic liquid catalyst by hydrogenation using metal and acid*, US Pat., US7727925 (2010).

90  Elomari, S., and Harris, T.V., *Regeneration of ionic liquid catalyst by hydrogenation using a supported catalyst*, US Pat., US7691771 (2010).

91  Elomari, S., and Harris, T.V., *Regeneration of ionic liquid catalyst by hydrogenation using a metal or metal alloy*, US Pat., US7651970 (2010).

92  Elomari, S., and Harris, T.V., *Regeneration of ionic liquid catalyst by hydrogenation using a homogeneous catalyst*, US Pat., US7678727 (2010).

93  Chauvin, Y. Hirschauer, A., Commereuc, D., Olivier-Bourbigou, H., Saussine, L., and Guibard, I., *Non-aqueous liquid composition with an ionic character and its use as a solvent*, US Pat., US5104840 (1992).

94  Commereuc, D., Forestière, A., Hugues, F., and Olivier-Bourbigou, H., *Sequence of processes for olefin oligomerization*, US Pat., US6646173 (2003).

95  Olivier-Bourbigou, H., Commereuc, D., and Harry, S., *Catalytic composition and process for the catalysis of dimerization codimerization and oligomerization of olefins*, US Pat., US6576724 (2003).

96  Lecocq, V., and Olivier-Bourbigou, H., *Catalyst composition containing an aluminoxane for dimerizing, co-dimerizing and oligomerizing olefins*, US Pat., US6911410 (2005).

97  Lecocq, V., and Olivier-Bourbigou, H., *Catalyst composition for dimerizing, co-dimerizing and oligomerizing and polymerizing olefins*, US Pat., US6951831 (2005).

98  Hillebrand, G., Hirschauer, A., Commereuc, D., Olivier-Bourbigou, H., and Saussine, L., *Process for hydroformylation using a catalyst based on cobalt and/or rhodium employed in a two-phase medium*, US Pat., US6469216 (2002).

99  Favre, F., Commereuc, D., Olivier-Bourbigou, H., and Saussine, L., *Hydroformylation process employing a catalyst based on cobalt and/or rhodium in a non-aqueous ionic solvent*, US Pat., US6410799 (2002).

100  Magna, L., Olivier-Bourbigou, H., Saussine, L., and Kruger-Tissot, V., *Hydroformylation process employing a cobalt-based catalyst in a non-aqueous ionic liquid with improved catalyst recycling*, US Pat., US7223374 (2007).

101  Olivier-Bourbigou, H., Saussine, L., Magna, L., and Proriol, D., *Hydroformylation method involving a cobalt-based catalyst in a non-aqueous ionic liquid*, US Pat., US7781621 (2010).

102  Bahrmann, H., and Bohnen, H., *Method for producing aldehydes*, US Pat., US6472565 (2002).

103  Bohnen, H., Herwig, J., Hoff, D., Wasserscheid, P., and van Hal, R., *Method for the production of aldehydes*, US Pat., US6995293 (2006).

104  Tustin, G.C., and Moncier, R.M., *Continuous carbonylation process*, US Pat., US6916951 (2005).

105  Magna, L., Olivier-Bourbigou, H., Harry, S., and Commereuc, D., *Process for carbonylating alcohols, employing a catalyst based on rhodium or iridium in a non-aqueous ionic liquid, with efficient catalyst recycling*, US Pat., US71157745 (2006).

106  Riisager, A., and Fehrmann, R., *A process for continuous carbonylation by supported ionic phase catalysis*, European Pat., EP 1 883 616B1 (2009).

107  Chauvin, Y., and Di Marco-Van Tiggelen, F., *Catalytic composition and olefin disproportion process*, US Pat., US5525567 (1996).

108   Gürtler, C., and Jautelat, M., *α,ω-Diene metathesis in the presence of ionic liquids*, US Pat., US6756500 (2004).

109   Schinski, W.L., and Driver, M.S., *Process for making fatty acid nitriles and fatty amines by cross-metathesis of normal alpha olefins*, US Pat., US6380420 (2002).

110   Olivier-Bourbigou, H., Hillion, G., and Vallée, C., *Process for co-producing olefins and diesters or diacids by homometathesis of unsaturated fats in non-aqueous ionic liquids*, US Pat., US7754904 (2010).

111   Hell, K., Hesse, U., and Weyershausen, B., *Verfahren zur Herstellung von organo-modifizierten Polysiloxanen unter VerwendungionischerFlüssigkeiten*, European Pat., EP1382630B1 (2006).

112   Harmer, M.A., and Junk, C.P., *Preparation of polytrimethylene ether glycol and copolymers thereof*, US Pat., US7238772 (2007).

113   Harmer, M.A., Junk, C.P., and Vickery, J., *Ionic liquids*, US Pat., US7544813 (2009).

114   Harmer, M.A., Junk, C.P., and Manzer, L.E., *Preparation of polytrimethylene ether glycol and copolymers thereof*, US Pat., US7405330 (2008).

115   Harmer, M.A., Junk, C.P., Vickery, J., and Miller, R., *Preparation of poly(tetramethylene) glycol*, US Pat., US7528287 (2009).

116   Harmer, M.A., Junk, C.P., and Manzer, L.E., *Preparation of poly(tetramethylene) glycol*, US Pat., US7402711 (2008).

117   Westman, J., *Preparation and use of ionic liquids in microwave-assisted chemical transformations*, US Pat., US6596130 (2003).

118   Li, Z., Tang, Y., and Cheng, J., *Use of ionic liquids as coordination ligands for organometallic catalysts*, US Pat., US7615644 (2009).

119   Garayt, M., Le Boulaire, V., Gree, D., Gree, R., Schanen, V., and Spindler, J-F., *Use of a composition of an ionic nature as a substrate reagent, a composition constituting a fluorination reagent and a method using same*, US Pat., US 7393980 (2008).

120   Mehnert, C.P., Dispenziere, N.C., and Schlosberg, R.H., *Process for conducting aldol condensation reactions in ionic liquid media*, US Pat., US6552232 (2003).

121   Company, C.E., Prats, L.G., and Boliart, J.C., *Friedel–Crafts acylation process in ionic liquids*, US Pat., US7595425 (2009).

122   Hommeltoft, S.I., *Process for the drying of a hydrocarbon stream*, US Pat., US6887442 (2005).

123   Fuchigami, T., Sekiguchi, K., and Masuda, G., *Method for photolyzing organic matter and method for treating wastewater*, US Pat., US7488425 (2009).

124   Couturier, J.L., and Guerret, O., *Method for preparing alkoxyamines from nitroxides*, US Pat., US6700007 (2004).

125   Graenacher, C., *Cellulose solution and cellulose derivative and process of making same*, US Pat., US1924238 (1933).

126   Swatloski, R.P., Rogers, R.D., and Holbrey, J.D., *Dissolution and processing of cellulose using ionic liquids*, US Pat., US6824599 (2004).

127   Varanasi, S., Schall, C.A., and Prasad Dadi, A., *Saccharifying cellulose*, US Pat., US7674608 (2010).

128  Maase, M., and Stegmann, V., *Solubility of cellulose in ionic liquids with addition of amino bases*, US Pat., US7754002 (2010).

129  Maase, M., and Stegmann, V., *Cellulose solutions in ionic liquids*, US Pat., US7749318 (2010).

130  Ebeling, H., and Fink, H.P., *Cellulose carbamate spinning solution, method for producing a cellulose carbamate nonowen, and use of the same*, European Pat., EP2110467 (2009).

131  Sumnicht, D.W., and Kokko, B.J., *Absorbent sheet having regenerated cellulose microfiber network*, US Pat., US7718036 (2010).

132  Scheibel, J.J., Keneally, C.J., Menkhaus, J.A., Seddon, K.R., and Chwala, P., *Method for modifying cellulosic polymers in ionic liquids*, US Pat., US7714124 (2010).

133  Myllymaki, V., and Aksela, R., *Starch esterification method*, European Pat., EP1664125 (2010).

134  Myllymaki, V., and Aksela, R., *Depolymerisation method*, European Pat., EP1704259 (2009).

135  Hecht, S.E., Niehoff, R.L., Narasimhan, K., Neal, C.W., Forshey, P.A., Phan, D.V., Brooker, A.D.M., and Combs, K.H., *Extracting biopolymers from a biomass using ionic liquids*, US Pat., US7763715 (2010).

# 2 Supercritical Fluids in Ionic Liquids

MAAIKE C. KROON

Department of Chemical Engineering and Chemistry, Eindhoven University of Technology, Eindhoven, The Netherlands

COR J. PETERS

Chemical Engineering Program, The Petroleum Institute, Abu Dhabi, United Arab Emirates
Department of Chemical Engineering and Chemistry, Eindhoven University of Technology, Eindhoven, The Netherlands

## ABSTRACT

Ionic liquids and supercritical fluids are both alternative environmentally benign solvents, but their properties are very different. Ionic liquids are non-volatile but often considered highly polar compounds, whereas supercritical fluids are non-polar but highly volatile compounds. The combination of these two types of solvents has some unique features. It has been discovered that the solubility of supercritical carbon dioxide in several ionic liquids is very high but that the solubility of ionic liquids in supercritical carbon dioxide is negligibly low. Therefore, organic solutes can be extracted from an ionic liquid using supercritical carbon dioxide without any contamination by the ionic liquid. The phase behaviour of many binary or ternary (ionic liquid + super-critical carbon dioxide) systems was subsequently studied. Combined with the fact that ionic liquids are excellent reaction media for catalysed reactions, this led to the development of chemical processes where the reaction was carried out in the ionic liquid and the product was extracted afterwards with super-critical carbon dioxide. Newest developments include the multi-functional use of supercritical carbon dioxide as extraction medium, transport medium, and miscibility controller in these processes, resulting in higher reaction and separation rates.

*Ionic Liquids Further UnCOILed: Critical Expert Overviews*, First Edition.
Edited by Natalia V. Plechkova and Kenneth R. Seddon.
© 2014 John Wiley & Sons, Inc. Published 2014 by John Wiley & Sons, Inc.

## 2.1 INTRODUCTION

Both ionic liquids and supercritical fluids have been described as alternative "green" solvents, which are highly tuneable. The properties of an ionic liquid can be tuned by the choice of the cation and the anion [1]. The properties of a supercritical fluid can be adjusted to be more "gas-like" (low solvency power) or "liquid-like" (high solvency power) by adjusting the pressure [2, 3]. Most commonly used supercritical fluids include carbon dioxide ($CO_2$), ethane, propane, ethane, and fluoroform (trifluoromethane, $CHF_3$) above their critical point. Supercritical $CO_2$, in particular, has attracted a lot of interest, because it is non-toxic, non-flammable, relatively inert, abundant, and inexpensive. Moreover, it is relatively easy to reach the critical conditions (304 K, 7.4 MPa) [3].

The properties of ionic liquids and supercritical $CO_2$ are very different. Ionic liquids are non-volatile but often considered as highly polar compounds, whereas $CO_2$ is a non-polar but highly volatile compound. The combination of these two types of solvents has some unique features. In 1999, it was reported that the solubility of supercritical $CO_2$ in [$C_4$mim][$PF_6$] was very high but that $CO_2$ is not able to dissolve these ionic liquids [4]. Therefore, it was found to be possible to extract a solute from an ionic liquid using supercritical $CO_2$ without any contamination by the ionic liquid [5]. The phase behaviour of many binary or ternary (ionic liquid + supercritical $CO_2$) systems was subsequently studied, and is addressed in Section 2.2. Combined with the fact that ionic liquids are excellent reaction media for catalysed reactions, this led to the development of chemical processes where the reaction was carried out in the ionic liquid and the product was extracted afterwards with supercritical $CO_2$. Newest developments include the multi-functional use of supercritical $CO_2$ as extraction medium, transport medium, and miscibility controller in these processes, resulting in higher reaction and separation rates. These applications of (ionic liquid + supercritical fluid) systems are described in Section 2.3. The chapter ends with some conclusions and an outlook with regard to (ionic liquid + supercritical fluid) systems.

## 2.2 PHASE BEHAVIOUR OF (IONIC LIQUID + SUPERCRITICAL FLUID) SYSTEMS

### 2.2.1 Experimental Methods

Different methods to determine the phase behaviour of (ionic liquid + supercritical fluid) systems are available. Synthetic methods are most commonly used to determine the phase behaviour of these systems [6–9], whereby mixtures of (ionic liquid + supercritical fluid) of known composition are prepared and the phase transitions within a certain pressure and temperature range are subsequently observed. The synthetic method is suitable over wide pressure and temperature ranges and very reliable, but does not allow the analysis of the phases in equilibrium. Another commonly used method is the static method [10–13], whereby the equilibrium cell at constant temperature is filled with a known amount of ionic liquid, which is brought into contact with a calibrated reservoir filled with supercritical fluid until equilibrium is reached as indicated by negligible pressure change. Dynamic methods are less commonly used and also less reliable, but allow analysis of the different phases [12]. However, the analysis of the supercritical phase is not always very useful, because many ionic liquids have negligible solubility (below detection limit) in the supercritical phase [4, 5]. The gravimetric balance, often used for determining gas solubilities in ionic liquids [14], is generally not suitable for measuring the phase behaviour of (ionic liquid + supercritical fluid) systems because of its low-pressure range ($<2$ MPa).

### 2.2.2    Phase Behaviour of Binary (Ionic Liquid + Supercritical Fluid) Systems

#### 2.2.2.1    *The Binary Ionic Liquid + Supercritical $CO_2$ System.*    The most widely investigated binary (ionic liquid + supercritical fluid) systems are the mixtures of 1,3-dialkylimidazolium ionic liquids with supercritical $CO_2$ [4–14]. A typical phase diagram of these systems is depicted in Figure 2.1.

From Figure 2.1, it can be concluded that the $CO_2$ solubility in a 1,3-dialkylimidazolium ionic liquid is high at lower pressures, but a nearly infinite bubble-point slope is present at a specific maximum concentration of $CO_2$, beyond which increasing the external pressure hardly increases the $CO_2$ solubility in the ionic liquid. According to Huang et al. [15], the reason for this sharp pressure increase at a certain maximum $CO_2$ concentration is that at this

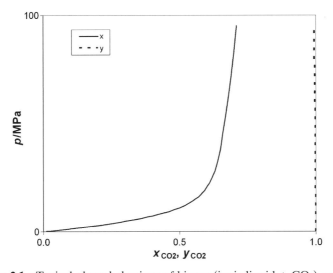

**Figure 2.1**    Typical phase behaviour of binary (ionic liquid + $CO_2$) systems.

point all cavities in the ionic liquid phase are occupied by $CO_2$, so that further insertion of $CO_2$ would require "breaking" the cohesive structure of the ionic liquid.

It was found that the anion predominantly determines the $CO_2$ solubility in 1,3-dialkylimidazolium ionic liquids [11]. Ionic liquids with anions containing fluoroalkyl groups, such as the $[NTf_2]^-$ anion, show highest $CO_2$ solubility [9, 11]. It was also observed that an increase in the alkyl chain length on the cation increases the $CO_2$ solubility in the ionic liquid [7, 8]. The solubility of $CO_2$ in an ionic liquid decreases with increasing temperature [16].

The extremely low solubility of 1,3-dialkylimidazolium ionic liquids in supercritical $CO_2$, as indicated by the straight dew point line at a $CO_2$ mole fraction of 100% in Figure 2.1, resulted in the use of supercritical $CO_2$ to extract products from these ionic liquids without solvent contamination [5]. However, some ionic liquids that do not incorporate a 1,3-dialkylimidazolium cation show completely different phase behaviour. For example, the ionic liquid $[P_{6\,6\,6\,14}]Cl$ was found to be able to dissolve in supercritical $CO_2$ up to a mass fraction of 7% [17], indicating that one has to be extremely cautious when stating that ionic liquids cannot dissolve in supercritical $CO_2$.

***2.2.2.2   The Binary (Ionic Liquid + Supercritical $ChF_3$) System.***   While supercritical $CO_2$ is not able to dissolve any 1,3-dialkylimidazolium ionic liquid [4], other supercritical fluids do. Ionic liquids are especially soluble in hydrocarbons that have a strong molecular interaction with the ionic liquid, such as supercritical $CHF_3$ [18, 19]. Figure 2.2 shows the general phase

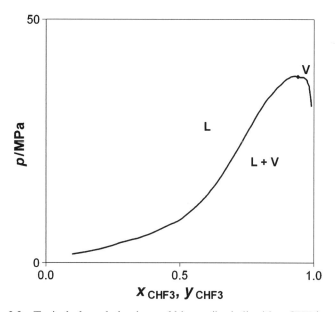

**Figure 2.2**   Typical phase behaviour of binary (ionic liquid + $CHF_3$) systems.

behaviour of binary (ionic liquid + CHF$_3$) systems [18–20]. This phase diagram is completely different from the phase diagram of binary (ionic liquid + CO$_2$) systems. The binary ionic liquid system with CHF$_3$ shows a closed phase envelope, including the occurrence of a critical point [18], whereas the CO$_2$ binary system with the same ionic liquid has an immiscibility gap between the CO$_2$ phase and the ionic liquid phase, even up to very high pressures. This has been attributed to the stronger molecular interactions between CHF$_3$ (with its strong permanent dipole moment) and the ionic liquid compared to those between CO$_2$ (no dipole moment) and the ionic liquid [8]. Again, it can be concluded that one has to be extremely cautious when stating that ionic liquids cannot dissolve in supercritical fluids. This is simply not true. In fact, the solubility of an ionic liquid in a supercritical phase depends on the curvature of the critical line in type III systems, according to the classification of Scott and van Konynenburg [21].

### 2.2.2.3  *Classification of Binary Ionic Liquid + Supercritical Fluid Systems.*
Scott and van Konynenburg [21] found that six different types of fluid phase behaviour exist, which are presented in Figure 2.3. With an exception of type VI, all types could be retrieved from the van der Waals equation of state. Although the original classification of Scott and van Konynenburg is still accepted, a detailed study on the occurrence of "holes" in ternary fluid multiphase systems [22], with CO$_2$ as one of the components, showed that acceptance of the existence of types I and V for binary CO$_2$ systems leads to inconsistency in the fluid phase transformations in ternary systems. For

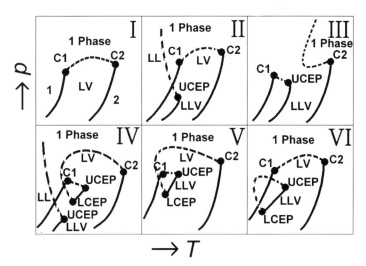

**Figure 2.3**  Classification of the liquid–vapour phase behaviour of binary systems according to Scott and van Konynenburg [21]. C, critical point; L, liquid; V, vapour; UCEP, upper critical end point; LCEP, lower critical end point.

instance, if in the ternary system ($CO_2$ + A + B), the binary system ($CO_2$ + A) has type III and the binary system ($CO_2$ + B) has type V fluid phase behaviour, a continuous transformation from type III into type V, by gradually replacing molecules A by molecules B, cannot be made. The same observation applies if type II or type IV is combined with type V or type I. These inconsistencies can only be overcome if it is assumed that type V in reality is type IV and, similarly, that type I in reality is type II, that is, both type I and type V must have a low-temperature liquid–liquid immiscibility region, as is the case in type IV. In all, this means that the original classification of Scott and van Konynenburg only comprises three independent types of fluid phase behaviour (II, III, and IV). From experiments in ternary $CO_2$ systems, it became apparent that between types II, III, and IV, continuous transformations are always possible, which is not the case if a type I or a type V is accepted to exist [22].

According to the foregoing discussion, the dispute whether binary (ionic liquid + $CO_2$) systems show type III or type V phase behaviour is not relevant and should be replaced by the question of whether we are dealing with type III or type IV [23]. As can be seen from Figure 2.3, type III has an upper critical end point (UCEP) of the nature $L_1 = V + L_2$, while type IV has, coming from higher to lower temperature, a UCEP ($L_1 = V + L_2$), followed by a lower critical end point (LCEP) of the nature ($L_1 = L_2 + V$) towards lower temperature and finally another UCEP ($L_1 = V + L_2$) as the beginning of the lower temperature branch of the three-phase equilibrium $L_1L_2V$. This means that if both a UCEP and an LCEP are present in the system, it will have type IV fluid phase behaviour. However, if only a UCEP can be identified in the system, it will have a type III fluid phase behaviour. An additional indication that we are dealing with type V phase behaviour is that this type has a critical line running from critical point 1 (C1) to critical point 2 (C2), which can be easily identified both experimentally and computationally, while type III phase behaviour shows a range of temperatures at which there are two immiscible phases up to infinite pressures. From the foregoing discussion, it follows that computational studies [14] suggesting that type V has to be assigned to (ionic liquid + $CO_2$) systems are not correct, and should be most likely type III [8, 24] or type IV, in case also an LCEP is present in the system. As it is experimentally observed that (1,3-dialkylimidazolium ionic liquid + $CO_2$) systems show the existence of two immiscible phases even up to extremely high pressures (>0.3 GPa) [5], binary (ionic liquid + $CO_2$) systems most likely will show a type III phase behaviour [8, 24]. In line with the foregoing discussion, (ionic liquid + $CHF_3$) systems, which show a critical line running from C1 to C2 as indicated in Figure 2.3 for type IV, most likely show a type IV fluid phase behaviour [25].

### 2.2.2.4 Modelling of Binary Ionic Liquid + Supercritical Fluid Systems.
Different types of equations of state have been used to model the phase behaviour of binary (ionic liquid + supercritical fluid) systems. Cubic equations of state such as the Peng–Robinson equation [18] and the Redlich–

Kwong equation [14, 25] have been used to describe the solubility of $CO_2$ and $CHF_3$ in ionic liquids. Because these cubic equations of state require the critical parameters of ionic liquids, which are unknown and have to be estimated by using group contribution methods [14], it is unreliable to apply cubic equations of state to ionic liquid systems. Moreover, cubic equations of state can only describe the $CO_2$ solubility in ionic liquids at low concentrations and pressures (below the critical pressure of $CO_2$), but cannot predict the dramatic increase in bubble-point pressure at higher $CO_2$ concentrations [18]. This is the reason why type V (type IV according to the previous discussion) phase behaviour was wrongly assumed for binary (ionic liquid + $CO_2$) systems on the basis of numerical calculations [23].

More reliable phase behaviour predictions for binary ionic liquid systems with $CO_2$ come from group contribution equations of state, such as the non-random lattice fluid equation of state [13] and the group contribution equation of state of Skjold–Jørgensen [26]. In group contribution methods, molecules are decomposed into groups which have their own parameters. Generally, ionic liquids are decomposed into a large group, consisting of the anion and the methylated (aromatic) ring of the cation, and a $CH_3$ group and various $CH_2$ groups that form the alkyl chain of the cation [13, 26]. For example, Figure 2.4 shows how the ionic liquid [$C_4$mim][$BF_4$] is decomposed into one $CH_3$ group, three $CH_2$ groups, and one [mim][$BF_4$] group. Pure group parameters are regressed from liquid density data [13]. Binary interaction parameters are fitted from infinite dilution activity coefficients and vapour–liquid equilibrium data of binary (ionic liquid + $CO_2$) systems [26]. In this way, the unknown critical parameters and vapour pressures of ionic liquids are not needed to determine group contribution equation of state parameters. Consequentially, phase equilibrium data can be predicted with higher accuracy [13, 26].

Statistical-mechanics-based equations of state are most predictive because they account explicitly for the microscopic characteristics of ionic liquids. The statistical association fluid theory models tPC-PSAFT [27, 28] and soft-SAFT [29] have successfully been used to model the phase behaviour of binary ionic liquid systems with $CO_2$ over a wide pressure range (0–40 MPa). These statistical mechanics-based equations of state consider the ionic liquids to be asymmetrical neutral ion pairs, either with a dipole moment to account for the charge distribution of the ion pair (for tPC-PSAFT) [27, 28] or with an

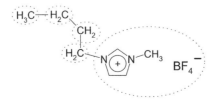

**Figure 2.4**   Decomposition of the ionic liquid [$C_4$mim][$BF_4$] into separate groups.

associating site mimicking the interactions between the cation and anion as a pair (in the case of soft-SAFT) [29]. Also, the associating interactions between ionic liquids and $CO_2$ are accounted for. All pure-component parameters for ionic liquids are calculated from available physicochemical data of the constituent ions, such as size, polarisability, and number of electrons [27]. This means that all parameters are physically meaningful. Only one binary interaction parameter for each possible binary pair is adjusted in order to fit the model to experimental vapour-liquid equilibrium data [26–28]. Statistical associating fluid theory models predict the phase behaviour of ionic liquid systems with $CO_2$ with high accuracy [27–29]. However, it is less suitable to predict the phase equilibria of ionic liquid systems with more polar compounds (e.g., $CHF_3$), because ionic liquid dissociation into its constituent ions is not taken into account [28].

### 2.2.3  Phase Behaviour of Ternary (Ionic Liquid + Supercritical Fluid) Systems

The number of different binary (ionic liquid + supercritical fluid) systems is already very large, but the number of possible ternary (ionic liquid + supercritical fluid) systems is orders of magnitude larger. However, the phase behaviour of only a few ternary (ionic liquid + supercritical fluid) systems has been investigated so far. Most studied ternary (ionic liquid + supercritical fluid) systems consist of an ionic liquid, supercritical $CO_2$, and an organic compound (*viz.* alkane, alcohol, ketone, ester, etc.) [30–40]. In some cases, the third compound is water [41–43].

Figure 2.5 shows the general phase behaviour of ternary (ionic liquid (liquid) + $CO_2$ (vapour) + organic (liquid)) mixtures. When the ionic liquid and the organic compound are completely miscible at ambient conditions (liquid + vapour), it is possible to induce the formation of a second liquid

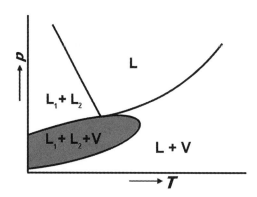

**Figure 2.5**  Phase behaviour of ternary (ionic liquid + supercritical $CO_2$ + organics) systems.

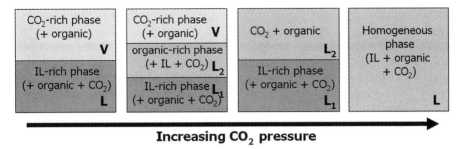

**Increasing CO$_2$ pressure**

**Figure 2.6**   Supercritical CO$_2$-induced "two-phase"–"three-phase"–"two-phase"–"one-phase" transition in ternary (ionic liquid + CO$_2$ + organics) systems.

phase by placing a pressure of CO$_2$ upon the mixture (liquid + liquid + vapour) [30–39]. The most dense phase is rich in ionic liquid, the newly formed liquid phase is rich in organics, and the vapour phase mostly contains CO$_2$ with some organics. Further pressurisation leads to expansion of the organic-rich phase with increased CO$_2$ pressure, while the ionic liquid-rich phase expands relative little. Eventually, this will lead to the disappearance of the vapour phase at the point where the organic-rich phase merges with the vapour phase [31–39]. At this moment, the last traces of ionic liquid that remained in the organic-rich liquid phase are expelled, and the resulting (CO$_2$ + organic phase) contains no detectable ionic liquid. Eventually, when the pressure is increased even further, one homogeneous liquid region is reached [36, 37].

Interestingly, it is thus possible to induce ternary (ionic liquid + CO$_2$) systems to undergo a "two-phase"–"three-phase"–"two-phase"–"one-phase" transition by only changing the CO$_2$ pressure (Fig. 2.6) [36]. Although the simple phase transition from two to three phases by addition of CO$_2$ was already known to occur in ternary CO$_2$ systems without an ionic liquid, it was discovered only recently to occur in ternary CO$_2$ systems in the presence of an ionic liquid [30]. Initially this phenomenon was wrongly identified as LCEP [30]. Thereafter, the transition from three to two phases at further CO$_2$ pressure increase was discovered, and also wrongly identified as K-point [31]. After all, both transitions are normal phase transitions without any criticality involved [40]. More recently, the formation of a homogeneous liquid phase at even higher CO$_2$ pressures was found [36, 37]. The location of this homogeneous liquid phase is hard to locate because it occurs in a relatively narrow range of CO$_2$ concentrations [36].

Ternary (ionic liquid + CO$_2$ + water) systems show similarities to ternary (ionic liquid + CO$_2$ + organic) systems. The supercritical CO$_2$ can cause liquid–liquid separation in hydrophilic (ionic liquid + water) mixtures [41–43].

It is more difficult to model the phase behaviour of ternary (ionic liquid + supercritical fluid) systems compared with the binary ones. The only equation of state that has been successfully applied to model ternary (ionic liquid + supercritical fluid) systems is the group contribution equation of state of Skjold–Jørgensen [44]. Excess Gibbs energy methods, commonly used to model ternary ionic liquid systems without any supercritical fluid added [45, 46], were not successful for this purpose.

## 2.3  CHEMICAL PROCESSING IN (IONIC LIQUID + SUPERCRITICAL FLUID) SYSTEMS

### 2.3.1  Separations in Ionic Liquid + Supercritical Fluid Systems

#### 2.3.1.1  *$CO_2$ Removal from Process Streams Using Ionic Liquids.*  Supercritical $CO_2$ has a high solubility in ionic liquids [4]. Other compounds show much lower solubilities in ionic liquids. After $CO_2$, the supercritical fluids $CHF_3$ and hydrogen sulfide have the highest solubilities and strongest interactions with the ionic liquid [18, 47], followed by methane [48, 49]. Carbon monoxide is less soluble [49, 50]. Dihydrogen ($H_2$) is the least soluble of all supercritical fluids studied [49].

These differences in solubility can be used to separate $CO_2$ from high-pressure streams by using ionic liquids as selective extractants [51] or in supported membranes [4, 52]. As opposed to conventional absorption techniques using amines, the lack of vapour pressure of ionic liquids minimises the loss of the capturing agent into the gas stream [52]. Examples include the separation of acid gases from natural gas [51, 53] or purifying the products from steam reforming or water gas-shift reactions using ionic liquids [49, 51]. For example, the production of $H_2$ from fossil fuels by steam reforming/water gas shift can be enhanced by simultaneous removal of the by-product $CO_2$ using an ionic liquid [49].

#### 2.3.1.2  *Recovery of Organic Compounds from Ionic Liquids with Supercritical $CO_2$.*  Because supercritical $CO_2$ is able to dissolve a wide range of organic compounds, but 1,3-dialkylimidazolium ionic liquids are not soluble in supercritical $CO_2$ [4], several studies have focussed on the recovery of organic compounds from these ionic liquids by using supercritical $CO_2$ as extractant [5, 54, 55]. Combined with the fact that ionic liquids are excellent reaction media for catalysed reactions [1], this led to the development of chemical processes where the reaction was carried out in the ionic liquid and the product was extracted afterwards with supercritical $CO_2$ [54, 55]. The main advantage is that the organic compound is recovered free of ionic liquid [5]. Disadvantages are the low extraction rate due to mass transfer limitations at the interface between the two phases, and the batch-wise operation of the process, making it is difficult to scale up [55].

A closer look at the phase behaviour of ternary (ionic liquid + supercritical $CO_2$ + organics) systems (Fig. 2.5 and Fig. 2.6) shows that this type of extraction is only possible in the "two-phase" region (liquid + supercritical fluid)

[36]. Here, the supercritical $CO_2$ phase does not contain any ionic liquid, while the solubility of the organic compound in $CO_2$ is sufficiently high [36]. The conditions under which the different phase transitions ("two-phase"–"three-phase"–"two-phase"–"one-phase") occur depend on the type of organic, the type of ionic liquid, and the concentrations [32, 56]. Stronger interaction between the ionic liquid and the organic compound makes it more difficult for $CO_2$ to induce the formation of a second liquid phase, and also to recover the organic compound [56]. This difference in affinity can be used for selective extraction of specific organics from ionic liquids by using $CO_2$ [57].

$CO_2$ at low concentrations was found to work as co-solvent (increasing the solubility of organics into the ionic liquid phase), while $CO_2$ at higher concentrations worked as an anti-solvent (decreasing the solubility of organics in the ionic liquid phase) [58]. The same type of phase behaviour was also observed for systems in which the organic is a solid instead of a liquid [59]. Therefore, it is also possible to recover an organic compound from an ionic liquid by crystallisation using supercritical $CO_2$ as anti-solvent [59].

### 2.3.2   Combined Reactions and Separations in Ionic Liquid + Supercritical Fluid Systems

*2.3.2.1   Continuous Biphasic Processes with Ionic Liquids and Supercritical $CO_2$.*   After reaction, the formed organic products can be separated batch-wise from ionic liquids by extraction with supercritical $CO_2$. It was found that continuous operation could be achieved when the supercritical $CO_2$ was used not only as an extraction medium, but also as a transport medium [60–65]. In this case, the supercritical $CO_2$ phase acts both as a reactant and as a product reservoir (Fig. 2.7). The reactants are transported into the reactor using supercritical $CO_2$ as the mobile phase. In the reactor, the reactants dissolve in the ionic liquid phase with immobilised catalyst, where the reaction takes place. The products are continuously extracted with the supercritical $CO_2$ stream.

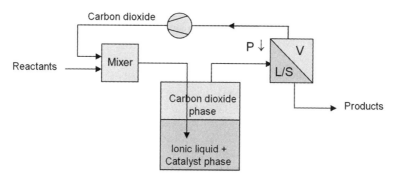

**Figure 2.7**   Continuous operation (reaction and separation) in biphasic (ionic liquid + supercritical $CO_2$) systems.

The product and $CO_2$ are separated downstream by controlled density reduction via pressure release or temperature increase. This biphasic process operation has been applied to hydrogenations [60], hydroformylations [61], dimerisations [62], (enzyme-catalysed) esterifications [63, 64], and the synthesis of cyclic carbonates (as $CO_2$ fixation method) [65].

Advantages of the biphasic operation are the ease of separation of the product and the catalyst, the enhanced stability and selectivity of the catalyst by the ionic liquid, and (in most cases) the increased reaction rate by adding supercritical $CO_2$ as compared to the biphasic operation without $CO_2$ [60–65]. However, the reported reaction rates in these biphasic systems are low compared with conventional catalytic single-phase processes, as a result of mass transfer limitations and low reactant solubilities. Moreover, mass transfer limitations also lead to low extraction rates [36]. The use of supported ionic liquid phase + supercritical $CO_2$ systems can overcome some of the mass transfer limitation problems [66].

In order to achieve high reaction rates, it is highly desirable to create a homogeneous liquid phase during reaction. In addition, instantaneous demixing into two phases, where the product is recovered from the phase that does not contain any ionic liquid, is desirable for a fast separation. In the next section, a continuous process that combines such features is presented.

### 2.3.2.2 Continuous Processes with Ionic Liquids, Supercritical $CO_2$, and the Miscibility Switch Phenomenon.

Figure 2.5 and Figure 2.6 show that it is possible reach a homogeneous phase in ternary (ionic liquid + supercritical $CO_2$ + organics) systems at high $CO_2$ pressures [36]. When the $CO_2$ pressure is subsequently lowered, the two-phase region (liquid + supercritical fluid) is reached again. This $CO_2$-induced switch in miscibility can be used to design a continuous process with high reaction and separation rates [67, 68].

The reaction is carried out in the homogeneous system, where the reactants as well as the catalyst dissolve in the ionic liquid [67]. The advantage of using an ionic liquid as reaction medium is that immobilised catalyst is stabilised against air and water oxidation by the ionic liquid, resulting in a longer lifetime of the catalyst without the need of regeneration [69]. The advantage of adding $CO_2$ to the reaction mixture is that the solubility of many reactants is increased (higher concentrations) and/or that reactants that are normally immiscible with pure ionic liquid can dissolve in (ionic liquid + $CO_2$) mixtures (the co-solvency effect) [70]. Therefore, it is possible to bring all components in high concentrations into one homogeneous phase. In this homogeneous system, the reaction takes place without any mass transfer limitations, which results in a high reaction rate [67]. Moreover, the addition of $CO_2$ to the reaction mixture leads to a lower viscosity of the reaction system and a higher diffusion rate of the reactants, resulting in a further increase in reaction rate [60]. The ionic liquid scarcely expands when $CO_2$ is dissolved because the $CO_2$ molecules occupy the cavities in the ionic liquid phase [15]. Therefore, the reaction volume can be kept small, leading to small equipment size.

The separation is carried out in the biphasic system [67]. Application of the miscibility switch (pressure release) results in the instantaneous formation of a second phase out of the homogeneous liquid system by spinodal demixing [36]. The light phase consists of supercritical $CO_2$ with dissolved products (and reactants in case of incomplete conversion), but does not contain any ionic liquid (because $CO_2$ does not usually dissolve ionic liquid) [4, 5]. The heavy phase consists of ionic liquid with dissolved catalyst and some remaining products (and some remaining reactants in the case of incomplete conversion). These phases can be separated from each other, and the pressure of the light phase is further decreased, leading to precipitation of the product (as a liquid or as a solid) out of the $CO_2$. In this way, pure product is obtained without any detectable ionic liquid or catalyst (and no reactants when the reaction is complete). The catalyst remains in the ionic liquid phase and can be easily recycled, without negatively affecting the activity and enantioselectivity. Also, the $CO_2$ can be recompressed and reused [67]. The essential advantage of using instantaneous demixing instead of conventional extraction with $CO_2$ is the higher rate of product separation from the ionic liquid (no mass transfer limitations) [67]. Another advantage of carrying out the separation in the biphasic system is that the energy consumption is low. Energy is only required for recompressing the $CO_2$, but no energy-intensive distillation step is needed. Compared with the conventional separation processes, the energy consumption in the novel process setup can be decreased by 50–80% [67].

The continuous process set-up in which reactions and separations are combined using ionic liquids, supercritical $CO_2$, and the miscibility switch phenomenon is schematically depicted in Figure 2.8. It should be noticed that the $CO_2$ has a multi-functional purpose as co-solvent in the reaction step, viscosity decreasing agent, miscibility controller, and separation medium. Since the principle of miscibility windows is a general phenomenon [36], it is likely that this process set-up is applicable to many industrial processes.

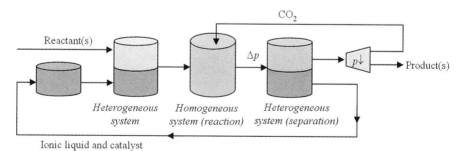

**Figure 2.8**  Continuous operation (reaction and separation) using ionic liquids, supercritical $CO_2$, and the miscibility switch phenomenon.

## 2.4  CONCLUSIONS AND OUTLOOK

Combining ionic liquids with supercritical fluids results in interesting phase behaviour. Binary (ionic liquid + supercritical $CO_2$) systems show a large immiscibility gap between the $CO_2$ phase and the ionic liquid phase up to very high pressures, and most likely have type III phase behaviour. Binary (ionic liquid + supercritical $CHF_3$) systems show a closed phase envelope and are therefore exhibit type IV phase behaviour. Supercritical $CO_2$ has high solubility in ionic liquids compared with other supercritical fluids. The solubility of 1,3-dialkylimidazolium ionic liquids in supercritical $CO_2$ is negligibly small, but this is not generally true for other ionic liquids. Ternary (ionic liquid + supercritical $CO_2$ + organics) systems can undergo a "two-phase"–"three-phase"–"two-phase"–"one-phase" transition by pressure increase.

Phase behaviour data of multi-component (ternary, quaternary, etc.) (ionic liquid + supercritical fluid) systems are relatively scarce. Future studies will therefore be directed to the phase behaviour of multi-component (ionic liquid + supercritical fluid) systems. Also, new modelling studies of the phase behaviour of multi-component (ionic liquid + supercritical fluid) systems are expected.

On the basis of the phase behaviour of multi-component (ionic liquid + supercritical fluid) systems, new applications will be sought. Combined (ionic liquid + supercritical fluid) systems will be used as "green" reaction and separation media, where the reaction product will be recovered from the supercritical phase (free of ionic liquid). A tendency to use the supercritical fluid for multiple purposes is observed. For example, supercritical $CO_2$ can act as extraction medium, transport medium, miscibility controller, and viscosity decreasing agent in these novel processes.

## REFERENCES

1   Wasserscheid, P., and Welton, T. (eds.), *Ionic Liquids in Synthesis*, 2nd ed. (Wiley-VCH Verlag, Weinheim, 2007).

2   Jessop, P.G., and Leitner, W. (eds.), *Chemical Synthesis Using Supercritical Fluids* (Wiley-VCH Verlag, Weinheim, 1999).

3   Beckman, E.J., Supercritical and near-critical $CO_2$ in green chemical synthesis and processing, *J. Supercrit. Fluids* **28**, 121–191 (2003).

4   Blanchard, L.A., Hancu, D., Beckman, E.J., and Brennecke, J.F., Green processing using ionic liquids and $CO_2$, *Nature* **399**, 28–29 (1999).

5   Blanchard, L.A., and Brennecke, J.F., Recovery of organic products from ionic liquids using supercritical carbon dioxide, *Ind. Eng. Chem. Res.* **40**, 287–292 (2001).

6   Kroon, M.C., Shariati, A., Costantini, M., Van Spronsen, J., Witkamp, G.J., Sheldon, R.A., and Peters, C.J., High-pressure phase behavior of systems with ionic liquids: part V. The binary system carbon dioxide + 1-butyl-3-methylimidazolium tetrafluoroborate, *J. Chem. Eng. Data* **50**, 173–176 (2005).

7  Shariati, A., and Peters, C.J., High-pressure phase equilibria of systems with ionic liquids, *J. Supercrit. Fluids* **34**, 171–182 (2005).

8  Shariati, A., Gutkowski, K., and Peters, C.J., Comparison of the phase behavior of some selected binary systems with ionic liquids, *AIChE J.* **51**, 1532–1540 (2005).

9  Schilderman, A.M., Raeissi, S., and Peters, C.J., Solubility of carbon dioxide in the ionic liquid 1-ethyl-3-methylimidazolium bis(trifluoromethylsulfonyl)imide, *Fluid Phase Equilib.* **260**, 19–22 (2007).

10  Kamps, A.P.S., Tuma, D., Xia, J., and Maurer, G., Solubility of $CO_2$ in the ionic liquid [bmim][$PF_6$], *J. Chem. Eng. Data* **48**, 746–749 (2003).

11  Aki, S.N.V.K., Mellein, B.R., Saurer, E.M., and Brennecke, J.F., High-pressure phase behavior of carbon dioxide with imidazolium-based ionic liquids, *J. Phys. Chem. B* **108**, 20355–20365 (2004).

12  Muldoon, M.J., Aki, S.N.V.K., Anderson, J.L., Dixon, J.K., and Brennecke, J.F., Improving carbon dioxide solubility in ionic liquids, *J. Phys. Chem. B* **111**, 9001–9009 (2007).

13  Kim, Y.S., Choi, W.Y., Jang, J.H., Yoo, K.P., and Lee, C.S., Solubility measurement and prediction of carbon dioxide in ionic liquids, *Fluid Phase Equilib.* **228–229**, 439–445 (2005).

14  Shiflett, M.B., and Yokozeki, A., Solubilities and diffusivities of carbon dioxide in ionic liquids: [bmim][$PF_6$] and [bmim][$BF_4$], *Ind. Eng. Chem. Res.* **44**, 4453–4464 (2005).

15  Huang, X., Margulis, C.J., Li, Y., and Berne, B.J., Why is the partial molar volume of $CO_2$ so small when dissolved in a room temperature ionic liquid? Structure and dynamics of $CO_2$ dissolved in [bmim$^+$] [$PF_6^-$], *J. Am. Chem. Soc.* **127**, 17842–17851 (2005).

16  Finotello, A., Bara, J.E., Camper, D., and Noble, R.D., Room-temperature ionic liquids: Temperature dependence of gas solubility selectivity, *Ind. Eng. Chem. Res.* **47**, 3453–3459 (2008).

17  Hutchings, J.W., Fuller, K.L., Heitz, M.P., and Hoffmann, M.M., Surprisingly high solubility of the ionic liquid trihexyltetradecylphosphonium chloride in dense carbon dioxide, *Green Chem.* **7**, 475–478 (2005).

18  Shariati, A., and Peters, C.J., High-pressure phase behavior of systems with ionic liquids: Measurements and modeling of the binary system fluoroform + 1-ethyl-3-methylimidazolium hexafluorophosphate, *J. Supercrit. Fluids* **25**, 109–117 (2003).

19  Shiflett, M.B., and Yokozeki, A., Solubility and diffusivity of hydrofluorocarbons in room-temperature ionic liquids, *AIChE J.* **52**, 1205–1219 (2006).

20  Shiflett, M.B., and Yokozeki, A., Binary vapor-liquid and vapor-liquid-liquid equilibria of hydrofluorocarbons (HFC-125 and HFC-143a) and hydrofluoroethers (HFE-125 and HFE-143a) with ionic liquid [emim][$Tf_2N$], *J. Chem. Eng. Data* **53**, 492–497 (2008).

21  Scott, R.L., and van Konynenburg, P.H., Static properties of solutions: Van der Waals and related models for hydrocarbon mixtures, *Discuss. Faraday Soc.* **49**, 87–97 (1970).

22  Peters, C.J., and Gauter, K., Occurrence of holes in ternary fluid multiphase systems of near-critical carbon dioxide and certain solutes, *Chem. Rev.* **99**, 419–431 (1999).

23  Raeissi, S., Florusse, L., and Peters, C.J., Scott-van Konynenburg phase diagram of carbon dioxide + alkylimidazolium-based ionic liquids, *J. Supercrit. Fluids* **55**,825–832 (2010).

24  Nwosu, S.O., Schleicher, J.C., and Scurto, A.M., High pressure phase equilibria for the synthesis of ionic liquids in compressed $CO_2$ for 1-hexyl-3-methylimidazolium bromide with 1-bromohexane and 1-methylimidazole, *J. Supercrit. Fluids* **51**, 1–9 (2009).

25  Yokozeki, A., and Shiflett, M.B., Global phase behaviors of trifluoromethane in ionic liquid [bmim][$PF_6$], *AIChE J.* **52**, 3952–3957 (2006).

26  Breure, B., Bottini, S.B., Witkamp, G.J., and Peters, C.J., Thermodynamic modeling of the phase behavior of binary systems of ionic liquids and carbon dioxide with the group contribution equation of state, *J. Phys. Chem. B* **111**, 14265–14270 (2007).

27  Kroon, M.C., Karakatsani, E.K., Economou, I.G., Witkamp, G.J., and Peters, C.J., Modeling of the carbon dioxide solubility in imidazolium-based ionic liquids with the tPC-PSAFT equation of state, *J. Phys. Chem. B* **110**, 9262–9269 (2006).

28  Karakatsani, E.K., Economou, I.G., Kroon, M.C., Peters, C.J., and Witkamp, G.J., tPC-PSAFT modeling of gas solubility in imidazolium-based ionic liquids, *J. Phys. Chem. C* **111**, 15487–15492 (2007).

29  Andreu, J.S., and Vega, L.F., Capturing the solubility behavior of $CO_2$ in ionic liquids by a simple model, *J. Phys. Chem. C* **111**, 16028–16034 (2007).

30  Scurto, A.M., Aki, S.N.V.K., and Brennecke, J.F., $CO_2$ as a separation switch for ionic liquid/organic mixtures, *J. Am. Chem. Soc.* **124**, 10276–10277 (2002).

31  Aki, S.N.V.K., Scurto, A.M., and Brennecke, J.F., Ternary phase behavior of ionic liquid (IL)—organic—$CO_2$ systems, *Ind. Eng. Chem. Res.* **45**, 5574–5585 (2006).

32  Mellein, B.R., and Brennecke, J.F., Characterization of the ability of $CO_2$ to act as an antisolvent for ionic liquid/organic mixtures, *J. Phys. Chem. B* **111**, 4837–4843 (2007).

33  Zhang, Z., Wu, W., Liu, Z., Han, B., Gao, H., and Jiang, T., A study of tri-phasic behavior of ionic liquid–methanol–$CO_2$ systems at elevated pressures, *Phys. Chem. Chem. Phys.* **6**, 2352–2357 (2004).

34  Zhang, Z., Wu, W., Wang, B., Chen, J., Shen, D., and Han, B., High-pressure phase behavior of $CO_2$/acetone/ionic liquid system, *J. Supercrit. Fluids* **40**, 1–6 (2007).

35  Fu, D., Sun, X., Qiu, Y., Jiang, X., and Zhao, S., High-pressure phase behavior of the ternary system $CO_2$ + ionic liquid [bmim][$PF_6$] + naphthalene, *Fluid Phase Equilib.* **251**, 114–120 (2007).

36  Kroon, M.C., Florusse, L.J., Kühne, E., Witkamp, G.J., and Peters, C.J., Achievement of a homogeneous phase in ternary ionic liquid/carbon dioxide/organic systems, *Ind. Eng. Chem. Res.* **49**, 3474–3478 (2010)

37  Kroon, M.C., Florusse, L.J., and Peters, C.J., Phase behavior of the ternary 1-hexyl-3-methylimidazolium tetrafluoroborate + $CO_2$ + methanol system, *Fluid Phase Equilib.* **294**, 84–88 (2010).

38  Chobanov, K., Tuma, D., and Maurer, G., High-pressure phase behavior of ternary systems (carbon dioxide + alkanol + hydrophobic ionic liquid), *Fluid Phase Equilib.* **294**, 54–66 (2010).

39  Ahn, J.Y., Lee, B.C., Lim, J.S., Yoo, K.P., and Kang, J.W., High-pressure phase behavior of binary and ternary mixtures containing ionic liquid [C$_6$-mim][Tf$_2$N], dimethyl carbonate and carbon dioxide, *Fluid Phase Equilib.* **290**, 75–79 (2010).

40  Kühne, E., Alfonsín, L.R., Mota Martinez, M.T., Witkamp, G.J., and Peters, C.J., Comment on "Characterization of the ability of CO$_2$ to act as an antisolvent for ionic liquid/organic mixtures," *J. Phys. Chem. B* **113**, 6579–6580 (2009).

41  Scurto, A.M., Aki, S.N.V.K., and Brennecke, J.F., Carbon dioxide induced separation of ionic liquids and water, *Chem. Commun.*, 572–573 (2003).

42  Zhang, Z., Wu, W., Gao, H., Han, B., Wang, B., and Huang, Y., Tri-phase behavior of ionic liquid–water–CO$_2$ system at elevated pressures, *Phys. Chem. Chem. Phys.* **6**, 5051–5055 (2004).

43  Bermejo, M.D., Montero, M., Saez, E., Florusse, L.J., Kotlewska, A.J., Cocero, M.J., Van Rantwijk, F., and Peters, C.J., Liquid-vapor equilibrium of the systems butyl-methylimidazolium nitrate-CO$_2$ and hydroxypropylmethylimidazolium nitrate-CO$_2$ at high pressure: Influence of water on the phase behavior, *J. Phys. Chem. B* **112**, 13532–13541 (2008).

44  Kühne, E., Martin, A., Witkamp, G.J., and Peters, C.J., Modeling the phase behavior of ternary systems ionic liquid + organic + CO$_2$ with a group contribution equation of state, *AIChE J.* **55**, 1265–1273 (2009).

45  Döker, M., and Gmehling, J., Measurement and prediction of vapor–liquid equilibria of ternary systems containing ionic liquids, *Fluid Phase Equilib.* **227**, 255–266 (2005).

46  Simoni, L.D., Lin, Y., Brennecke, J.F., and Stadtherr, M.A., Modeling liquid-liquid equilibrium of ionic liquid systems with NRTL, electrolyte-NRTL, and UNIQUAC, *Ind. Eng. Chem. Res.* **47**, 256–272 (2008).

47  Pomelli, C.S., Chiappe, A., Vidis, A., Laurenczy, G., and Dyson, P.J., Influence of the interaction between hydrogen sulfide and ionic liquids on solubility: Experimental and theoretical investigation, *J. Phys. Chem. B* **111**, 13014–13019 (2007).

48  Raeissi, S., and, Peters, C.J., High pressure phase behaviour of methane in 1-butyl-3-methylimidazolium bis(trifluoromethylsulfonyl)imide, *Fluid Phase Equilib.* **294**, 67–71 (2010).

49  Raeissi, S., and Peters, C.J., A potential ionic liquid for CO$_2$-separating membranes: Selection and gas solubility studies, *Green Chem.* **11**, 185–192 (2009).

50  Kumełan, J., Pérez-Salado Kamps, A., Tuma, D., and Maurer, G., Solubility of CO in the ionic liquid [bmim][PF$_6$], *Fluid Phase Equilib.* **228–229**, 207–211 (2005).

51  Anthony, J.L., Aki, S.N.V.K., Maginn, E.J., and Brennecke, J.F., Feasibility of using ionic liquids for carbon dioxide capture, *Int. J. Environ. Tech. Manag.* **4**, 105–115 (2004).

52  Fortunato, R., Afonso, C.A.M., Reis, M.A.M., and Crespo, J.G., Supported liquid membranes using ionic liquids: Study of stability and transport mechanisms, *J. Membrane Sci.* **242**, 197–209 (2004).

53  Shiflett, M.B., and Yokozeki, A., Separation of CO$_2$ and H$_2$S using room-temperature ionic liquid [bmim][PF$_6$], *Fluid Phase Equilib.* **294**, 105–113 (2010).

54 Brown, R.A., Pollet, P., McKoon, E., Eckert, C.A., Liotta, C.L., and Jessop, P.G., Asymmetric hydrogenation and catalyst recycling using ionic liquid and supercritical carbon dioxide, *J. Am. Chem. Soc.* **123**, 1254–1255 (2001).

55 Dzyuba, S.V., and Bartsch, R.A., Recent advances in applications of room temperature ionic liquid/supercritical $CO_2$ systems, *Angew. Chem., Int. Ed.* **42**, 148–150 (2003).

56 Kühne, E., Perez, E., Witkamp, G.J., and Peters, C.J., Solute influence on the high-pressure phase equilibrium of ternary systems with carbon dioxide and an ionic liquid, *J. Supercrit. Fluids* **45**, 27–31 (2008).

57 Bogel-Łukasik, R., Najdanovic-Visak, V., Barreiros, S., and Nunes da Ponte, M., Distribution ratios of lipase-catalyzed reaction products in ionic liquid supercritical $CO_2$ systems: Resolution of 2-octanol enantiomers, *Ind. Eng. Chem. Res.* **47**, 4473–4480 (2008).

58 Kroon, M.C., Van Spronsen, J., Peters, C.J., Sheldon, R.A., and Witkamp, G.J., Recovery of pure products from ionic liquids using supercritical carbon dioxide as co-solvent in extractions or as anti-solvent in precipitations, *Green Chem.* **8**, 246–249 (2006).

59 Kroon, M.C., Toussaint, V.A., Shariati, A., Florusse, L.J., Van Spronsen, J., Witkamp, G.J., and Peters, C.J., Crystallization of an organic compound from an ionic liquid using carbon dioxide as anti-solvent, *Green Chem.* **10**, 333–336 (2008).

60 Liu, F., Abrams, M.B., Baker, R.T., and Tumas, W., Phase-separable catalysis using room temperature ionic liquids and supercritical carbon dioxide, *Chem. Commun.* **433–434** (2001).

61 Webb, P.B., Sellin, M.F., Kunene, T.E., Williamson, S., Slawin, A.M.Z., and Cole-Hamilton, D.J., Continuous-flow hydroformylation of alkenes in supercritical fluid—ionic liquid biphasic systems, *J. Am. Chem. Soc.* **125**, 15577–15588 (2003).

62 Ballivet-Tkatchenko, D., Picquet, M., Solinas, M., Franciò, G., Wasserscheid, P., and Leitner, W., Acrylate dimerization under ionic liquid—supercritical carbon dioxide conditions, *Green Chem.* **5**, 232–235 (2003).

63 Reetz, M.T., Wiesenhöfer, W., Franciò, G., and Leitner, W., Continuous flow enzymatic kinetic resolution and enantiomer separation using ionic liquid/ supercritical carbon dioxide media, *Adv. Synth. Catal.* **345**, 1221–1228 (2003).

64 Zhang, Z., Wu, W., Han, B., Jiang, T., Wang, B., and Liu, Z., Phase separation of the reaction system induced by $CO_2$ and conversion enhancement for the esterification of acetic acid with ethanol in ionic liquid, *J. Phys. Chem. B* **109**, 16176–16179 (2005).

65 Xiao, L.F., Li, F.W., Peng, J.J., and Xia, C.G., Immobilized ionic liquid/zinc chloride: Heterogeneous catalyst for synthesis of cyclic carbonates from carbon dioxide and epoxides, *J. Mol. Catal. A* **253**, 265–269 (2006).

66 Hintermair, U., Höfener, T., Pullmann, T., Franciò, G., and Leitner, W., Continuous enantioselective hydrogenation with a molecular catalyst in supported ionic liquid phase under supercritical $CO_2$ flow, *Chem. Cat. Chem.* **2**, 150–154 (2010).

67 Kroon, M.C., Shariati, A., Florusse, L.J., Peters, C.J., Van Spronsen, J., Witkamp, G.J., Sheldon, R.A., and Gutkowski, K.E., *Process for carrying out a chemical reaction*, World Pat., WO 2006/088348 A1 (2006).

68 Kühne, E., Santarossa, S., Perez, E., Witkamp, G.J., and Peters, C.J., New approach in the design of reactions and separations using an ionic liquid and carbon dioxide

as solvents: Phase equilibria in two selected ternary systems, *J. Supercrit. Fluids* **46**, 93–98 (2008).

69  Guernik, S., Wolfson, A., Herskowitz, M., Greenspoon, N., and Geresh, S., A novel system consisting of Rh-DuPHOS and ionic liquid for asymmetric hydrogenations, *Chem. Commun.* 2314–2315 (2001).

70  Solinas, M., Pfaltz, A., Cozzi, P.G., and Leitner, W., Enantioselective hydrogenation of imines in ionic liquid/carbon dioxide media, *J. Am. Chem. Soc.* **126**, 16142–16147 (2004).

# 3 The Phase Behaviour of 1-Alkyl-3-Methylimidazolium Ionic Liquids

KEIKO NISHIKAWA

Division of Nanoscience, Graduate School of Advanced Integration Science, Chiba University, Japan

## ABSTRACT

Some of the unique properties of room temperature ionic liquids are remarkably manifest in their thermal behaviour, viz. they exhibit low melting points despite being salts, difficulty in crystallisation, premelting over a wide temperature range, excessive supercooling, and complex thermal histories. In addition, 1-alkyl-3-methylimidazolium ionic liquids show peculiar thermal behaviour at the phase changes, namely "rhythmic melting and crystallisation," "intermittent crystallization," in the premelting regions, and ultra-slow phase changes. These thermal behaviours could be detected only by the use of an ultra-sensitive differential scanning calorimeter with nano-Watt stability, and simultaneous measurements of calorimetry and Raman scattering. This unique thermal behaviour is attributed to the flexibility of the alkyl chains bonded to the imidazolium ring, and linkage of the phase transition or structural relaxation with the conformational changes.

## 3.1 PHASE TRANSITIONS LINKED WITH CONFORMATIONAL CHANGES OF CATIONS

Ionic liquids are attracting much attention because of their characteristic properties [1–4] and potential utilities as functional liquids [5–10]. Their unique properties are particularly exhibited in their thermal behaviour, such as

*Ionic Liquids Further UnCOILed: Critical Expert Overviews*, First Edition.
Edited by Natalia V. Plechkova and Kenneth R. Seddon.
© 2014 John Wiley & Sons, Inc. Published 2014 by John Wiley & Sons, Inc.

premelting over a wide temperature range, excessive supercooling, various crystal–crystal phase transitions, and the existence of complex thermal histories [11–21]. Researchers have performed thermal analyses of ionic liquids and their crystals using differential scanning calorimetry (DSC), thermogravimetric analysis (TGA), and so on [11–21]. These studies have provided useful information regarding their thermal properties. In this chapter, we focus specifically on the thermal behaviour of 1-alkyl-3-methylimidazolium ionic liquids and their crystals, especially their phase behaviour.

Many experimental results have suggested that the existence of multiple conformers for a constituent ion seriously affects the structural and thermal properties of the ionic liquid. For ionic liquids, the importance of a variety in conformational structures of constituent ions was first recognised by Holbrey et al. [22] and by the Hamaguchi group [23, 24] independently. The two groups pointed out the existence of polymorphism in $[C_4mim]Cl$ crystals. In the former paper, Holbrey et al. concluded that the multiple conformers inhibit the easy crystallisation of ionic liquids [22]. Berg reviewed the existence of various conformers in liquids and their crystals for many ionic liquids [25]. Raman spectroscopic studies backed up by density functional theory (DFT) calculations revealed that the occurrence of multiple conformers of 1-alkyl-3-methylimidazolium ions and their populations are different in the liquid and crystalline states [1, 26–30]. With an apparatus for ultra-sensitive DSC [17, 31], and simultaneous measurements of DSC and Raman spectroscopy [29], my group also confirmed that the melting and crystallisation of $[C_4mim]Br$ and $[C_4mim]Cl$ occurred directly linked with the cooperative conformational change of the butyl group in the cation [17, 29]. Moreover, not limited to $[C_4mim]Br$ and $[C_4mim]Cl$, we affirm that most curious thermal behaviour of 1-alkyl-3-methylimidazolium ionic liquids is due to the link between the phase transitions and the conformational changes of the alkyl chains bonded to the imidazolium ring [17, 18, 20, 21, 29, 32, 33].

Some examples on the varieties of conformers of 1-alkyl-3-methylimidazolium cations, and the relationships between conformational changes and phase behaviour, are presented in the following sections. Prior to their detailed discussion, the conformers of the three cations, $[C_2mim]^+$, $[^iC_3mim]^+$ {1-(2-propyl)-3-methylimidazolium cation}, and $[C_4mim]^+$, are outlined briefly here.

### 3.1.1  Conformers of 1-Alkyl-3-Methylimidazolium Cations

Stable and possible conformational structures for the three cations $[C_2mim]^+$, $[^iC_3mim]^+$, and $[C_4mim]^+$ in gaseous state can be obtained by DFT calculation using the Gaussian03 program package [34]. The quantum mechanical calculations suggest that there exist several rotational isomers for each of these cations: $[C_2mim]^+$ [27], $[^iC_3mim]^+$ [30], and $[C_4mim]^+$ [35, 36]. Taking into account both energy differences among isomers, and results from single crystal structure analyses of the corresponding salts, however, we can pick up a limited number of rotational isomers as possible conformers for each cation in their

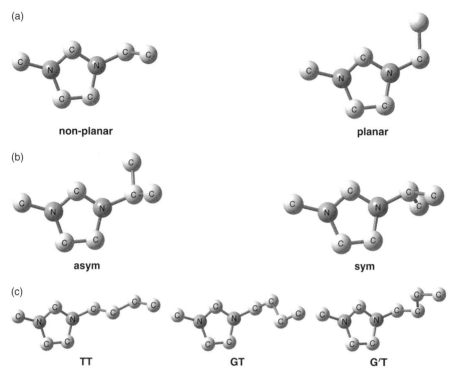

**Figure 3.1**   Stable conformers for (a) $[C_2mim]^+$, (b) $[^iC_3mim]^+$, and (c) $[C_4mim]^+$.

crystal or liquids states. The conformers to be borne in mind are shown in Figure 3.1. For $[C_2mim]^+$, Figure 3.1a, two rotational isomers around the N1-C1′ axis are stable, namely **planar** and **non-planar** conformers [27]. The change of potential energy accompanying the rotation around the N1-C1′ axis is shown in Figure 3.2a. It is seen that the **non-planar** conformer is about 2 kJ mol$^{-1}$ more stable than the **planar** conformer in the free cation. In fact, the **non-planar** conformation is adopted in most crystals of 1-alkyl-3-methylimidazolium salts, including for the simple $[C_2mim]^+$ cation. It is characteristic that the energy barrier from non-planar to planar is very small.

For the $[^iC_3mim]^+$ ion [30], there exist two possible conformers: the symmetric form (**sym**) and the asymmetric form (**asym**). As shown in Figure 3.1b, **sym** refers to the arrangement where two methyl groups in the isopropyl group are positioned symmetrically with respect to the imidazolium ring. For the **asym** conformer, the plane formed by N1-C1′-C2′ is almost perpendicular to the plane of the imidazolium ring. Figure 3.2b shows the potential energy for the rotation of the isopropyl group of $[^iC_3mim]^+$ around the N1-C1′ axis, as calculated by DFT. The dihedral angle of C5-N1-C1′-H is defined as the torsion angle. The two local minima, at torsion angles of *ca.* 30° and 180°, correspond to asym and sym, respectively.

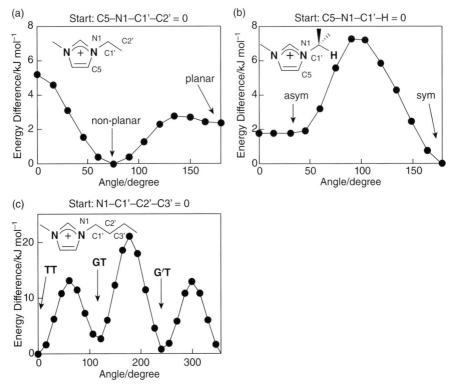

**Figure 3.2**  Potential energies as functions of torsion angles: (a) [C₂mim]⁺, (b) [ⁱC₃mim]⁺, and (c) [C₄mim]⁺.

There are three rotational isomers as possible conformers for [C₄mim]⁺ in the crystal or liquid states among nine or more conformers [35, 36], that is, the *gauche trans* (**GT**), *trans trans* (**TT**), and *gauche' trans* (**G′T**) conformations. They are caused by the rotation of the butyl group around the C1′-C2′ axis, keeping the arrangement for the C2′-C3′ axis in the *trans* conformation. These three isomers are shown in Figure 3.1c. Figure 3.2c shows the potential energy change accompanied with rotation around the C1′-C2′ axis.

### 3.1.2   DSC Measurements of the Bromides of [C₂mim]⁺, [ⁱC₃mim]⁺, and [C₄mim]⁺

The DSC traces for [C₂mim]Br [37], [ⁱC₃mim]Br [18], and [C₄mim]Br [17] are shown in Figure 3.3a,b,c, respectively. The bromides were selected as the standard samples to focus solely on the conformations of 1-alkyl-3-methylimidazolium cations. To avoid contamination, such as from unreacted precursor materials, the salts were recrystallised, and single crystals were selected for starting samples. The crystals were dried sufficiently under a vacuum of about 10⁻² Pa.

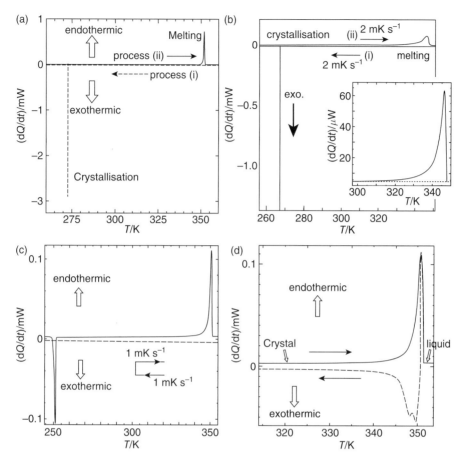

**Figure 3.3**  Overall DSC traces of (a) [C$_2$mim]Br, (b) [$^i$C$_3$mim] Br, and (c) [C$_4$mim]Br. (d) DSC curves for [C$_4$mim]Br around the melting point. The broken curve shows the trace where the sample is cooled on the way to the premelting process. The heating and cooling rate are 1 mK s$^{-1}$.

The DSC traces of [C$_2$mim]Br are shown in Figure 3.3a [37]. The sample was cooled from 360 K (liquid state) to 223 K, and then heated from 223 to 360 K at the scanning rate of 2 mK s$^{-1}$. The salt crystallised at 273 K in the cooling process (broken line) and melted at 352 K in the heating process (solid curve). Raman spectroscopic measurements revealed that [C$_2$mim]$^+$ in the crystalline state of the bromide is in the non-planar form, while both planar and non-planar conformers are mixed in the liquid and supercooled liquid states [27].

Figure 3.3b indicates the DSC traces of [$^i$C$_3$mim]Br obtained at the cooling and heating rates of 2 mK s$^{-1}$ [18]. The exothermic crystallisation peak appeared at about 267 K in the cooling process, Process (i), and the endothermic melting peak appeared at about 346 K in the heating process, Process (ii). As found for [C$_2$mim]Br, [$^i$C$_3$mim]Br crystallises in the cooling process and melts in the heating process. From this viewpoint, these two salts are more "normal" than [C$_4$mim]Br (discussed later). The crystallisation of [$^i$C$_3$mim]Br occurs at about 80 K lower than melting, indicating the stable existence of a supercooled state and that it is hard to crystallise. The same characteristic feature has been reported for many ionic liquids [11–17, 20, 21]. As shown in Figure 3.3b, the melting peak has a very broad range, extending beyond 20 K, while the crystallisation peak is sharp. The magnified drawing around the melting peak is illustrated in the inset in Figure 3.3b. Contamination of a sample commonly causes broadening of its DSC melting curve. To avoid the effect of contamination, several different single crystals were selected for study—the sample was pure enough to crystallise as single crystals. However, the peak is too broad to regard this melting as a simple process of loosening of the crystalline lattices. We conclude that this broadening is due to premelting, characteristic of these samples, as well as other ionic liquids. In comparison with a similar broad temperature range observed during the premelting of [C$_4$mim]Br [17], we assumed that the structural changes of the [$^i$C$_3$mim]$^+$ cations occurred during both the melting and crystallisation processes. Our Raman spectroscopic experiments [30] confirmed that the conformation of the [$^i$C$_3$mim]$^+$ ions in the crystalline state is **asym**, while **sym** and **asym** coexist in the liquid and supercooled liquid states and that a conformational change between the two occur in the premelting region, as expected.

The DSC traces for [C$_4$mim]Br are shown in Figure 3.3c [17]. In the measurement, crystalline [C$_4$mim]Br was first melted completely by heating, cooled down to 223 K as shown by broken line, and then heated up to 360 K as shown by a solid curve. The cooling and heating rates were 1 mK s$^{-1}$. No peak was found in the cooling process. However, in the heating process exothermic and endothermic peaks were found at about 250 and 350 K, respectively. That is, during the cooling process from the liquid, the liquid or supercooled liquid state continued down to the lowest temperature of the experiment (223 K) and crystallisation did not occur. In the heating process, crystallisation was observed at about 250 K and the crystal melted at about 350 K.

It is noted that the melting peak is broad, ranging over several Kelvin. This peak is too broad to regard this melting as a simple process. We thus suggest that this broadness is due to premelting, characteristic of the present sample. [C$_4$mim]Br appears as only one crystal form, in which the C1′-C2′-C3′ conformation of the butyl group is in **GT** [22], while both of the **GT** and **TT** conformations coexist in liquid and supercooled liquid states [26, 29]. As a result, some of the cations must change their conformation from **GT** to **TT** upon melting, and vice versa at the crystallisation. We consider that this premelting behaviour is due to the increase of some fluctuation in the libration around the C-C bonds (especially C1′-C2′) within the potential minimum. This triggers the cooperative changes of conformation in the butyl chain and leads to melting. Our Raman spectroscopic experiments confirmed that a part of the $[C_4mim]^+$ ions with **GT** conformation change in the premelting region [29].

From the magnified DSC curve around the crystallisation point (not shown here), it was noted that the crystallisation peak is divided into two. In Figure 3.3c, the specimen in the temperature region below the crystallisation peak is in a supercooled liquid phase, in which both of the **GT** and **TT** conformers coexist [26, 29]. On the other hand, $[C_4mim]Br$ in the temperature region above that of this peak is in the crystalline phase, where only the **GT** conformer exists [22, 26, 29]. Taking into consideration the structural change of the cation mentioned above, we interpret the observed split in the crystallisation peak as being due to the dynamics of crystallisation, namely the $[C_4mim]^+$ ions with the **GT** conformation and $Br^-$ anions crystallise initially, while the $[C_4mim]^+$ ions with the **TT** conformation do not form crystals until after they have changed conformation from **TT** to **GT**. The lower-temperature component of the peak can be assigned to the direct crystallisation of **GT** conformers, and the component at the higher temperature to the crystallisation of **TT** conformers after the change of conformation from **TT** to **GT**. Of course, these changes of the conformation must be cooperative to form a crystal. A large number of cations must cooperatively change their conformations. As a result, the dynamics of crystallisation of $[C_4mim]^+$ ions with the **TT** configuration is slow enough to be detected separately. As the experiment was performed at a constant heating rate of $1$ mK s$^{-1}$, the temperature difference of the two separate components corresponds to the time required for the cooperative change of conformers from **TT** to **GT**; it was about 5 minutes. We anticipated that the same change in the conformation occurs in the melting process. Indeed, when we reversed the heating process to cooling in the premelting region, we observed the split in the peak. Figure 3.3d shows the trace of the experiments, where the sample was first heated to a point in the premelting region (solid line), and then was cooled (broken line). The heating and cooling rates for the traces were $1$ mK s$^{-1}$. As is evident in the figure, the splitting of the peak was also observed. This splitting was more remarkable for the sample which was heated up nearly to the top of the melting curve. The peak of higher temperature is assigned to the direct crystallisation of the portion which locally melts with **GT** conformers and that of lower temperature corresponds to the crystallisation of the portion with **TT** conformers after the change of conformation to **GT**.

## 3.2  SUITABLE EQUIPMENT FOR THE THERMAL ANALYSIS OF IONIC LIQUIDS

Many types of equipment for thermal analyses are commercially available and give useful information regarding the thermal properties of samples. If the sample is an ionic liquid, however, the closest attention must be paid to the sensitivity/stability of the equipment, and to the capability of setting experimental conditions. This is because the thermal behaviour of ionic liquids is generally very complex, and thermal histories are frequently exhibited in their responses. Some phenomena occur at an unbelievably slow speed, which is thought to be related to the complex thermal histories of the ionic liquids. Without recognising the extremely slow changes, some thermal measurements of ionic liquids were performed at scanning speeds, which were much more rapid than their relaxation rates. This is thought to be one of the origins of confusion in thermal data reported for ionic liquids. Here, two laboratory-constructed scanning instruments are introduced, which are suitable for the thermal analyses of ionic liquids.

The first example is a nano-Watt-stabilised DSC instrument of the heat-flux type, which was designed and constructed by Tozaki [17, 31]. The design is based on the following concepts: thermoelectric modules are utilised for a heat-flux sensor (the Seebeck effect) and heat pumps (the Peltier effect), and the system adopted for temperature regulation is a type of predictive controller [38] instead of a proportional integral derivative (PID) controller. In this system, the amount of the current passed into the Peltier elements to supply or remove heat is predicted both from the difference between the actual sample temperature and the set temperature, and from the temperature history up to the starting point of regulation. We have programmed this procedure to repeat every 2 seconds. This controller causes no ripples because no on–off switching is used. By setting up multiple adiabatic boxes around a sample, this DSC instrument can measure a heat flow with a baseline fluctuation within $\pm 3$ nW. This baseline stability, which corresponds to the sensitivity of the DSC, is $10^2$–$10^3$ times higher than that of commercially available DSC equipment [31]. This apparatus is also designed to scan as slowly as $\sim 0.01$ mKs$^{-1}$, enabling us to mimic a nearly quasi-static process. This scanning speed corresponds to taking 28 hours to raise the temperature by 1 K. As the relaxation time of the apparatus is estimated to be $\sim 2$ seconds, the dynamics of a thermal process can be traced for a relaxation time exceeding 2 seconds. Measurements can be made during either process, heating or cooling, by changing the direction of the electric current through the thermoelectric modules. This performance is essential for thermal measurements on ionic liquids, since many of them often display different behaviour during the cooling and heating processes. The measurable temperature region of the DSC instrument is from 220 to 453 K, and the rate of cooling or heating is controllable in the range from 0.01 to 10 mK s$^{-1}$. The temperature of the sample was measured using a platinum resistance thermometer. The heat flow measurement was performed with a 1% margin. The temperature was controlled within $\pm 0.15$ mK.

The second example is the equipment for simultaneous measurements of DSC and Raman spectroscopy [29]. The outline of the apparatus is as follows. The laboratory-made calorimeter is combined with the commercially available Raman spectrometer. The spectrometer is a fibre optically coupled Raman spectrometer (Hololab, Kaiser Optical Systems) equipped with a GaAlAs diode laser (wavelength: 785 nm). The optical resolution is 4 cm$^{-1}$, and a spectrum in the range of 100 to 3450 cm$^{-1}$ can be measured simultaneously due to the adoption of a multiplex grating system backed up by a charge coupled device (CCD) camera. As for the calorimeter, thermoelectric modules are used both as a heat flux sensor and as a heat pump. A free piston stirring cooler is applied to obtain low temperatures, down to 153 K. Temperature is measured using a platinum resistance thermometer. The baseline stability of the calorimetry trace is measured as being 5 $\mu$W. One of the characteristics of the calorimeter is its capability of carrying out heating and/or cooling experiments at an extremely slow rate, sufficient for mimicking a quasi-static condition. From the standpoint of Raman spectroscopy, this apparatus can be regarded as a sample holder by which spectra at various temperatures ranging from 153 to 403 K can be obtained under sufficiently stabilised thermal conditions. From the standpoint of calorimetry, the apparatus gives us direct information on the structure change of the sample during accompanying thermal phenomena. Laser power was set at less than 10 mW for simultaneous measurements to observe the *in situ* thermal behaviour of ionic liquids.

## 3.3   THE PHASE BEHAVIOUR OF [C₄mim][PF₆]

The phase behaviour of [C₄mim][PF₆] is introduced here as an example of where complicated crystal–crystal transitions occur accompanied with various conformational changes of the butyl group in the [C₄mim]$^+$ cation [20]. [C₄mim][PF₆] is one of the most popular ionic liquids because it consists of representative cations and anions. There are hence reports on the liquid [39–41] and crystalline [42, 43] structures for [C₄mim][PF₆]. For the crystal structure, two groups independently obtained crystals belonging to the same crystalline phase, and reported that the butyl group of the cation is in the **G′T** conformation, as shown in Figure 3.1c [42, 43]. From thermodynamic studies, some groups indicated that [C₄mim][PF₆] has two crystal polymorphs [42, 44–48] and complex phase transition behaviour [42, 46, 48]. However, there was some confusion about the phase behaviour in previous reports.

We investigated the phase transition behaviour and cation structures of [C₄mim][PF₆] by calorimetry and Raman spectroscopy equipped with a precise temperature-control stage [20]. Consequently, we found that [C₄mim][PF₆] has three crystal polymorphs, and all the phase transitions except the glass transition occur accompanied by conformational changes of the butyl group around the C1′-C2′ axis.

### 3.3.1   Phase Transitions

Measurements were carried out using the apparatus described in Section 3.2, which made it possible to perform simultaneous measurements of calorimetry and Raman spectroscopy [29].

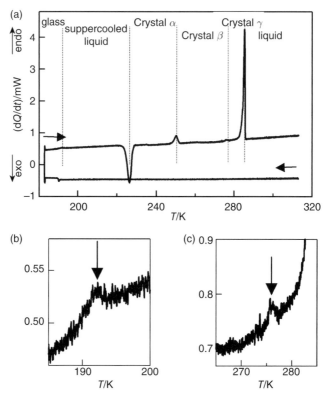

**Figure 3.4**    (a) Calorimetric curve for [C$_4$mim][PF$_6$] at a scanning rate of 5 mK s$^{-1}$ for the whole temperature range. Expanded curves: (b) around the glass transition and (c) around the phase transition from **Crystal** $\beta$ to **Crystal** $\gamma$.

Figure 3.4 shows calorimetric curves for [C$_4$mim][PF$_6$] in the range of 183–313 K. To follow and match the relatively slow thermodynamic changes of ionic liquids, the measurements were performed at a scan rate of 5 mK s$^{-1}$, which was much lower than the rates in typical calorimetric measurements. In the cooling process, no phase transition peak appeared except for the glass transition. On the other hand, several phase transition peaks were observed (Fig. 3.4) in the heating process, such as the glass transition (192 K), crystallisation (226.5 K), two crystal–crystal transitions (250.3 K, 276 K), and melting (285.3 K). The temperature values in parentheses are peak-top values. The temperature for the glass transition during heating is in good agreement with that obtained by adiabatic calorimetry, at 190.6 K [44]. We called the three observed crystalline phases **Crystal** $\alpha$, **Crystal** $\beta$, and **Crystal** $\gamma$ in order of increasing temperature.

We focus now on the crystal–crystal phase transitions. As mentioned earlier, there are two endothermic peaks at 250.3 and 276 K in the calorimetric curve between the crystallisation and melting peaks. Most other groups detected only the signal corresponding to our larger peak at 250.3 K, and they reported that [C$_4$mim][PF$_6$] had two crystalline phases [42, 47]. We think that the peak at 276 K is difficult to detect by DSC measurements with typical scanning rates because it is a very small peak that appears close to the melting peak, and the phase change seems to occur very slowly. Although the signal at 276 K is very small, we can clearly distinguish two phases under our experimental conditions, and the two phases have different structures, as shown in the next section.

### 3.3.2   Cation Structure in Each Phase

The structure of the [C$_4$mim]$^+$ ion in each crystalline phase was studied by Raman spectroscopy [20]. The results are shown in Figure 3.5a. Raman bands in the range of 580–640 cm$^{-1}$ are known as marker bands for rotational isomers of the butyl group in the [C$_4$mim]$^+$ ion [23, 26]. The existence of nine rotational isomers of [C$_4$mim]$^+$ was demonstrated by quantum chemical calculations [35, 36]. Taking into account both energy differences among the isomers and results from single crystal structure analyses for [C$_4$mim]$^+$ salts [20, 24, 42, 43, 49, 50], we can pick three rotational isomers as possible conformers for [C$_4$mim]$^+$ in the crystal or liquid states, that is, **GT**, **TT**, and **G′T** conformations caused by the rotation of the butyl group around the C1′-C2′ axis. The three isomers are shown in Figure 3.1c, and the corresponding active Raman bands from DFT calculations are displayed in Figure 3.5b,c,d.

   **Crystal** $\beta$ has a distinguishable band at 624 cm$^{-1}$, which can be assigned to the **TT** conformation of [C$_4$mim]$^+$, Figure 3.5c [23, 26]. In the region of 580–640 cm$^{-1}$, the Raman spectrum of **Crystal** $\alpha$ is similar to that of **Crystal** $\gamma$. They have characteristic peaks at about 500, 600, and 700 cm$^{-1}$, suggesting that the cation structures in these crystalline phases will be assigned to **GT**, Figure 3.5b [23], or **G′T**, Figure 5d, conformations. The difference between **Crystal** $\alpha$ and **Crystal** $\gamma$ is found in the region of 300–350 cm$^{-1}$. From the DFT calculations for [C$_4$mim]$^+$, the Raman bands observed at higher frequency (338 cm$^{-1}$) and lower frequency (326 cm$^{-1}$) are attributed to the **GT** and **G′T** conformations, respectively. As a result, **Crystal** $\alpha$ can be assigned to **GT**, and **Crystal** $\gamma$ to **G′T**.

   It should be noted that a small component due to the **G′T** conformation exists in **Crystal** $\alpha$. We studied the temperature dependence of the Raman scattering intensities of **Crystal** $\alpha$. The upper spectrum in Figure 3.5a was measured at 229 K after crystallisation during heating. The second spectrum was obtained from **Crystal** $\alpha$ at 193 K during cooling. As expected, no phase transition occurred in the cooling process. If **GT** and **G′T** conformers coexist in an asymmetric unit in **Crystal** $\alpha$, the intensity ratio of the bands at higher (338 cm$^{-1}$) and lower (326 cm$^{-1}$) frequencies will not change. However, the band intensity at 326 cm$^{-1}$ originating from **G′T** clearly decreases with falling temperature. This indicates that the most stable conformer in **Crystal** $\alpha$ is **GT**, and we conclude that the conformation of the cation in **Crystal** $\alpha$ is **GT**. We

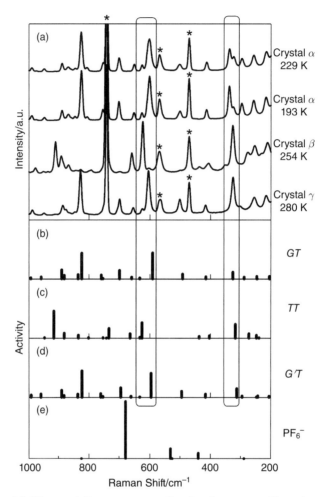

**Figure 3.5** (a) Observed Raman spectra for the three crystalline phases (asterisks indicate the anion bands). (b)–(d) Raman active bands, calculated with DFT, for the **GT**, **TT**, and **G′T** conformations of [C₄mim]⁺, respectively. (e) Raman active bands, calculated with DFT, for [PF₆]⁻.

consider two possibilities for the origin of the appearance of **G′T**: (1) the rotational isomerism reaction between **GT** and **G′T** easily occurs in **Crystal α** (i.e., there is a low activation energy for rotational isomerism), or (2) part of the **G′T** conformer was left as amorphous solid or liquid without crystallisation. We often observe the coexistence of crystals and supercooled liquid in ionic liquids. Although not reported yet, we have recently found from a combination measurements of nuclear magnetic resonance (NMR) and Raman spectroscopy that **Crystal α** changes to **Crystal γ** even at *ca.* 227 K (crystallisation temperature from the supercooled liquid) when **Crystal α** is kept for a long time (such as one day). This observation supports possibility (1) for the origin. This phenomenon shows that the phase transitions of ionic liquids are in general extremely slow and that our observation of their thermal behaviour is strongly dependent on the thermal histories of the samples.

The $[C_4mim]^+$ ion in the $[C_4mim][PF_6]$ crystalline phase was reported to take the **G′T** conformation from single crystal structure analyses at 173 K [43] and 193 K [42]. These results seem to be inconsistent with our finding because the **G′T** conformer appears in **Crystal** γ, which is the highest-temperature crystalline phase. However, this apparent inconsistency can be resolved by the complex thermal history of $[C_4mim][PF_6]$ [46]. Moreover, we measured Raman spectra of **Crystal** γ at 193 K (not shown), which were obtained by cooling from **Crystal** γ with no phase transition. Two research groups grew single crystals by different methods [42, 43]. We suppose that the crystals they first made were **Crystal** γ and that their structures were retained as the temperatures were lowered to the measurement temperatures without a phase transition.

Figure 3.6 shows the temperature dependence of Raman spectra during melting. This region includes the phase transition from **Crystal** γ (**G′T**) to liquid. The band at 624 cm$^{-1}$ that is assigned to the **TT** conformation arises through melting, and behaviour such as the appearance or disappearance of bands characteristic to each conformation were observed in other phase transition regions. Therefore, we can conclude that all phase transitions of $[C_4mim]$ $[PF_6]$, except for the glass transition, accompany conformational change of the butyl group. It is noted that structural change in the cation becomes abruptly active just near the peak-top temperature of melting, although the premelting

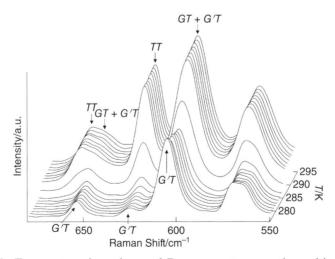

**Figure 3.6** Temperature dependence of Raman spectra near the melting point of $[C_4mim][PF_6]$. The scanning rate was 5 mK s$^{-1}$, and each spectrum was obtained every 1 K.

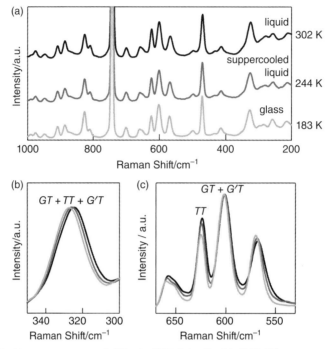

**Figure 3.7**    Raman spectra of $[C_4mim][PF_6]$ in liquid (302 K), supercooled liquid (244 K), and glass states (183 K) in the ranges of (a) 200–1000 cm$^{-1}$, (b) 300–350 cm$^{-1}$, and (c) 530–670 cm$^{-1}$.

phenomenon starts at a temperature 4–5 K lower. The same tendencies were observed in the melting processes of $[C_4mim]Br$ [29] and $[^iC_3mim]Br$ [30]. In our reported results on ionic liquids [18, 29, 30], the conformational changes of constituent ions seem to be delayed after the start of the thermal signal of the phase transitions.

Figure 3.7a shows the Raman spectra of $[C_4mim][PF_6]$ in the liquid (at 302 K), supercooled liquid (at 244 K in the cooling process), and glass (at 183 K) states [20]. These spectra resemble each other, and this demonstrates that the three states are composed of cations with similar structure. According to the Raman spectra in the ranges of 300–350 cm$^{-1}$, (Fig. 3.7b) and 530–670 cm$^{-1}$ (Fig. 3.7c), the **GT**, **TT**, and **G'T** conformers are considered to be mixed in these states, as in other 1-alkyl-3-methylimidazolium ionic liquids.

Many crystal structures have been reported for a range of $[C_4mim]^+$ salts [22, 24, 42, 43, 49–51]. Some typical data are summarised in Table 3.1. The cation conformations in these crystal structures are concentrated in the **GT** conformation. Our Raman scattering study shows the existence of three types of crystalline phase for $[C_4mim][PF_6]$, where conformations of the cation differ. No report has been made of 1-butyl-3-methylimidazolium ionic liquids with

**TABLE 3.1    Cation Conformations in Crystal Structures of [C₄mim]X Obtained by X-ray Diffraction.**

| Anion | $Cl^{22, 24}$ | $Br^{22}$ | $I^{50}$ | $Tf_2N^{49}$ | $CH_3SO_3^{51}$ | $CH_3SO_4^{51}$ | $PF_6^{42, 43}$ |
|---|---|---|---|---|---|---|---|
| Conformation | **TT, GT** | **GT** | **GT** | **GT + G′T** | **GT** | **GT** | **G′T** |

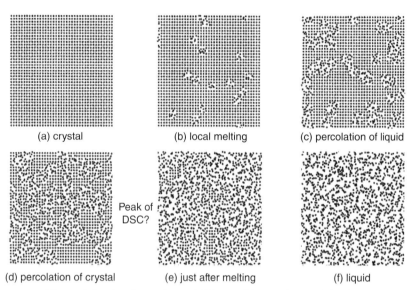

(a) crystal          (b) local melting          (c) percolation of liquid

Peak of DSC?

(d) percolation of crystal          (e) just after melting          (f) liquid

**Figure 3.8**    A schematic model for melting and crystallisation.

all three conformations, **GT**, **TT**, and **G′T**, in their measured crystal strucutures. [C₄mim][PF₆] is one of the most representative examples showing the conformational variety of ionic liquids. At the same time, the complex isomerisation of the constituent cations has caused some confusion regarding the phase behaviour and thermodynamic properties of [C₄mim][PF₆].

## 3.4    NOVEL PHASE TRANSITION BEHAVIOURS OF ROOM TEMPERATURE IONIC LIQUIDS

### 3.4.1    A Model for Melting and Crystallisation

Following our results on [C₄min]Br [17, 29] and [ⁱC₃mim]Br [18, 30], we have modelled the melting and crystallisation behaviour of ionic liquids, and the schematic diagram of the model is shown in Figure 3.8. In a crystalline state, ions with a certain specific conformation are arranged regularly (Fig. 3.8a). In the heating process, small domains in a crystalline area start to melt first, accompanied by conformational changes of the ions, probably at the regions with crystalline defects or at the surface/boundary of the crystals (Fig. 3.8b); specifically, local melting occurs and the regions are composed of mixtures of different conformers. These premelting domains grow as the temperature is increases (Fig. 3.8c) until percolation of the crystal areas finally breaks off (Fig. 3.1d). Up to such a temperature, local melting and crystallisation in the pre-melting region can occur reversibly by heating or cooling [17, 18]. We may assume that many small crystalline domains remain even after the melting (Fig. 3.8e), where the DSC trace falls to the baseline level of the liquid. The reverse phenomena will be observed in the crystallisation process from the liquid.

The above-mentioned processes are considered to occur not only in ionic liquids, but also in "normal" substances. However, the phenomena in ionic liquids can be detected with the supersensitive DSC equipment because their phase transitions are extremely slow.

### 3.4.2    Rhythmic Melting and Crystallisation

We performed detailed measurements, focussing on the melting process of [C$_4$mim]Br [32]. As shown in Figure 3.3c, the salt has a wide premelting region. The single crystal melted perfectly, cooled down to room temperature, and in turn heated at the rate of 0.02 mK s$^{-1}$ from room temperature to 355 K. This heating rate corresponds to 13.9 hours to raise the temperature of the sample by 1 K.

The melting trace of [C$_4$mim]Br is shown in Figure 3.9a, where the data collected every 40 seconds are plotted as dots. As shown in the figure, the DSC curve is very noisy. Selecting five representative regions in the curve, we num-bered the divisions around 344.5, 345.9, 349.6, 351.2, and 353.2 K as (1), (2), (3), (4), and (5), respectively. Divisions (1) and (2) correspond to the regions near and just on the starting point of premelting, respectively.Division (3) cor-responds to the middle point of the premelting curve, and the trace is noisiest in this region. Division (4) is the point where the crystal just melts. Division (5) corresponds to the region where the sample is stabilised as the liquid state. Figure 3.9b and Figure 3.9c are expanded representations of Divisions (3) and (5). Each dot in the expanded plots refers to the measuring point, which is carried out every 4 seconds. Therefore, the lateral axes correspond to time axes as well as temperature axes.

We have made frequency analyses for 5000 points in each Division, by drawing the baseline and measuring the deviation of each peak-top from the base line. As for Divisions (1) and (5), the distribution curves by the frequency analysis are regarded almost perfectly as Gaussian functions. And the half widths at 1/e of the peak-tops are about 3–4 nW. Thus, the deviations of Divi-sions (1) and (5) represent statistical thermal/electronic noise of the apparatus, and just refer to the sensitivity or resolution of the DSC equipment [31]. On the other hand, other curves slightly differ from the Gaussian distributions,

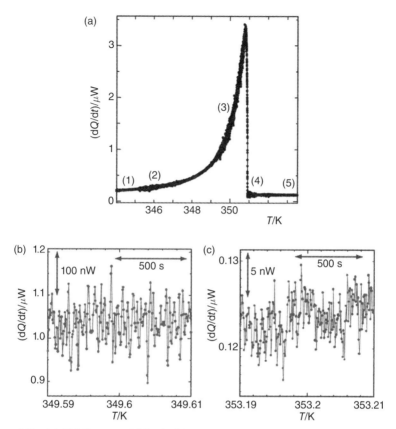

**Figure 3.9**   (a) DSC trace of [C$_4$mim]Br near the melting point. The heating rate is 0.02 mK s$^{-1}$. Five divisions are selected and numbered as (1)–(5). (b) Expanded figure of the division (3). (c) Expanded figure of the division (5). Dots in (b) and (c) refer to the measuring points of every 4 seconds.

and the widths are wider than the width due to intrinsic noises. Therefore, we can conclude that the states of Divisions (2), (3), and (4) are not in thermo-dynamic equilibrium, and the deviation is not noise, and has some physical meanings.

To understand the present results, the two following facts should be borne in mind. First, the premelting with over 10 K accompanies the conformational change of the butyl group from **GT** to **TT**, as well as the changes in the crystal-line lattice [17, 29]. Second, *ab initio* calculations showed that the energy dif-ference between **GT** and **TT** conformations is within ±1 kJ mol$^{-1}$ [35, 36]. Such a small energy difference implies that the conformational change for a free ion occurs easily. Therefore, the Divisions (2) and (3) will denote that, by the local temperature increase due to the fluctuation of the heating, a part of the

$[C_4mim]^+$ ions change their conformation from **GT** to **TT** in a domain and the domain melts, corresponding to Figure 8b. As only a small thermal energy is supplied because of the extremely slow heating rate, the thermal energy is absorbed by the melting domain itself, and the neighbouring one, which is just going to melt. The crystallisation of the domain then occurs. In turn, thermal energy is released by the crystallisation. Using this energy and the supplied energy from the thermoelectric modules, a part of the crystalline domain melts. These processes can be repeated rhythmically. The rhythmic melting and crystallisation appear as "noise" on the DSC trace, as shown in Figure 3.9a,b. Just after melting, namely in Division (4) in Figure 3.9a, the aggregation domains where most of the ions are in the **GT** conformation are going to crystallise again, or a great number of ions are going to change their conformations between **TT** and **GT**. In other words, the "noisy" trace of Division (4) indicates that the state is non-equilibrium. Ions in the domains are going to rhythmically crystallise or melt even at higher temperature than the melting one. We should note that the DSC pattern (frequency or heat flow magnitude) of the change at just the melting point, Division (4), is almost the same as the pattern at the starting point of the premelting shown in Division (2).

The number of ion pairs of $[C_4mim]Br$ in the domains can be estimated from the frequency distribution curves. The total enthalpy of the melting is reported to be $23.0-23.6$ kJ mol$^{-1}$ [17]. If we assume that each peak or valley in the expanded figures refers to the melting or crystallising of a domain, and regard the half-width of the distribution curve at $1/e$ as the averaged value of the heat flow, the number of ion pairs of $[C_4mim]^+$ and $Br^-$ in a domain is $2.0 \times 10^{12}$ for Divisions (2) and (4), and $1.8 \times 10^{13}$ for Division (3). This number of ion pairs corresponds to a radius of $4.9 \times 10^3$ nm for Divisions (2) and (4), and $1.0 \times 10^4$ nm for Division (3), if the domains are assumed to be spheres. This study was the first to estimate the domain size relating to the melting or crystallisation of ionic liquids [32].

From the expanded curves, the timescale of melting/crystallisation of the domain can be obtained, because the interval of measured points in the expanded figures corresponds to 4 seconds and 0.08 mK. For Divisions (1) and (5), the widths of the oscillatory peaks are randomly distributed in the range of $4-60$ seconds. On the other hand, the widths of the corresponding peaks of Divisions (2), (3), and (4) are similar to each other; the half-width values of the equivalent peaks in (2), (3), and (4) are $12-16$ seconds, $12-20$ seconds, and $8-12$ seconds, respectively. These values correspond to the timescales of the rhythmic transition of the domains. It is reported that several minutes are necessary to achieve equilibrium for the portions of the two conformers of **GT** and **TT** in $[C_4mim]Cl$ by the instant melting [1]. The long relaxation time of the present phase transition could be due to the accompanying cooperative conformational change of a large number of ions, of the order of $10^{12}-10^{13}$. As shown here, a supersensitive thermodynamic measurement is a very effective method to detect such slow dynamic behaviour involving heat flow. The present heating rate matches well to the slow dynamics.

This curious phenomenon was named "rhythmic melting and crystallisation" [32], which was observed in the premelting region in the heating process of [C$_4$mim]Br. Melting and crystallisation are periodically repeated with the endothermic and exothermic heat transfer cycles that alternately trigger subsequent changes. A rhythmic change of volume phase transition in polymer gels is well known [52]. Because the structures of the ions are relatively simple compared with polymers, the observed phase transition in an ionic liquid is unique. We think that the main origin of the rhythmic melting and crystallisation of the ionic liquid is a coupling of the melting/crystallisation and cooperative conformational change of the ions.

### 3.4.3  Intermittent Crystallisation

The next sample to be considered is [$^i$C$_3$mim]Br [18, 33], the overall DSC traces of which are shown in Figure 3.3b. To investigate the details of the premelting phenomenon of the sample, we performed cyclic experiments. Namely, the sample was heated at the rate of 2 mK s$^{-1}$ from room temperature to the preset temperature, Process (1), and then cooled down, Process (2). The temperature, set at a specific point in the premelting region, is hereafter denoted as the "returning temperature." We performed numerous experiments varying the returning temperature and the cooling rate, and some novel phenomena were observed depending on the experimental conditions. Two of them are introduced here.

When the returning temperature was the peak-top temperature of the melting DSC trace and the cooling rate was set at 0.05 mK s$^{-1}$, a very curious phenomenon was observed. The result is shown in Figure 3.10a, where the

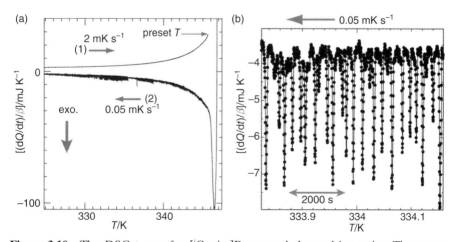

**Figure 3.10**  The DSC traces for [$^i$C$_3$mim]Br around the melting point. The curve shows the trace wherein the sample was cooled down from 346.0 K. (a) Full-scale results, and (b) a magnified section of data around 334 K in the cooling run.

values of the vertical axis are divided by the heating or cooling rate ($\beta$) to normalise the rate differences. The DSC signals of the reversible crystallisation traced simple curves with accompanying fine exothermic spikes. The magnified drawing around 334 K is shown in Figure 3.10b. The small peaks were sharp, and were separated piece by piece. Moreover, all of them were exothermic peaks. In the latter figure, the data are plotted as dots every 4 seconds. As emphasised in our papers [18, 32, 33], these notched peaks are not due to electronic noise, but intrinsically arise from the present sample. We have named this phenomenon "intermittent crystallisation" [33]. The present DSC instrument, with super-high sensitivity and quick response times, has enabled us to detect this phenomenon. These results reveal that there exist local melting domains in the premelting region, as indicated in the schematic model, shown in Figure 3.8b, and that the domains with almost the same order in size are spatially separated from each other at about 334 K. The spikes refer to the crystallisation of the domains.

For [$^i$C$_3$mim]Br, more than 90% of the crystalline area seemed to melt up to the returning temperature, as estimated from the amount of absorbed heat. Being cooled down suddenly, most of the [$^i$C$_3$mim]$^+$ and Br$^-$ ions appear to be smoothly crystallised, because the DSC trace is smooth just after returning, as shown in Figure 3.10a. The exothermic spikes appear after 70–80% release of the total enthalpy of crystallisation. This behaviour implies that a large number of [$^i$C$_3$mim]$^+$ ions in the premelting region maintained an **asym** conformation (i.e., the conformation in the crystalline phase) that leads to smooth crystallisation. This speculation is supported by our experimental studies involving simultaneous measurements of Raman spectroscopy and calorimetry [29]. The results for [$^i$C$_3$mim]Br [30] and [C$_4$min]Br [29] showed that a major portion of the conformational changes occurred at the temperature at the peak-top of the melting DSC trace, or just after this point. In the cooling process of [$^i$C$_3$mim]Br, these domains seemed to remain where **sym** and **asym** conformers are mixed together, suggesting that **sym** isomers cannot easily turn to an **asym** conformation.

We estimated the order of enthalpy for a domain in the intermittent crystallisation to be $0.7 \times 10^{-5}$ J. Comparing with the molar crystallisation enthalpy of 13.9 kJ mol$^{-1}$, the average number of ion pairs in the domains was then estimated to be $\sim 3 \times 10^{14}$. The timescale of crystallisation can be estimated from Figure 3.10b. The interval of intermittent crystallisation was about 200–230 s, and about 100 s is required for each domain to crystallise under the present experimental conditions. These values are much longer than the relaxation time of the apparatus.

Once the returning temperature exceeded the peak-top temperature of the DSC trace, the observed phenomena altered drastically. First, we confirmed that the sample cooled from the temperature exceeding the peak-top did not entirely crystallise in the premelting region but overall crystallisation occurred at about 267 K, as shown in Figure 3.3b. This demonstrates that the upper limit for the reversible melting and crystallisation is nearly at the peak-top temperature.

We discovered another unique phenomenon in the measurement performed at the returning temperature of 346.5 K and with a cooling rate of 0.05 mK s$^{-1}$. This temperature was just 0.2 K higher than the peak-top temperature. As shown in Figure 3.3b, at this temperature the melting curve did not drop to the baseline of the liquid state, meaning that the sample did not become liquid completely. The overall DSC traces of this experiment are shown in Figure 3.11a. As shown in this figure, a small exothermic peak first appeared in the cooling process. This seems to be assigned to the crystallisation of a very small portion. Crystalline grains remain at this temperature, and **asym** conformers, in a quantity enough to form crystals, remain [30]. However, the crystalline grains cannot grow up to become an overall crystalline state. We consider that this is because the percolation of the crystalline area breaks off — see the schematic model shown in Figure 3.8d. It is speculated that the flat region of the baseline in the cooling process refers to the coexisting state of supercooled liquid and crystalline grains. Although this may be thought to be a strange phenomenon, we often observe the stable coexistence of supercooled liquid and crystals for many ionic liquids.

Although overall crystallisation did not occur in the premelting region, the DSC trace was very noisy in some temperature ranges, as shown in Figure 3.11a. The magnified drawing around 340.52 K is shown in Figure 3.11b, where data for every 4 seconds are plotted as dots. The noise intrinsic to the apparatus was $\pm 3$ nW. The observed spikes were about 10 times larger than the 6-nW

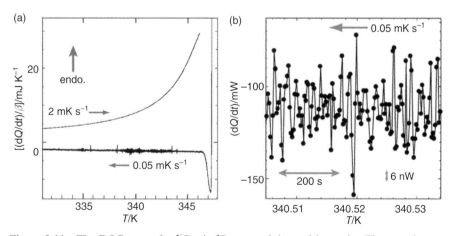

**Figure 3.11**    The DSC traces for [$^i$C$_3$mim]Br around the melting point. The sample was cooled down with a cooling rate of 0.05 mK s$^{-1}$ from a temperature of 0.2 K higher than the peak-top of the melting trace. **(a)** Full-scale data set, and **(b)** magnified section around 340.52 K.

margin shown by the arrow, and indicated that they were attributed to a thermal phenomenon characteristic of the sample. It is clear that the pattern of periodic endothermic and exothermic heat transfers is different from the pattern of intermittent exothermic heat transfers. The notched peaks are rhythmic crystallisation and melting, and we observed the same phenomenon for [$C_4$mim]Br in the premelting region and in the liquid region just after melting during the heating process, as described in Section 3.4.2. At the temperature slightly exceeding the peak-top for the case of [$^i$C$_3$mim]Br, it is thought that crystalline domains remain surrounded by a large amount of liquid areas and that the percolation of the crystalline area breaks off. However, it is supposed that thermal fluctuations easily occur around the remaining crystalline domains. The melted region around these remaining crystalline grains crystallises as cooling occurs. The heat released during crystallisation melts the neighbouring domains. Due to the endothermic process of melting, crystallisation occurs in turn. Crystallisation and melting are thereby repeated rhythmically. These were observed as the periodic spikes in the DSC trace, as shown in Figure 3.11a,b.

In the domains where the rhythmic crystallisation and melting occurred, $5–9 \times 10^{12}$ ion pairs were contained. The timescale of the phenomenon was estimated to be 10–20 seconds. Compared with the intermittent crystallisation where the returning temperature was preset at peak-top temperature (Fig. 3.10b), the released or absorbed heat in the rhythmic crystallisation and melting was smaller, and the timescale was correspondingly shorter. These results may be contrary to the expectation that the premelting domains should become larger in the returning experiment from 346.5 K, as seen in Figure 3.11a. However, as for the sample heated up to the temperature exceeding the peak-top, a large number of [$^i$C$_3$mim]$^+$ ions changed to a **sym** conformation [30], and they mixed with the **asym** ions. As a result, it seems to be more difficult for larger domains to crystallise during the extremely slow cooling process. On the other hand, even a small amount of the heat released during local crystallisation will easily melt the small crystallising domains because of the large mixture of **sym** and **asym** conformations. This seems to be the reason for rhythmic crystallisation and melting occurring in the smaller size regions and with a shorter periodicity. By repeating rhythmic crystallisation and melting, the whole sample seems to relax into the supercooled liquid state.

## 3.5  CONCLUDING REMARKS

The thermal behaviour of some 1-alkyl-3-methylimidazolium ionic liquid is presented, with a focus on their phase transitions. In addition to the common thermal phenomena for most ionic liquids, such as premelting over a wide temperature range and excessive supercooling, the following unique thermal phenomena were observed:

1. Existence of thermal histories affecting their thermal behaviours.
2. Frequent existence of complex crystal–crystal phase transitions and/or polymorphism.
3. Extremely slow phase transitions.
4. Curious phenomena in phase transitions such as "rhythmic melting and crystallisation" and "intermittent crystallisation."

These phenomena are generally observed for most ionic liquids, so the following discussion can be applied to most ionic liquids. Although the above-mentioned phenomena (1–4) are different, the origins of these behaviours must be closely related; that is, the conformational changes in the dense field are likely to affect the phase transitions and cause these phenomena. Multiple conformers for a constituent ion commonly exist in ionic liquids. Energies of some of the conformers are almost the same, and the small energy differences among them are strongly influenced by the types of counter ion and by interactions resulting from the relative positions and orientations of ions. In the liquid state, coexistence of multiple conformers is entropically profitable. Therefore, most phase transitions of ionic liquids link with the conformational changes of constituent ions. This link causes complex and curious phase transitions. Due to the Coulomb interactions, the density of an ionic liquid is generally 10–20% higher than that of a molecular substance with similar components. The decrease in the free volume for conformational changes is probably the major origin of the retarded phase transition for ionic liquids.

## REFERENCES

1  Hamaguchi, H., and Ozawa, R., Structure of ionic liquids and ionic liquid compounds: Are ionic liquids genuine liquids in the conventional sense?, *Adv. Chem. Phys.* **131**, 85–104 (2005).

2  Wishart, J.F., and Castner, E.W. Jr. (eds.), *Physical Chemistry of Ionic Liquids*, Special Issue of *J. Phys. Chem. B* **111** (18), 4639–5029 (2007).

3  Rogers, R.D., and Voth, G.A. (eds.), *Ionic Liquids*, Special Issue of *Acc. Chem. Res.* **40** (11), 1077–1236 (2007).

4  Castner, E.W. Jr., and Wishart, J.F., Spotlight on ionic liquids *J. Chem. Phys.* **132**, Article ID 120901 (2010).

5  Welton, T., Room-temperature ionic liquids. Solvents for synthesis and catalysis, *Chem. Rev.* **99**, 2071–2083 (1999).

6  Rogers, R.D., and Seddon, K.R. (eds.), *Ionic Liquids — Industrial Applications for Green Chemistry*, ACS Symposium Ser., Vol. 818, American Chemical Society, Washington DC (2002).

7  Wasserscheid, P., and Welton, T. (eds.), *Ionic Liquids in Synthesis*, VCH-Wiley, Weinheim (2003).

8  Ohno, H. (ed.), *Electrochemical Aspects of Ionic Liquids*, John Wiley & Sons, Hoboken (2005).

9   Plechkova, N.V., and Seddon, K.R., Applications of ionic liquids in the chemical industry, *Chem. Soc. Rev.* **37**, 123–150 (2008).

10  Greaves, T.L., and Drummond, C.J., Protonic ionic liquids: Properties and applications, *Chem. Rev.* **108**, 206–237 (2008).

11  Holbrey, J.D., and Seddon, K.R., The phase behaviour of 1-alkyl-3-methylimidazolium tetrafluoroborates; ionic liquids and ionic liquid crystals, *J. Chem. Soc., Dalton Trans.* **13**, 2133–2139 (1999).

12  Ngo, H.L., LeCompte, K., Hargens, L., and McEwen, A.B., Thermal properties of imidazolium ionic liquids, *Thermochim. Acta* **357–358**, 97–102 (2000).

13  Huddleston, J.G., Visser, A.E., Reichert, W.M., Willauer, H.D., Broker, G.A., and Rogers, R.D., Characterization and comparison of hydrophilic and hydrophobic room temperature ionic liquids incorporating the imidazolium cation, *Green Chem.* **3**, 156–164 (2001).

14  Marsh, K.N., Boxall, J.A., and Lichtenthaler, R., Room temperature ionic liquids and their mixtures—a review, *Fluid Phase Equilibr.* **219**, 93–98 (2004).

15  Nishida, T., Tashiro, Y., and Yamamoto, M., Physical and electrochemical properties of 1-alkyl-3-methylimidazolium tetrafluoroborate for electrolyte, *J. Fluorine Chem.* **120**, 135–141 (2003).

16  Fredlake, C.P., Crosthwaite, J.M., Hert, D.G., Aki, S.N.V.K., and Brennecke, J.F., Thermophysical properties of imidazolium-based ionic liquids, *J. Chem. Eng. Data* **49**, 954–964 (2004).

17  Nishikawa, K., Wang, S., Katayanagi, H., Hayashi, S., Hamaguchi, H., Koga, Y., and Tozaki, K., *J. Phys. Chem. B* **111**, 4894–4900 (2007).

18  Nishikawa, K., Wang, S., Endo, T., and Tozaki, K., *Bull. Chem. Soc. Jpn.* **82**, 806–812 (2009).

19  Triolo, A., Russina, O., Fazio, B., Appetecchi, G.B., Carewska, M., and Passerini, S., Nanoscale organization in piperidinium-based room temperature ionic liquids, *J. Chem. Phys.* **130**, Article ID 164521 (2009).

20  Endo, T., Kato, T., Tozaki, K., and Nishikawa, K., Phase behaviours of room temperature ionic liquid linked with cation conformational changes: 1-Butyl-3-methylimidazolium hexafluorophosphate, *J. Phys. Chem. B* **114**, 407–411 (2010).

21  Endo, T., Kato, T., and Nishikawa, K., Effects of methylation at the 2 position of the cation ring on phase behaviours and conformational structures of imidazolium-based ionic liquids, *J. Phys. Chem. B* **114**, 9201–9208 (2010).

22  Holbrey, J.D., Reichert, W.M., Nieuwenhuyzen, M., Johnston, S., Seddon, K.R., and Rogers, R.D., Crystal polymorphism in 1-butyl-3-methylimidazolium halides: Supporting ionic liquid formation by inhibition of crystallization, *Chem. Commun.* 1636–1637 (2003).

23  Hayashi, S., Ozawa, R., and Hamaguchi, H., Raman spectra, crystal polymorphism, and structure of a prototype ionic-liquid [bmim]Cl, *Chem. Lett.* **32**, 498–499 (2003).

24  Saha, S., Hayashi, S., Kobayashi, A., and Hamaguchi, H., Crystal structure of 1-butyl-3-methylimidazolium chloride. A clue to the elucidation of the ionic liquid structure, *Chem. Lett.* **32**, 740–741 (2003).

25  Berg, R.W., Raman spectroscopy and *ab-initio* model calculations on ionic liquids, *Mon. Chem.* **138**, 1045–1075 (2007), and references cited therein.

26  Ozawa, R., Hayashi, S., Saha, S., Kobayashi, A., and Hamaguchi, H., Rotational Isomerism and structure of the 1-butyl-3-methylimidazolium cation in the ionic liquid state, *Chem. Lett.* **32**, 948–949 (2003).

27  Umebayashi, Y., Fujimori, T., Sukizaki, T., Asada, M., Fujii, K., Kanzaki, R., and Ishiguro,S.,Evidence of conformational equilibrium of 1-ethyl-3-methylimidazolium in its ionic liquid salts: Raman spectroscopic study and quantum chemical calculations, *J. Phys. Chem. A* **109**, 8976–8982 (2005).

28  Berg, R.W., Deetlefs, M., Seddon, K.R., Shim, I., and Thompson, J.M., Raman and ab initio studies of simple and binary 1-alkyl-3-methylimidazolium ionic liquids, *J. Phys. Chem. B* **109**, 19018–19025 (2005).

29  Endo,T.,Tozaki, K., Masaki, T., and Nishikawa, K., Development of apparatus for simultaneous measurements of Raman spectroscopy and high-sensitivity calorimetry, *Jpn. J. Appl. Phys.* **47**, 1775–1779 (2008).

30  Endo, T., and Nishikawa, K., Isomer populations in liquids for 1-isopropyl-3-methyl-imidazolium bromide and its iodide, and their conformational changes accompanying the crystallizing and melting processes, *J. Phys. Chem. A* **112**, 7543–7550 (2008).

31  Wang, S., Tozaki, K., Hayashi, H., and Inaba, H., Nano-watt stabilized DSC and ITS applications, *J. Therm. Anal. Cal.* **79**, 605–613 (2005).

32  Nishikawa, K., Wang, S., and Tozaki, K., Rhythmic melting and crystallizing of ionic liquid 1-butyl-3-methylimidazolium bromide, *Chem. Phys. Lett.* **458**, 88–91 (2008).

33  Nishikawa, K., and Tozaki, K., Intermittent crystallization of an ionic liquid: 1-Isopropyl-3-methylimidazolium bromide, *Chem. Phys. Lett.* **463**, 369–372 (2008).

34  Frisch, M.J., Trucks, G.W., Schlegel, H.B., Scuseria, G.E., Robb, M.A., Cheeseman, J.R., Montgomery, J.A. Jr., Vreven, T., Kudin, K.N., Burant, J.C., Millam, J.M., Iyengar, S.S., Tomasi, J., Barone, V., Mennucci, B., Cossi, M., Scalmani, G., Rega, N., Petersson, G.A., Nakatsuji, H., Hada, M., Ehara, M., Toyota, K., Fukuda, R., Hasegawa, J., Ishida, M., Nakajima, T., Honda, Y., Kitao, O., Nakai, H., Klene, M., Li, X., Knox, J.E., Hratchian, H.P., Cross, J.B., Adamo, C., Jaramillo, J., Gomperts, R., Stratmann, R.E., Yazyev, O., Austin, A.J., Cammi, R., Pomelli, C., Ochterski, J.W., Ayala, P.Y., Morokuma, K., Voth, G.A., Salvador, P., Dannenberg, J.J., Zakrzewski, V.G., Dapprich, S., Daniels, A.D., Strain, M.C., Farkas, O., Malick, D.K., Rabuck, A.D., Raghavachari, K., Foresman, J.B., Ortiz, J.V., Cui, Q., Baboul, A.G., Clifford, S., Cioslowski, J., Stefanov, B.B., Liu, G., Liashenko, A., Piskorz, P., Komaromi, I., Martin, R.L., Fox, D.J., Keith, T., Al-Laham, M.A., Peng, C.Y., Nanayakkara, A., Challacombe, M., Gill, P.M.W., Johnson, B., Chen, W., Wong, M.W., Gonzalez, C., and Pople, J.A., *Gaussian 03* (2004).

35  Turner, E.A., Pye, C.C., and Singer, R.D., Use of ab initio calculations toward the rational design of room temperature ionic liquids, *J. Phys. Chem. A* **107**, 2277–2288 (2003).

36  Tsuzuki, S., Arai, A.A., and Nishikawa, K., Conformational analysis of 1-butyl-3-methylimidazolium by CCSD(T) level ab initio calculations: Effects of neighboring anions, *J. Phys. Chem. B* **112**, 7739–7747 (2008).

37   Imanari, M., Uchida, K., Miyano, K., Seki, H., and Nishikawa, K., NMR study on relationships between reorientational dynamics and phase behaviour of room-temperature ionic liquids: 1-alkyl-3-methylimidazolium cations, *Phys. Chem. Chem. Phys.* **12**, 2959–2967 (2010).

38   Kojima, A., Yoshimura, Y., Iwasaki, H., and Tozaki, K., Millikelvin-stabilized cell for the precise study of phase transitions, in D.C. Ripple (ed.), *Temperature: Its Measurement and Control in Science and Industry*, Vol. 7, AIP Conference Proceedings, American Institute of Physics, **684**, 921–926 (2003).

39   Morrow, T.I., and Maginn, E.J., Molecular dynamics study of the ionic liquid 1-*n*-butyl-3-methylimidazolium hexafluorophosphate, *J. Phys. Chem. B* **106**, 12807–12813 (2002).

40   Antony, J.H., Mertens, D., Breitenstein, T., Dölle, A., Wasserscheid, P., and Carper, W.R., Molecular structure, reorientational dynamics, and intermolecular interactions in the neat ionic liquid 1-butyl-3-methylimidazolium hexafluorophosphate, *Pure Appl. Chem.* **76**, 255–261 (2004).

41   Urahata, S.M., and Ribeiro, M.C.C., Single particle dynamics in ionic liquids of 1-alkyl-3-methylimidazolium cations, *J. Chem. Phys.* **122**, 024511–024519 (2005).

42   Choudhury, A.R., Winterton, N., Steiner, A., Cooper, A.I., and Johnson, K.A., In situ crystallization of low-melting ionic liquids, *J. Am. Chem. Soc.* **127**, 16792–16793 (2005).

43   Dibrov, S.M., and Kochi, J.K., Crystallographic view of fluidic structures for room-temperature ionic liquids: 1-Butyl-3-methylimidazolium hexafluorophosphate, *Acta Crystallogr.* **C62**, o19–o21 (2006).

44   Kabo, G.J., Blokhin, A.V., Paulechka, Y.U., Kabo, A.G., Shymanovich, M.P., and Magee,J.W., Thermodynamic properties of 1-butyl-3-methylimidazolium hexafluoro-phosphate in the condensed state, *J. Chem. Eng. Data* **49**, 453–461 (2004).

45   Jin, H., O'Hare, B., Dong, J., Arzhantsev, S., Baker, G.A., Wishart, J.F., Benesi, A.J., and Maroncelli, M., Physical properties of ionic liquids consisting of the 1-butyl-3-methylimidazolium cation with various anions and the bis(trifluoromethylsulfonyl) imide anion with various cations, *J. Phys. Chem. B* **112**, 81–92 (2008).

46   Triolo, A., Mandanici, A., Russina, O., Rodriguez-Mora, V., Cutroni, M., Hardacre, C., Nieuwenhuyzen, M., Bleif, H.-J., Keller, L., and Ramos, M.A., Thermodynamics, structure, and dynamics in room temperature ionic liquids: The case of 1-butyl-3-methyl imidazolium hexafluorophosphate ([bmim][PF$_6$]), *J. Phys. Chem. B* **110**, 21357–21364 (2006).

47   Domańska, U., and Marciniak, A., Solubility of 1-alkyl-3-methylimidazolium hexafluoro-phosphate in hydrocarbons, *J. Chem. Eng. Data* **48**, 451–456 (2003).

48   Troncoso, J., Cerdeiriña, C.A., Sanmamed, Y.A., Romaní, L., and Rebelo, L.P.N., Thermodynamic properties of imidazolium-based ionic liquids: Densities, heat capacities, and enthalpies of fusion of [bmim][PF$_6$] and [bmim][NTf$_2$], *J. Chem. Eng. Data* **51**, 1856–1859 (2006).

49   Paulechka, Y.U., Kabo, G.J., Blokhin, A.V., Shaplov, A.S., Lozinskaya, E.I., Golovanov, D.G., Lyssenko, K.A., Korlyukov, A.A., and Vygodskii, Y.S., IR and X-ray study of polymorphism in 1-alkyl-3-methylimidazolium bis(trifluoromethane sulfonyl)imides, *J. Phys. Chem. B* **113**, 9538–9546 (2009).

50 Nakakoshi, M., Shiro, M., Fujimoto, T., Machinami, T., Seki, H., Tashiro, M., and Nishikawa, K., Crystal structure of 1-butyl-3-methylimidazolium iodide, *Chem. Lett.* **35**, 1400–1401 (2006).

51 Santos, C.S., Rivera-Rubero, S., Dibrov, S., and Baldelli, S., Ions at the surface of a room-temperature ionic liquid, *J. Phys. Chem. C* **111**, 7682–7691 (2007).

52 Yoshida, R., Takahashi, T., Yamaguchi, T., and Ichijo, H., Self-oscillating gel, *J. Am. Chem. Soc.* **118**, 5134–5135 (1996).

# 4  Ionic Liquid Membrane Technology

JOĂO G. CRESPO

REQUIMTE/CQFB, Department of Chemistry, Faculdade de Ciencias e Tecnologia, Universidade Nova de Lisboa, Caparica, Portugal

RICHARD D. NOBLE

Alfred T. & Betty E. Look Professor, Chemical Engineering Department, University of Colorado, Boulder, Colorado, USA

## ABSTRACT

Ionic liquids have some unique physical/chemical properties that make them excellent materials to be used in various morphologies as membranes. These morphologies include ionic liquids impregnated into the pores of supports (supported liquid membrane), polymeric versions of ionic liquids, composites of the polymer with ionic liquid, and three-component systems where there is also a solid phase such as a zeolite, and finally gelled versions. These membranes have been demonstrated for various gas and liquid phase separations. For gases, the most studied separation is $CO_2$ from $N_2$ or $CH_4$. Separation of organics from water is the most common liquid phase application. One additional application is as barrier materials where the membrane acts to protect against chemical warfare agents (CWAs) or other toxic chemicals while allowing water vapour to be transported. These membranes can also be used for electrochemical applications due to the ionic nature of the material.

## 4.1  IONIC LIQUIDS: DEFINITIONS AND PROPERTIES

Ionic liquids are molten salts composed of a bulky and asymmetric organic cation and organic or inorganic anions, which create a low lattice energy in the

*Ionic Liquids Further UnCOILed: Critical Expert Overviews*, First Edition.
Edited by Natalia V. Plechkova and Kenneth R. Seddon.
© 2014 John Wiley & Sons, Inc. Published 2014 by John Wiley & Sons, Inc.

crystalline structure. This structure lowers their melting point, allowing these salts to be in the liquid state at room temperature (also sometimes known has room temperature ionic liquids [RTILs]). Ionic liquids generally exhibit negligible vapour pressure [1], high thermal stability, and good solvating capacity for both organic and inorganic compounds, among many other unique properties. Given the very large number of possible ionic liquids, their properties can be fine-tuned by adequate selection of specific ions and/or functional groups, making them "tailored solvents" that can be designed to fit the requirements of a specific process.

Their unique properties allow for their application in numerous chemical and industrial processes which include chemical, catalytic, and biological reactions; organic/inorganic synthesis; separation processes; separation and purification of gases; removal of contaminants; and the replacement of conventional organic solvents. There are relevant industrial processes that use ionic liquids, such as the BASIL (Biphasic Acid Scavenging utilizing Ionic Liquids) process developed by BASF™ [2] and the Difasol process developed by the Institut Français du Pétrole, which is an improvement of the traditional Dimersol process [3].

In order to identify the best ionic liquids for a particular application, it is necessary to know their thermophysical properties and to understand the phase behaviour of the systems containing ionic liquids. This task cannot be accomplished using only the available experimental data due to the very large number of possible combinations of ionic liquids, solvents, and target solutes. It is therefore necessary to develop predictive models able to describe the behaviour of these systems, based on a few selected experimental measurements.

To model the phase behaviour, a number of excess Gibbs free energy models have been applied to mixtures containing ionic liquids and different solvents. Some classical local composition models, such as the non-random two liquid (NRTL) and UNIversal QUAsiChemical (UNIQUAC), were also applied with success to the description of these systems [4]. The modified Flory–Huggins equation and a lattice model based on polymer-solution models have also been applied with good results [5]. Although the models employed in these studies provide good correlations, they have a very limited predictive capability since they require parameters fitted to experimental data when considering the ionic liquid complex groups. A less rigorous, but more predictive, alternative is the use of the COnductor-likeScreeningMOdel for Real Solvents (COSMO-RS), proposed by Klamt and Eckert [6], which does not require adjustable parameters being applicable to a large number of possible ionic liquids and solvents [7].

Gas solubility is one of the most important thermophysical properties when considering gas–ionic liquid systems. One approach is based on regular solution theory (RST) [8–11]. In this theory, the solubility is related to the difference in the solubility parameter for the solute and solvent squared. The solubility parameter for each component is an independent value. The

**TABLE 4.1    Carbon Dioxide Solubility Data for the $[C_2mim][NTf_2]$–$[C_2mim][BF_4]$ System [13]**

| Ionic Liquid Mixture | $H_{CO_2}$ / atm | $cm^3$ gas/g IL |
|---|---|---|
| $[C_2mim][NTf_2]$ | $50 \pm 1$ | 0.41 |
| 25 mol% $[C_2mim][BF_4]$ | $58 \pm 3$ | 0.35 |
| 50 mol% $[C_2mim][BF_4]$ | $65 \pm 1$ | 0.43 |
| 75 mol% $[C_2mim][BF_4]$ | $85 \pm 5$ | 0.43 |
| 90 mol% $[C_2mim][BF_4]$ | $91 \pm 1$ | 0.46 |
| 95 mol% $[C_2mim][BF_4]$ | $94 \pm 1$ | 0.46 |
| $[C_2mim][BF_4]$ | $100 \pm 2$ | 0.50 |

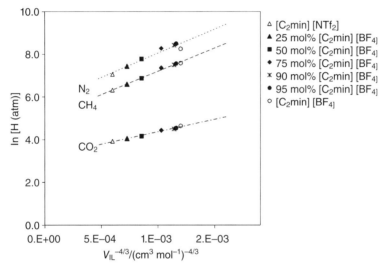

**Figure 4.1**    Average natural logarithm of the Henry's constant versus average measured mixture molar volume to the $-4/3$ power at $40\,°C$. The lines represent the RST models for each gas [13].

solubility parameters of many gases and vapours have been tabulated (Table 4.1). The temperature dependence can also be evaluated [12].

Figure 4.1 demonstrates that the RST is valid for ionic liquid mixtures. The linear relationship shows that RST can well describe the behaviour of gases in ionic liquids.

The negligible vapour pressure of ionic liquids, and the possibility to adjust and control their solubility in various solvents, make them particularly interesting for a large number of industrial processes, minimising or avoiding their loss to the environment, with the resulting economic and ecological benefits. Also, the fact that they can be designed in order to promote a high affinity to target compounds is extremely attractive and opens perspectives for selective separation processes in the gas and liquid phases, as will be discussed in this chapter.

The integration of ionic liquids, and/or their chemistry, in synthetic membranes has been discussed since the late 1990s [14, 15]. Membranes can supply an adequate environment for ionic liquids by providing conditions for their containment and use in reduced amounts. Some properties that are usually negatively associated with ionic liquids, such as high viscosity and corrosiveness, do not represent a relevant problem if short transport paths and appropriate micro-environments are provided, respectively, as in the case in integrated membrane–ionic liquid systems. Additionally, cost issues that are also associated with some ionic liquids may be circumvented by the use of membrane systems, where a reduced amount of ionic liquid is employed and, in many cases, under regenerative conditions.

Although commonly regarded as "green solvents," many ionic liquids described in the literature and used in numerous processes have been characterised as toxic to humans and other biological entities [16, 17]. In many cases, the "green character" of ionic liquids results mostly from their ease of confinement due to their negligible volatility and adjustable solubility. Membrane confinement can contribute significantly to reinforce their sustainable use.

## 4.2 STRUCTURE AND MORPHOLOGY OF IONIC LIQUID MEMBRANES

Ionic liquids can be used in various morphologies or configurations as membranes. Three intrinsic properties of ionic liquids that differentiate them from common organic solvents and water are non-volatility, thermal stability, and tuneable chemistry. These characteristics make ionic liquid membranes promising in industrial applications, particularly in gas (i.e., $CO_2$) separation and sequestration. The incorporation of these ionic liquids in a membrane can improve membrane separation performance, adding increased "liquid-like" gas solubility behaviour to the membrane. In addition, the ionic liquid can be functionalised to provide increased solubility and selectivity.

### 4.2.1 Supported Liquid Membranes (SLMs) and Contactors

Membrane contactors are devices that promote the non-dispersive contact between two different phases (gas–liquid or liquid–liquid) in order assure a high interfacial area for mass and heat transfer [18, 19]. The membranes may be porous or non-porous. In this section, we will focus our attention on porous membrane contactors that separate two different phases. The transport of target solutes from a feed to a receiving stream takes place through the pores of the membrane, which are wetted by the phase with the highest affinity towards the membrane (hydrophobic–hydrophobic or hydrophilic–hydrophilic

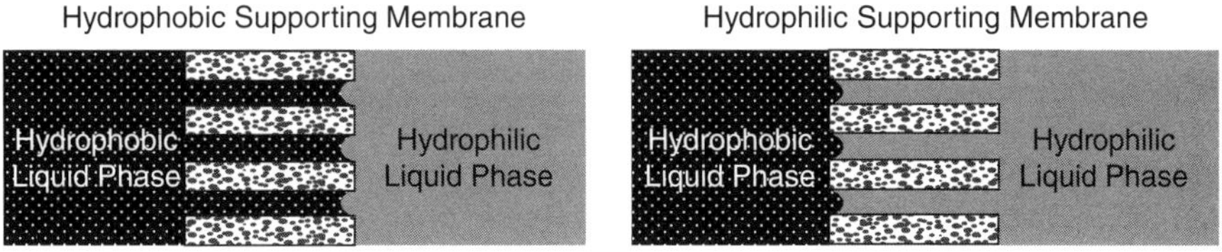

**Figure 4.2**   Wetting of contactor membranes according with hydrophobic–hydrophobic or hydrophilic–hydrophilic affinity.

**TABLE 4.2   Specific Area of Contact of Mass Transfer Equipment**

| Equipment | Specific Area/$m^{-1}$ |
|---|---|
| Settler/decanter | 3–30 |
| Plate and packing columns | 30–300 |
| Rotating disc columns | 150–500 |
| Membrane contactors | 1500–7000 |

affinity). Figure 4.2 shows a scheme of a hollow fibre membrane contactor, where the hydrophobic membranes are preferentially wetted by a hydrophobic solvent, and the opposite situation where a hydrophilic membrane contactor wetted by a hydrophilic liquid phase.

This equipment has several important advantages over conventional equipment used for mass transfer processes, such as a very high specific interfacial area (up to $\sim$7000 $m^2$ $m^{-3}$, see Table 4.2 for comparison with traditional equipment), which translates into high mass transfer rates, and no risk of phase dispersion if an appropriate pressure difference is adjusted between the two sides of the membrane. This avoids the need for subsequent phase separation and has a number of additional advantages in terms of fluid selection (no need for a density difference between the two phases) and operating conditions (no risk of flooding or short circuiting).

Both polymeric and ceramic membranes may be used, based on the required conditions in terms of chemical and thermal stability. This equipment has been progressively adopted in industrial and medical applications for gas absorption and stripping, gas humidification and drying, and liquid–liquid extraction [20].

The concept of a membrane contactor has been extended to the development of SLMs where a selected solvent is immobilised and retained inside the porous structure of the membrane material by capillary forces. In this case, an integrated absorption/stripping process can take place in the same equipment unit, which also applies to integrated extraction/re-extraction (see Fig. 4.3). This configuration is particularly attractive because very small amounts of solvent are required within the porous structure and because the solvent is

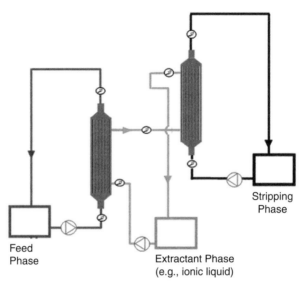

**Figure 4.3** Diagram of membrane contactor integrated extraction/re-extraction system.

continuously "regenerated" when using the integrated absorption/stripping or extraction/re-extraction approach. Additionally, the transport path through the immobilised solvent phase can be greatly reduced to a few micrometres, when compared with the situation of transport in a typical membrane contactor where transport through the bulk solvent phase also has to be considered.

The assembly and operation of membrane contactors in bench and industrial scales is relatively simple but the preparation of SLMs, although simple on a laboratory scale [21–23], still remains a technical challenge for large-scale applications.

The major drawback of membrane contactors and SLMs is the fact that most solvents are volatile and/or may be solubilised in the contacting phase(s). These features translate into a loss of solvent, with consequent failure of stability and selectivity, and contamination of the contacting phase(s). Due to their extremely low volatility, ionic liquids represent an interesting alternative, assuring no solvent loss during operation of gas–liquid processes. If liquid–liquid systems are considered, ionic liquids may also be applied advantageously through modulation of their solubility in the contacting liquid phases, by judicious selection of their cation and anion constituents.

The use of membrane contactors and SLM systems employing ionic liquids has been widely reported in the literature for gas and vapour absorption, where feed streams containing target solutes are processed with ionic liquids that exhibit high selectivity and good chemical and thermal stability. The recent development of task-specific ionic liquids [24], notably $CO_2$ task-specific ionic liquids [25], opens very interesting perspectives for selective gas and vapour transport and capture. The good thermal stability of many ionic liquids at temperatures of up to ~250 °C opens also the possibility for operation at process temperatures, with the corresponding advantages in terms of mass transport rates, process simplicity, and economy.

Room temperature ionic liquids on a micro-porous support have shown promising $CO_2$ separation performance in the laboratory. Various 1-alkyl-3-methylimidazolium ionic liquids on a micro-porous support or substrate exhibit $CO_2$ permeabilities of 1000 Barrers with a $CO_2/N_2$ selectivity of 21. However, when the pressure drop across the membrane exceeds values as low as 2 atm, the liquid is dislodged from the membrane, destroying its functionality [26].

Research at the University of Colorado has demonstrated that SLMs containing ionic liquids (SILMs) can be prepared and used for gas separations. Relatively stable SILMs can be made by impregnating commercially available micro-porous hydrophilic polymeric (polyethersulfone [PES]) or ceramic (Alumina Anodisc®) substrates. Various ionic liquids such as [$C_2$mim] [$N(SO_2CF_3)_2$], [$C_2$mim][$CF_3SO_3$], [$C_2$mim][$N(CN)_2$], [$P_{6\ 6\ 6\ 14}$]Cl, and [$C_6$mim] [$N(SO_2CF_3)_2$] were used to prepare these SILMs, which exhibited a combination of high permeability and selectivity for carbon dioxide.

Another type of supported membrane involves incorporation of ionic liquids into a conventional polymer membrane [27–29]. In this configuration, the material will maintain the solubility selectivity while enhancing the diffusion rate across the membrane since the material now has more liquid-like behaviour. The ionic liquid is contained in the membrane due to the strong electrostatic interactions with both the liquid and the ionic polymer. The development of systems for separation of mixed gases and removal/recovery of vapours will be addressed in detail later in this chapter.

With the use of membrane contactors, and SILMs for liquid–liquid contact and transport, the challenge remains in finding the adequate compromise between highly selective systems for target solutes, while assuring no loss of the ionic liquid phase to the contacting phase(s). Successful applications for solute transport between organic solvent phases and ionic liquid phases have been reported [30], although transport between an aqueous phase and an ionic liquid phase is more problematic, even when the solubility of the ionic liquid in aqueous media is extremely reduced, due to the progressive formation of aqueous microdomains inside the ionic liquid phase, leading to an incremental loss of selectivity and stability [31, 23]. Separations in the liquid phase will be also discussed in detail later in this chapter.

### 4.2.2  Polymer Ionic Liquid Membranes

*4.2.2.1  Ionic Liquid as a Polymerised Membrane (poly[RTILs]).*  Room temperature ionic liquids (RTILs) can be made in a polymer form. If the cation contains a polymerisable group, it can be readily converted into solid, dense polymers—poly(RTILs)—for use as gas separation membranes [32]. The cation can be readily functionalised to incorporate unsaturated carbon–carbon bonds that can form polymers [32, 33] to be used as a membrane. This configuration eliminates the issues of ionic liquid displacement due to a pressure gradient, as in the case of a SILM using a solid micro-porous support where the liquid is held in the pores by relatively weak capillary forces.

The cation–anion constituents of a polymerised ionic liquid can be varied to achieve an application-specific performance, a so-called tuning of the RTIL membrane to be task specific. Bara et al. [32] synthesised and converted a series of ionic liquid monomers with varying length of $N$-alkyl substituents to form polymer films. These membranes were tested for their performance in separations involving $CO_2$, $N_2$, and $CH_4$. The $CO_2$ permeability was observed to rise in a non-linear fashion as the $N$-alkyl substituent was increased. Specific performance differences will be discussed in more detail in a subsequent discussion of gas separation using RTIL membranes.

Poly(RTILs) are now a reality (see Figure 4.4). Imidazolium materials are versatile building blocks, with modular or "snap together" chemistry, which allows for the incorporation of a wide variety of functional groups in the RTILs. Imidazolium cations have been tailored to form "task-specific" ionic liquids for gas separations and other applications [25, 32, 34].

Cross-linkable gemini RTILs (GRTILs) have been synthesised into thin film polymer sheet membranes, poly(GRTILs) [33]. These membranes were found to have relatively low permeability to various gas constituents—$CO_2$, $N_2$, $CH_4$, and $H_2$—when compared to other poly(RTIL) membranes, due to highly restricted diffusion through the membrane's more contorted molecular structure. These membranes have a potential application as barriers, preventing the transport of a given gas constituent. Such an application would be in protective garment materials for protection against hazardous vapours such as CWAs and toxic industrial chemicals (TICs). As with other RTIL membranes, these could be "tuned" to a specific task or application.

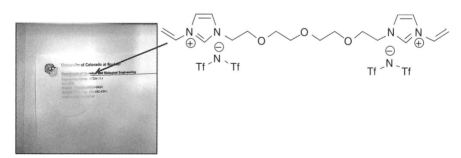

**Figure 4.4**  A semi-transparent disc of a poly(RTIL), containing no free ionic liquid, prepared from the illustrated vinylimidazolium ionic liquid on a polyethersulfone (PES) support.

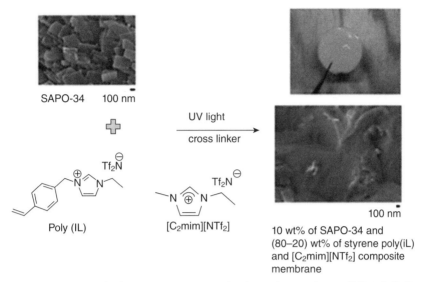

**Figure 4.5**   A typical three-component mixed-matrix membrane (MMM) [36].

### 4.2.2.2   *Three-Component Composite Membranes (MMMs).*

More recently, a three-component MMM has been developed that alleviates the problem associated with a lack of adhesion between the polymer and the solid phase (Figure 4.5). Free ionic liquid is added to the material that wets both the polymer and the solid. This wetting phase provides a selective layer between the solid and polymer. Since the IL is non-volatile, it remains in the membrane (see, for example, Figure 4.6) [35, 36]. The three components of a composite membrane are a poly(RTIL), a free RTIL, and a micro-porous solid, such as a zeolite. This is also referred to as an MMM.

### 4.2.3   Gelled Ionic Liquids

A recent morphology advancement is the formation of composite liquid–polymer structures [27, 29]. These materials have improved mechanical properties in comparison with the liquid (in an SLM), but improved diffusion properties compared with the polymer phase. This structure is achieved by the blending of a "free" ionic liquid with poly(RTILs) to form a homogeneous composite membrane. Technically, this is a composite membrane. However, with enough free ionic liquid, it can be considered as a gelled membrane.

Voss et al. [37] fabricated and analysed the performance of a true gelled RTIL, 1-hexyl-3-methylimidazolium bis{(trifluoromethyl)sulfonyl}amide, [$C_6$mim] [$NTf_2$]. The gelled RTIL is produced using the low molecular weight gelator (LMOG), 12-hydroxystearic acid. Here the gelled structure is a gelled RTIL with no poly(RTIL) (see Fig. 4.7).

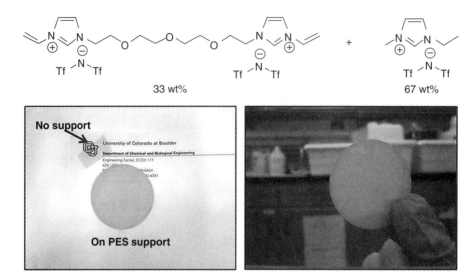

**Figure 4.6** A composite of a poly(RTIL) with incorporated free [C$_2$mim][NTf$_2$], supported on a polyethersulfone (PES) support.

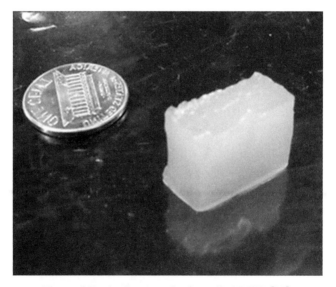

**Figure 4.7** A photograph of a gelled RTIL [37].

The gel would normally require a support since it is not as mechanically stable as a regular polymer film membrane, such as poly(RTIL). The preparation procedure is independent of how it would be used.

## 4.3   CHARACTERISATION OF IONIC LIQUID MEMBRANES

The characterisation of ionic liquids is widely discussed in the literature covering all relevant aspects, from purity issues, structural aspects, thermodynamic, thermophysical and transport properties, electrical and optical properties, and ecotoxicity. The variety of techniques used for such extensive characterisation include, in a non-exhaustive list, elemental analysis, nuclear magnetic resonance (NMR), mass spectrometry, Fourier transform infrared (FTIR) and X-ray photoelectron spectroscopy (XPS) techniques, small-angle X-ray scattering, thermogravimetry and calorimetry, conductivity, cyclic voltammetry and impedance spectroscopy, rheology studies, and toxicity evaluation with various *in situ* and *in vivo* model systems [38–40].

A detailed presentation and discussion of these studies is outside the scope of this chapter. In the following, particular attention will be given to specific techniques and studies which helped to characterise the structure and performance of ionic liquid membranes. In order to facilitate this discussion, SLMs and contactors will be examined separately from dense membranes incorporating ionic liquids chemistry.

When dealing with SILMs and contactors, the first concern is the assurance of the quality of the process of ionic liquid immobilisation inside the porous membrane structure. A good immobilisation procedure implies that all pores at the membrane surface are filled with ionic liquid, assuring that no solute transport takes place through "empty" pores. The most common evaluation procedure uses a simple gas permeation test under controlled pressure difference between the feed and the downstream circuits, where the pressure increase in the downstream circuit is monitored online with a pressure transducer. A sudden increase in pressure corresponds to a situation where the transport of gas starts to take place by convective transport through unfilled pores. This simple test can determine the operating pressure difference range for a given SILM. Pressure differences of up to 2 atm can be achieved, without breakthrough, if an adequate selection of the membrane material and the immobilised ionic liquid is adopted [23].

For some gas–vapour permeation applications, this range of pressure differences may be sufficient but, for other specific applications, higher operating pressure differences are required. In those cases, the use of modified dense membranes may constitute a better approach. Stability of SLMs and contactors for liquid phase separations has been also evaluated by checking periodically the elemental composition of the membrane near its interface using XPS. By varying the incidence angle of the X-ray beam, it is possible to obtain information about the elemental composition and electronic state of the elements at different depths (up to 10 nm) inside the membranes and know if the ionic liquid initially immobilised inside the membrane has been progressively removed [23]. Scanning electron microscopy (SEM) has been also proposed for monitoring the retention of ionic liquid within the membrane porous structure [40]. Dynamic performance tests are also essential to characterising the behaviour of ionic liquid membrane contactors and ionic liquid supported membranes. The selectivity and flux of target solutes, during operation under defined conditions, characterise the membrane behaviour and provide understanding if relevant changes in the overall membrane system (supporting membrane and ionic liquid) are taking place. This comment is also valid when studying modified dense membranes with inclusion of ionic liquids.

When dense membranes are modified by inclusion of "ionic liquid chemistry," other complementary characterisation techniques may be used. As an example, the introduction of ionic liquid cations in proton exchange membranes such as Nafion® may be followed off-line by Raman spectroscopy [38] and online, in real time, by confocal Raman spectroscopy. This technique can provide concentration profiles of the ionic liquid cation and other compounds present across the membrane (see Figure 4.8).

The swelling behaviour of these membranes when exposed to different solvents [42], and their ability to retain solvent molecules (namely, water), under controlled temperature conditions are also relevant characteristics that have been studied by optical microscopy and thermogravimetry [41], respectively. The electrical properties of modified dense membranes with inclusion

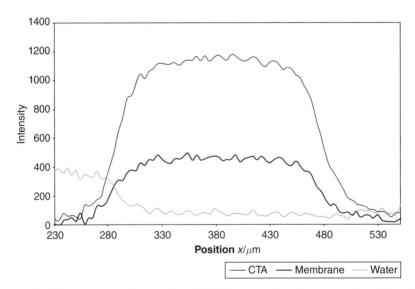

**Figure 4.8**   The concentration profile of [CTA]$^+$ (cetyltrimethylammonium, [N$_{1\,1\,1\,16}$]$^+$) and water inside a Nafion® membrane, obtained by online confocal Raman spectroscopy (unpublished data by L. Neves, I.R. Coelhoso J.G. Crespo, P. Huguet, and P. Sistat).

of ionic liquid chemistry have been widely studied by impedance spectroscopy techniques, aiming at understanding the impact of introducing such chemistry on the properties of these membranes in with respect to the transport of charged species [41]. Proton mobility, as well as water structuring in these membranes, has also been studied by low-field NMR spectroscopy [43].

## 4.4    RECENT APPLICATIONS OF IONIC LIQUID MEMBRANES

### 4.4.1    Separation of Gases and Vapours: $CO_2$ Separation

#### *4.4.1.1    Definitions.*

*4.4.1.1.1    Permeability.*    The permeability, $P$, is the product of the solubility, $S$, and the diffusivity, $D$. It is a measure of the volume flow rate of a gas per unit surface area through a membrane, normalised by the pressure gradient. It is also the product of the gas flux and membrane thickness divided by the pressure drop across the membrane. The units of permeability are Barrers, where 1 Barrer $= 10^{-10}$ cm$^3$ (STP) (cm cm$^{-2}$ s$^{-1}$ (cm Hg)$^{-1}$).

*4.4.1.1.2    Selectivity.*    The separation selectivity of a membrane for a combination of gases is the ratio of the permeabilities of two gases in a mixture, $P_1/P_2$. That is, then, equivalent to the ratio of the products of the solubility and diffusivity for each of the two gases, $S_1D_1/S_2D_2$. It turns out that for conventional, non-RTIL polymer membranes, the selectivity is driven by the diffusion component, while for RTIL membranes (and poly(RTILs)), that selectivity is driven by the solubility component. This is due to the charged nature of the RTIL and the structure generated in poly(RTILs). RTIL "tuning" (modification of cation and or anion and alkyl group) can affect both solubility and diffusivity. The changes in diffusivity are usually minor for the polymer phase, but the solubility can change based on free volume, charge, and hydrophilic versus hydrophobic behaviour.

Solubility is obtained by measuring the gas flux through a membrane and then calculating the permeability of that gas through the membrane [44]. The permeability is determined from the measured gas flux normalised by the membrane thickness and pressure drop across the membrane. The mode of transport through the membrane is assumed to be solution diffusion, the permeability being the product of these two properties. Diffusion through the membrane can be calculated from the lag time measured before steady state is achieved in the gas flux determination [44]. The solubility is then calculated from the permeability and diffusivity.

In the case of $CO_2$ separations with RTIL membranes, the $CO_2$ solubility is enhanced over that of other gases by its quadrupole moment. The $CO_2$ quadrupole interacts with the charge of the ionic liquid to achieve the improved solubility and higher selectivity values with respect to other gases.

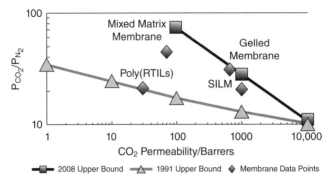

**Figure 4.9**  Robeson plot showing both 1991 and 2008 upper bounds.

*4.4.1.1.3  Robeson Plot: A Metric for Polymer Gas Separation Membrane Performance.*  The Robeson plot [45] is a log–log plot of selectivity versus permeability. Specifically, it is a plot of the selectivity between two gases, $P_1/P_2$, on the $y$-axis against the permeability of the more permeable gas, $P_1$, on the $x$-axis. Typical non-RTIL membranes hit a so-called upper bound [46] on the Robeson plot. Currently two so-called upper bounds exist; the 1991 upper bound [45] and the 2008 upper bound [47]. The 1991 "upper bound" separates non-RTIL polymers and SLMs from membrane configurations utilising RTILs; and the 2008 upper bound defines the limit of the new RTIL-based membranes. Conventional polymer membranes used for gas separation exhibit a performance trade off. The more permeable a membrane is to a given gas, the less selective it is for that gas, and vice versa [46]. This characteristic explains the positioning of the so-called upper bounds on the Robeson plot (Figure 4.9).

*4.4.1.2  CO₂ Separation.*  Supported ionic liquid membranes show promise in gas phase separation. Much effort is continuing in developing new SILMs for this area of application, particularly with respect to $CO_2$ gas separations. Industrial applications of $CO_2$ gas separation include natural gas sweetening, respiratory gas enrichment in life-support systems, as well as $CO_2$ scrubbing from power plant combustion exhaust.

Currently, the industrial methods used for $CO_2$ separations exact a significant energy cost, either for refrigeration in cryogenic condensation separations or regeneration of the separation medium in a bulk fluid separation, for example, aqueous amine $CO_2$ removal in natural gas sweetening or scrubbing of power plant flue gas.

Current $CO_2$ separation technology is dominated by pressure and temperature swing adsorption using liquid solvents. However, membrane-based separations have evolved and are becoming commercially viable. SILM membranes may represent an attractive, cost-effective alternative to such industrial separation applications, as the amount of material required to process the same amount of gas is greatly reduced, and regeneration process steps may be eliminated. Such membranes have been used in natural gas sweetening and have been proposed for use in $CO_2$ capture from flue gases. RTILs present a highly versatile and tuneable platform for the development of new materials and processes aimed at the capture of $CO_2$ from power plant flue gas and in natural gas sweetening.

Membranes, both conventional and RTIL, have potential for separation of $CO_2$ from $N_2$ and $CH_4$ in industrial processes. High-throughput, highly selective polymer membranes may prove to be an alternative to amine-based $CO_2$ separation in natural gas sweetening [48, 49]. A small, but growing, segment of natural gas commercial sweetening operations with high $CO_2$ concentrations and low gas throughput do utilise polymer membranes. In applications with both high $CO_2$ concentration and high throughput, a combined amine-membrane process is being used.

Potential exists for polymer membranes in $CO_2$ capture in fossil fuel burning power plants, separating $CO_2$ from $N_2$ in the flue gas [50]. Membranes could also be used in oxycombustion, separating $O_2$ from $N_2$ in air to produce an oxygen-enriched environment in which to burn coal. The subsequent products of the combustion, $CO_2$ and $H_2O$, could be processed by condensation and $CO_2$ sequestration.

### 4.4.1.3  *SILMs in CO₂ Separations.*

*4.4.1.3  SILMs in CO₂ Separations.*  Evaluation of several different imidazolium-based RTILs revealed that SILMs fabricated with those RTILs possessed permeability and selectivity properties for $CO_2/N_2$ that were superior to both SLMs and conventional polymer membranes, when viewed in the context of a "Robeson plot." Indeed, SILMs of $[C_2mim][NTf_2]$ on a microporous support exhibited an ideal $CO_2$ permeability in excess of 1000 Barrers with a selectivity for $CO_2/N_2 = 21$ [51]. However, the SILM liquid component is still subject to displacement from the support if the pressure drop across the membrane exceeds the capillary forces or electrostatic forces holding the RTIL in its matrix.

### 4.4.1.4  *Poly(RTILs).*

*4.4.1.4  Poly(RTILs).*  A possible solution to the RTIL displacement under higher differential pressure would be to formulate a polymerised RTIL-based monomer to create a poly(RTIL) [11]. Indeed, RTILs with polymerisable groups can be readily converted into solid, dense poly(RTILs) for use as gas separation membranes.

The tuneable chemistry of imidazolium-based RTILs allows for the generation of new types of $CO_2$-selective polymer membranes. A series of imidazolium-based RTILs bearing various length linear alkyl substituents and polymerisable units composed of a styrene or acrylate groups was evaluated [32]. By coating these monomers (mixed with a small amount of photo-initiator and cross-linker) on a porous support, and subsequent photo-polymerisation, Bara et al. formed poly(RTIL) materials as thin ($\sim$150 $\mu$m) films and studied their properties and performances relating to permeability to, and separation of, the gases $CO_2$, $N_2$, and $CH_4$ [32]. They measured the permeability, $P$, the solubility,

$S$, and the diffusivity, $D$, of the three gases for each of both styrene- and acrylate-based poly(RTILs) [32]. $CO_2$ permeability was observed to rise in a non-linear fashion as the alkyl substituent was increased. $CO_2/N_2$ separation performance was relatively unaffected as $CO_2$ permeability increased.

The correlation of increase in permeability of these gases through these poly(RTILs) with increasing length of the alkyl substituent may be due to the creation of more free space in the polymer, the free space being formed by an inefficient packing of the $n$-alkyl groups. Poly(RTILs) contain large ions, one tethered to a backbone and one free. Free space may also be generated by local repulsion between immobilised ions and so-called ion-phobic alkyl chains. It is worth noting here that poly(RTILs) differ from liquid phase RTILs in that the cations in the polymer are fixed (the anions can move very slowly), whereas there is more movement or rearrangement of ions in liquid phase RTILs.

The results from the poly(RTILs) tested by Bara et al. [32] yielded points on the Robeson plot on or slightly above that upper bound (1991 upper bound), indicating that these poly(RTILs) possess as good as or better separation properties for the gases considered ($CO_2/N_2$) than those of conventional, non-RTIL membranes. This is an encouraging result that should stimulate additional evaluation of new and different RTILs fabricated into poly(RTILs).

Poly(RTILs) give superior separation results for $CO_2/N_2$ separations when compared with conventional polymer membranes. They also perform as well as or better in $CO_2/N_2$ separations than their liquid RTIL counterparts containing $[NTf_2]^-$ anions [44] due to added diffusion selectivity imparted by the poly(RTIL) relative to the RTIL liquid. In addition, poly(RTILs) dissolve about twice as much $CO_2$ volume per cubic centimetre of material as their liquid counterpart.

### 4.4.1.5 MMMs in CO₂ Separation.

Mixed-matrix membranes are now showing promise with respect to $CO_2$ separation, combining the selectivity and transport properties of inorganic membrane materials with the fabrication properties of organic polymers. Two-component solid–solid (inorganic membrane material and organic polymer) MMMs exhibit improved catalytic, optical, and electrical properties. However, such two component membranes have limitations in light gas separation due to inadequate adhesion between the polymer and the inorganic component. This can lead to membrane failure.

In an attempt to solve this adhesion problem, or shortcoming, Hudiono et al. [35] fabricated and evaluated the light gas transport properties of a three-component composite MMM consisting of poly(RTIL), RTIL, and zeolite materials. As well as improving the adhesion of the polymer-solid membrane components, the addition of an RTIL, due to its inherent solubility and transport properties, would likely improve the gas separation properties of the membrane, making it well suited for light gas separation.

In their evaluation, Hudiono et al. prepared and tested three different composite membranes. The membrane configurations were

1. RTIL ($[C_2mim][NTf_2]$)–zeolite
2. poly(RTIL)–zeolite
3. poly(RTIL)–RTIL ($[C_2mim][NTf_2]$)–zeolite

Note that the zeolite used was SAPO-34, the RTIL was 1-ethyl-3-methylimidazolium bis{(trifluoromethyl)sulfonyl}amide ($[C_2mim][NTf_2]$), and the RTIL monomer (poly(RTIL)) was a styrene-based imidazolium RTIL monomer with methyl group substitution and $[NTf_2]^-$ anions. The poly(RTIL) monomer and RTIL were chosen for their ability to absorb $CO_2$.

The $CO_2$ separation results from the three-component membrane were superior to those from the solid–solid, zeolite–poly(RTIL), membrane, $CO_2$ permeability going from 13.9 to 72 Barrers, and $CO_2/N_2$ selectivity going from 35 to 44 [34]. One drawback was a drop in $CO_2/CH_4$ selectivity from 35 to 30 with the three-component membrane.

The $CO_2$ separation results from the membranes agreed with Maxwell equation predictions involving calculations of overall conductance (permeability). This indicates that there were no defects in the fabricated membranes.

The adhesion of the RTIL to both the polymer and inorganic zeolite was also investigated using electron microscopy and contact angle measurement. The work of adhesion, the work per unit area required to separate two materials, was calculated using contact angle, surface tension, and the Young–Dupre equation.

Hudiono et al. [35] concluded that the RTIL addition improves the structural integrity of the membrane by its action as a wetting agent to improve the interfacial adhesion between the poly(RTIL) and the inorganic zeolite. The charged nature of the RTIL may also contribute to that improved interfacial adhesion/interaction.

This improved interfacial adhesion and improved $CO_2/N_2$ selectivity indicate that such a three-component composite membrane would be viable for gas separations, increasing the mechanical stability of the membrane. It is also likely that, by altering the relative amounts of zeolite and RTIL in this three-component configuration, additional separation performance could be achieved. Additional performance improvement might also come from selection of different RTILs.

### 4.4.1.6 *Gelled Membranes in $CO_2$ Separation.* Supported liquid membranes possess superior transport properties compared with solid polymer membranes, as the diffusivity in the liquid is greater than that in the polymer. However, those liquid membranes are subject to loss due to evaporation into the gas phase as well as displacement or the liquid phase by differential pressures. Use of RTILs in SLMs eliminates the loss due to evaporation, having a negligible vapour pressure. The use of the RTIL in an SLM preserves the superior liquid transport properties.

However, the problem of RTIL displacement due to pressure differential remains. This mechanical deficiency can be overcome by use of a gelled membrane structure. The gelled membrane is a blend of free liquid RTIL and poly(RTIL) to form a homogeneous composite or gelled membrane [34]. This provides improved mechanical stability over the supported liquid option while retaining much of the performance of the SLM. Technically, this is a composite membrane, but as the liquid portion increases it becomes a gelled structure. The inclusion of the free liquid RTIL provides increased diffusivity in this gelled configuration. Effectively, the gelled membrane provides a solid polymer-like membrane structure with liquid-like gas transport.

Voss et al. [37] fabricated and tested a gelled membrane for $CO_2$ gas separation and compared the results obtained with other membrane configurations. They used an imidazolium-based RTIL, $[C_6mim][NTf_2]$, in their work [37]. The gel was fabricated using an LMOG (12-hydroxystearic acid). In general, imidazolium-based RTILs have excellent solubility and solubility selectivity for $CO_2$ over $N_2$. The gelled RTIL membrane yielded gas separation performance properties on a par with those obtained from a "neat" RTIL membrane (a supported RTIL membrane). The $CO_2$ gas permeabilities obtained were 650 Barrers for the gelled membrane and 700 Barrers for the neat RTIL SILM. Selectivities for $CO_2/N_2$ were 31 and 33, respectively, for the gel and the neat SILM, where "neat" refers to a pure liquid RTIL.

This is a significant result in that it demonstrates that the gas separation properties of the gel are nearly as good as those of the supported RTIL membrane, and possess superior mechanical properties. In addition, these results are similar to results for other RTIL membrane configurations that are potentially viable candidates for industrial gas separations.

### 4.4.2   Barrier Materials

Currently, protective garment materials for protection against hazardous vapours such as CWAs and TICs are either dense, completely impermeable polymers (i.e., cross-linked butyl rubber) or composite membrane systems that rely on an active sorbent, such as activated carbon, to remove hazardous penetrants. The first type of protective material quickly generates heat stress in the wearer because evaporative cooling is blocked, and the second type is often bulky, heavy, and has a limited capacity and shelf life.

New functionalised poly(RTIL) composite membrane materials that allow facile water vapour transport for evaporative cooling while completely blocking the transport of the blister CWA simulant, CEES (2-chloroethyl ethyl sulfide; $C_2H_5SCH_2CH_2Cl$), have recently been reported. The first system is based on a new hydrophilic poly(diol-RTIL) that readily transports water vapour [52], blended with a commercial organic amine or copolymerised with an amine-containing co-monomer. For example, an approximately 150-$\mu$m-thick film of a 9:1 composite of poly(diol-RTIL)/$N,N,N',N'$-tetramethylhexane-1,6-diamine, Figure 4.12(a), was found to exhibit a water vapour flux at room temperature of 3770 g m$^{-2}$·day while completely blocking CEES vapour transport. The desired water vapour flux for military protective garments is 1500–

2000 g m$^{-2}$·day$^{-1}$ [53]. In comparison, a pure poly(diol-RTIL) film of the same approximate thickness, but without added amine, shows a water vapour flux of 6200 g m$^{-2}$·day$^{-1}$ and 100% CEES penetration. Also, a regular non-hydroxylated poly(RTIL) film with added amine is able to completely block CEES penetration but has essentially no measurable water vapour transport.

A second system is based on the same hydrophilic poly(diol-RTIL) that exhibits high water vapour breathability, but it is blended with basic zeolite particles (e.g., Na–zeolite Y) instead of amine additives. This is the same approach as previously reported for gas separations [34]. An approximately 150-$\mu$m-thick film of this material containing 20 wt% Na–zeolite Y was found to have a water vapour flux of 2870 g m$^{-2}$·day$^{-1}$, while completely blocking CEES vapour transport. Mechanistic studies on these two breathable poly(RTIL) composite systems for blister agent (i.e., mustard gas (HD); bis(2-chloroethyl)sulfide; $S(CH_2CH_2Cl)_2$) protection are currently in progress with CEES, in order to determine their mechanisms of action.

Bara et al. [33] reported permeability results from membranes fabricated from cross-linked GRTILs. They found that the permeabilities to the gases $CO_2$, $N_2$, $CH_4$, and $H_2$ were much lower when compared with previously studied poly(RTIL) membranes. They concluded that further investigation of these types of membranes was warranted for possible application in barrier materials.

### 4.4.3 Separations in the Liquid Phase

Considering the possibility of designing extremely hydrophobic ionic liquids, such as imidazolium-based ionic liquids with long alkyl side chains (e.g., [C$_8$mim]$^+$-based ionic liquids), the potential use of these ionic liquids in aqueous/IL membrane contactors and/or aqueous/IL/aqueous SLMs seemed rather attractive. Actually this, and other hydrophobic ionic liquids, proved to be rather selective for the extraction of target hydrophobic solutes from aqueous media, such as esters of amino acids and other organic compounds [15, 54, 40]. When these ionic liquids were tested in SILMs, for integrated extraction/re-extraction of hydrophobic solutes from/to aqueous environments, the initial target solute selectivity was observed to decrease progressively until there was a total loss of selectivity. This disappointing behaviour was understood when transport studies using tritiated water, added to the feed compartment, demonstrated clearly that water transport occurs between the two aqueous compartments through the hydrophobic ionic liquid. This behaviour was also confirmed by the transport of sodium chloride, insoluble in very hydrophobic ionic liquids, whose transport through this type of SILMs was explained assuming that water clusters develop inside the hydrophobic ionic liquid, providing the necessary micro-environment for solubilising sodium chloride. The formation of dynamic water clusters inside ionic liquids was also confirmed experimentally and by recent dynamic simulation studies.

As a consequence of this behaviour, the practical application of SILMs for solute recovery from aqueous environments is limited, although several strategies have been developed in order to hinder and retard the development of water clusters inside the ionic liquid phase. In situations where a prolonged operation is not required, it may be possible, by using these methodologies, to operate during a time window where a desirable selectivity is maintained [42].

This behaviour under the presence of water also raises important questions when operating with vapour streams containing a non-negligible amount of water vapour. Under these circumstances, strategies for controlling the presence of water vapour in the vapour phase have to be taken into consideration. In specific cases, the ability to develop water clusters inside ionic liquids, even if they are extremely hydrophobic, may have an advantage. This is the case of enzymatic catalysis in ionic liquids, where a controlled water activity inside the ionic liquid is necessary in order to assure an optimal enzyme conformation and corresponding activity and selectivity.

The use of ionic liquid membrane contactors and SILMs for solute recovery and transport from/to organic solvents has revealed to be less prone to stability problems. A large number of successful applications have been referred to in the literature for the selective recovery of target solutes from organic media, their transport through the ionic liquid phase, and recovery in a receiving organic media. One of the first applications described in the literature [15] refers to the resolution of the isomeric amines diisopropylamine (DIIPA) and triethylamine (TEA), which were continuously fractionated over two weeks without any observable decrease in selectivity, using a SLM that contained the RTIL 1-butyl-3-methylimidazolium hexafluorophosphate immobilised in the porous structure of a hydrophilic polyvinylidene fluoride membrane. A higher affinity and transport rate was observed for the secondary amine, which establishes stronger hydrogen bonding with the imidazolium cation. The reason for a stronger hydrogen bonding for the secondary amine DIIPA over that for the tertiary amine TEA may be attributed to the combination of the effect of the higher basicity of DIIPA ($pK_b = 2.95$), over that of TEA ($pK_b = 3.35$), and the steric hindrance of TEA observed on formation of the H–C(2) hydrogen bonding with the imidazolium cation (see Figure 4.10).

These studies demonstrate the feasibility of using ionic liquids as a new kind of solvent in SLMs for selective transport of organic molecules. Systematic experiments have been performed and reported in the literature [55–58] with different mixtures of compounds with representative organic functional groups. From these results, it can be concluded that the appropriate combination of selected ionic liquids and supporting membranes is crucial for achieving good selectivity in a given separation problem.

The emergence of new ionic liquids and the high variety of commercial supporting membranes will enable the design of IL/supporting membrane systems that allow one to obtain the desired selectivity for a specific substrate mixture. The high selectivity obtained for the separation of mixtures with very

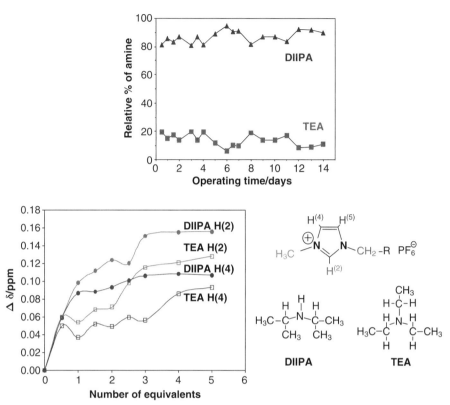

**Figure 4.10** Selective transport of diisopropylamine (DIIPA) over triethylamine (TEA), using a 1-butyl-3-methylimidazolium hexafluorophosphate SLM.

similar boiling points demonstrates the potential for the use of this methodology for continuous separation of compounds from complex mixtures, namely separations difficult to achieve by using traditional distillation methods.

Finally, it is also worth mentioning the integration of catalysis in ionic liquid media with membrane processing, for catalyst and solvent reuse. Actually, the post-reaction separation of ionic liquids and catalysts from reaction products is an unresolved challenge in the application of ionic liquids to organometallic catalysis. Recent reports [59–61] addressed this challenge using organic solvent nanofiltration technology. As an example [62], Suzuki reactions were carried out in a homogeneous solution, comprising 50:50 wt% ethyl ethanoate and ionic liquid. The post-reaction mixture was diluted further with ethyl ethanoate and then separated by nanofiltration into a permeate fraction and a retained (retentate) fraction. The product was recovered in the nanofiltration permeate, while the ionic liquid and palladium catalyst were retained by the membrane and recycled into subsequent consecutive reactions. The organic solvent nanofiltration was able to separate the Suzuki reaction product from both catalyst and ionic liquid. All the ionic liquids screened showed positive effects on the catalytic stability, significantly reducing the formation of palladium black and providing high reaction yields over consecutive recycles.

### 4.4.4 Fuel Cells and Electrochemical Applications

Although polymer-electrolyte membrane (PEM) fuel cells have been known for a long time, they have not yet reached large-scale development as some issues are still unresolved. These are mainly related to limited functional characteristics of the perfluorosulfonic acid membrane electrolyte, whose conductivity relies on the level of water content. Consequently, its stability is limited to temperatures below ~100 °C. Proton exchange membranes that operate at temperatures above 120 °C are needed to avoid catalyst poisoning, enhance electrochemical reactions, simplify the design, and reduce the cost of fuel cells. In order to design new membranes for elevated temperature operation, it is necessary to understand the chemistry, morphology, and dynamics of protons and water molecules in current membranes. Based on such fundamental understanding, membranes can be modified by controlling the polymer chemistry and architecture, or by adding inorganic fillers that can retain water under relatively low humidity conditions.

In this line of thought, Nafion® membranes have been modified by partially replacing protons for ionic liquid cations. This modification aims at promoting the water solvation of these cations, with a consequent increase in water structuring and retention even at higher operating temperatures [41]. Other approaches involve the development of anhydrous membranes based on phosphoric acid doped polymers, ionic liquid-infused polymer gels, and solid acids which enable fuel cell operation above 150 °C [63, 64]. Considerable work remains to be done to identify proton transport mechanisms in novel membranes and evaluate membrane stability under real operating conditions.

This issue may be also addressed by switching to ionic liquid-based polymer membranes. Various strategies have been tested for the preparation of these membranes, including polymerisation of the components with the formation of polycations, polyanions, copolymers and special, double-ion structures, as well as blends with a neutral macromolecule, typically polyvinylidene fluoride [65]. These architectures offer a water-independent proton conductivity associated with a high thermal stability, which allows high operational temperatures, as well as basic pHs. These approaches open new possibilities in terms of catalyst design, with the ultimate goal of replacing platinum, which is impractically rare.

In this case, the proton relay molecule, instead of being $H_2O$, is an imidazole or an amine. The operating temperature is higher, and the environmental pH increases from 7 to 11. Neutral to basic pHs may open new possibilities in terms of catalyst design, the media being far less aggressive than the highly acidic perfluorosulfonic membrane, with the ultimate goal of replacing platinum [66]. The investigation of ionic liquid membranes is still in its infancy but their potential is important.

## 4.5  FUTURE DIRECTIONS

### 4.5.1  Bioreactive Systems

The literature describes a large number of studies where enzymatic catalysis was accomplished with advantage in ionic liquid media [55, 67]. As ionic liquids can be designed with different cation and anion combinations, allowing for tailoring reaction solvents with specific desired properties, they provide the opportunity to carry out many important biocatalytic reactions that are difficult to perform in traditional solvents. As compared with those observed in conventional organic solvents, the use of enzymes in ionic liquids may present significant advantages, such as high conversion rates, high enantioselectivity, better enzyme stability, as well as better recoverability and recyclability [68]. To date, a wide range of approaches has been taken to further improve the performance of enzymes in ionic liquids.

The use of enzymatic membrane reactors, where the membrane may play multiple roles—support for interface stabilisation if a biphasic system is used (membrane contactor) and support for enzyme immobilisation (or, at least, micro-environment for enzyme accommodation)—has been also widely described in the literature [69]. Identification of the best strategy for solubilising enzymes into ionic liquids while keeping or improving their, activity, selectivity, and stability remains the main challenge. Mixtures of ionic liquids with a controlled amount of water, called hydrated ionic liquids, may strongly influence the protein conformation and ultimately, its function. It is important to note that the chemical activity of water remains very low because water molecules are strongly involved in the solvation of the ions. Cholinium dihydrogenphosphate ($[N_{1\,1\,1\,2OH}][H_2PO_4]$), composed of a chaotropic cation and a kosmotropic anion, is a typical example. Hydrated (20% water:$[N_{1\,1\,1\,2OH}]$$[H_2PO_4]$) was shown to be an excellent solvent for cytochrome c and other proteins. In this liquid, cytochrome c was reported to show electron transfer activity even 18 months after storage at room temperature [70]. The small amount of water provides the required hydrogen-bonding environment for the protein, assuring that it can adopt a conformation that guarantees its activity, selectivity, and stability.

Another area of research refers to the development of enzymatic fuel cells, which have already been demonstrated using aqueous electrolyte solutions. The use of ionic liquids instead of aqueous salt solutions prevents loss of enzymatic activity, as a result of the local environment that they provide [71]. Considering the large variety of substrates and enzymes available for biofuel cell applications, the use of ionic liquids as electrolytes paves the way for the development of novel bio-energy conversion systems, making biofuel cells attractive next generation energy-conversion devices.

### 4.5.2 Ionic Liquid Systems with Enhanced Selectivity

Since the first example of a chiral ionic liquid, 1,1-bis{(2S)-2-methylbutyl} imidazolium bromide, containing a chiral cation, was reported in 1997 [72], the number of publications dealing with chiral ionic liquids grew rapidly and, nowadays, a large pool bearing either chiral cations or anions (but seldom both) and a wide range of functionalities is available [73]. Despite the rapid design of new chiral ionic liquids, successful applications remained hidden for some time. Actually, nine years after the original publication, the use of a chiral ionic liquid in a synthesis with an enantiomeric excess (>90%) was reported for the first time [74]. Nevertheless, this field is growing rapidly, and applications are generally divided into three different groups: chiral ionic liquids in asymmetric synthesis, spectroscopic applications of chiral ionic liquids, and chromatographic applications of chiral ionic liquids.

The combination of chiral ionic liquids with membranes has been scarcely reported in the literature although the potential for its use has been recognised [75]. Developments achieved with chiral ionic liquids as stationary phases for gas chromatography may be adapted with advantage using membrane contactor technology. Ephedrinium-based* chiral ionic liquids constitute a good example. In 2004, the group of Armstrong [76] published the first direct enantiomeric separation of compounds by using an ephedrinium-based stationary phase in gas chromatography. A *N,N*-dimethylephedrinium-based chiral ionic liquid, previously described by Wasserscheid, was coated on a fused-silica capillary column with a brown polyimide layer to generate a new chiral stationary phase, which was able to successfully separate a range of chiral alcohols and diols, chiral sulfoxides, and some chiral epoxides and acetamides.

Chiral ionic liquids may be associated in membranes through the development of SLMs, where an ionic liquid phase is immobilised inside the membrane porous support or through the development of membrane polymers integrating the chemistry of selected chiral ionic liquids (see previous discussion in this chapter). Making use of one of the most interesting advantages of membranes—contrarily to chromatographic processes, where particles are surrounded by a single environment, membranes can face two different environments simultaneously—integration with chiral ionic liquids can allow for a simultaneously enrichment of each contacting phase in a selected enantiomeric form. Alternatively, chiral ionic liquid-modified membranes can also be used in membrane chromatography processes.

### 4.5.3 Stimuli-Responsive Ionic Liquid Systems

Many ionic liquids contain transition metals. Therefore, it is not surprising that ionic liquids with pronounced magnetic properties have been described. Among the first examples were those reported by Okuno et al. [77], who

---

* Ephedrine = (R*,S*)-2-(methylamino)-1-phenylpropan-1-ol.

described the ionic liquid $[C_4mim][FeCl_4]$ as a material with a very large magnetic susceptibility and demonstrated that non-magnetic materials can be transported and separated readily in this ionic liquid with the help of magnetic-field gradients. Recently, ionic liquids responsive to magnetic-field gradients have been used in SLM systems showing adjustable transport properties for target solutes, based on the applied external magnetic field.

The facile tuneability of the physicochemical properties of ionic liquids can be used to modulate the wetting behaviour of polymer films and membranes. Recently [78], modified negatively charged polyimide films have been prepared using ionic liquids based on electrostatic self-assembly rather than using covalent links. To reveal the effect of anions on surface wettability, the authors screened a series of anions including $Br^-$, $[BF_4]^-$, $[PF_6]^-$, $[NTf_2]^-$, $[OTf]^-$, $[ClO_4]^-$, and $[NO_3]^-$. This was the first time that the reversible wettability of polymer surfaces was modulated using ionic liquids.

Like anions, the change in cations in ionic liquids also has the ability to modulate surface wettability, which was demonstrated by using the series $[C_2mim]Br$, $[C_4mim]Br$, $[C_6mim]Br$, and $[C_8mim]Br$ [78]. Using ionic liquids to modify surfaces provides a new strategy for reversible switching between hydrophobicity and hydrophilicity. However, the maximum contact angle can only reach up to $95°$, and the highest change of contact angle is less than $45°$. No switching between superhydrophobicity and superhydrophilicity via the exchange of counterions of ionic liquids has been reported yet.

## REFERENCES

1   Earle, M.J., Esperança, J.M.S.S., Gilea, M.A., Lopes, J.N.C., Rebelo, L.P.N., Magee, J.W., Seddon, K.R., and Widegren, J.A., The distillation and volatility of ionic liquids, *Nature* **439**, 831–834 (2006).

2   Plechkova, N.V., and Seddon, K.R., Applications of ionic liquids in the chemical industry, *Chem. Soc. Rev.* **37**, 123–150 (2008).

3   Chauvin, Y., Hirschauer, A., and Olivier, H., Alkylation of isobutane with 2-butene using 1-butyl-3-methylimidazolium chloride aluminum-chloride molten salts as catalysts, *J. Mol. Catal.* **92**, 155–165 (1994).

4   Domańska, U., Laskowska, M., and Marciniak, A., Phase equilibria of (1-ethyl-3-methylimidazolium ethylsulfate plus hydrocarbon, plus ketone, and plus ether) binary systems, *J. Chem. Eng. Data* **53**, 498–502 (2008).

5   Bendova, M., and Wagner, Z., Thermodynamic description of liquid-liquid equilibria in systems 1-ethyl-3-methylimidazolium ethylsulfate+c7-hydrocarbons by polymer-solution models, *Fluid Phase Equilibr.* **284**, 80–85 (2009).

6   Klamt, A., and Eckert, F., COSMO-RS: A novel and efficient method for the a priori prediction of thermophysical data of liquids, *Fluid Phase Equilibr.* **172**, 43–72 (2000).

7   Domańska, U., Pobudkowska, A., and Eckert, F., Liquid-liquid equilibria in the binary systems (1,3-dimethylimidazolium, or 1-butyl-3-methylimidazolium methyl-sulfate plus hydrocarbons), *Green Chem.* **8**, 268–276 (2006).

8  Scovazzo, P., Camper, D., Kieft, J., and Noble, R.D., Regular solution theory and $CO_2$-gas solubility in room temperature ionic liquids, *Ind. Eng. Chem. Res.* **43**, 6855–6860 (2004).

9  Camper, D., Scovazzo, P., and Noble, R.D., Gas solubilities in room temperature ionic liquids, *Ind. Eng. Chem. Res.* **43**, 3049–3054 (2004).

10  Camper, D., Becker, C., Koval, C., and Noble, R.D., Diffusion and solubility in room temperature ionic liquids, *Ind. Eng. Chem. Res.* **45**, 445–450 (2006).

11  Camper, D., Bara, J.E., Koval, C.A., and Noble, R.D., Bulk fluid solubility and membrane feasibility of RMIM-based room temperature ionic liquids, *Ind. Eng. Chem. Res.* **45**, 6279–6283 (2006).

12  Finotello, A., Bara, J.E., Camper, D., and Noble, R.D., Room temperature ionic liquids: Temperature dependence of gas solubility selectivity, *Ind. Eng. Chem. Res.* **47**, 3453–3459 (2008).

13  Finotello, A., Bara, J.E., Camper, D., and Noble, R.D., Ideal gas solubilities and solubility selectivities in a binary mixture of ionic liquids, *J. Phys. Chem. B* **112**, 2335–2339 (2008).

14  Scovazzo, P., Visser, A.E., Davis, J.H., Rogers, R.D., Noble, R.D., and Koval, C., Supported ionic liquid membranes and facilitated ionic liquid membranes, *Abstr. Pap. Am. Chem. S.* **221**, U558, 2001.

15  Branco, L.C., Crespo, J.G., and Afonso, C.A.M., Studies on the selective transport of organic compounds using ionic liquids as novel supported liquid membranes, *Chemistry* **8**, 3865–3871 (2002).

16  Luís, P., Garea, A., and Irabien, A., Quantitative structure-activity relationships (QSARs) to estimate ionic liquids ecotoxicity EC50 (*Vibrio ficheri*), *J. Mol. Liq.* **152**, 28–33 (2010).

17  Frade, R.F.M., and Afonso, C.A.M., Impact of ionic liquids in environment and humans: An overview, *Hum. Exp. Toxicol.* **29**, 1038–1054 (2010).

18  Prasad, R., and Sirkar, K.K., Microporous membrane solvent extraction, *Separ. Sci. Technol.* **22**, 619–640 (1987).

19  Zhang, Q., and Cussler, E.L., Hollow fiber gas membranes, *AIChE Journal* **31**, 1548–1553 (1985).

20  Sirkar, K.K., Membranes, phase interfaces, and separations: Novel techniques and membranes—An overview, *Ind. Eng. Chem. Res.* **47**, 5250–5266 (2008).

21  Cho, T.H., Fuller, J., and Carlin, R.T., Catalytic hydrogenation using supported ionic liquid membranes, *High Temp. Mater. P.-US* **2**, 543–558 (1998).

22  Rogers, R.D., Visser, A.E., Davies, J.H., Koval, C., DuBois, D.L:, Scovazzo, P., and Noble, R.D., Choosing ionic liquids for supported ionic liquid membranes, *Abstr. Pap. Am. Chem. S.* **223**, U647–U647 (2002).

23  Fortunato, R., Afonso, C.A.M., Benavente, J., and Crespo, J.G., Stability of supported ionic liquid membranes as studied by X-ray photoelectron spectroscopy, *J. Membr. Sci.* **256**, 216–223 (2005).

24  Visser, A.E., Swatloski, R.P., Reichert, W.M., Mayton, R., Sheff, S., Wierzbicki, A., Davies, J.H., and Rogers, R.D., Task-specific ionic liquids for the extraction of metal ions from aqueous solutions, *Chemical Commun.* **1**, 135–136 (2001).

25  Bates, E.D., Mayton, R.D., Ntai, I., and Davis, J.H., $CO_2$ capture by a task-specific ionic liquid, *J. Am. Chem. Soc.* **124**, 926–927 (2002).

26  Scovazzo, P., Kieft, J., Finan, D.A., Koval, C., DuBois, D., and Noble, R., Gas separations using non-hexafluorophosphate [PF$_6$]$^-$ anion supported ionic liquid membranes, *J. Membr. Sci.* **238**, 57 (2004).

27  Bara, J.E., Hatakeyama, E.S., Gin, D.L., and Noble, R.D., Improving CO$_2$ permeability in polymerized room-temperature ionic liquid gas separation membranes through the formation of a solid composite with a room-temperature ionic liquid, *Polymer. Adv. Tech.* **19**, 1415–1420 (2008).

28  Bara, J.E., Gin, D.L., and Noble, R.D., Effect of anion on gas separation performance of polymer room temperature ionic liquid composite membranes, *Ind. Eng. Chem. Res.* **47**, 9919–9924 (2008).

29  Bara, J.E., Noble, R.D., and Gin, D.L., Effect of "free" cation substituent on gas separation performance of polymer–room-temperature ionic liquid composite membranes, *Ind. Eng. Chem. Res.* **48**, 4607–4610 (2009).

30  Branco, L.C., Crespo, J.G., and Afonso, C.A.M., High selective transport of organic compounds using supported liquid membranes based on ionic liquids, *Angew. Chem. Int. Edit.* **41**, 2771–2773 (2002).

31  Fortunato, R., Afonso, C.A.M., Reis, M.A.M., and Crespo, J.G., Supported liquid membranes using ionic liquids: Study of stability and transport mechanisms, *J. Membr. Sci.* **242**, 197–209 (2004).

32  Bara, J.E., Lessmann, S., Gabriel, C.J., Hatakeyama, E.S., Noble, R.D., and Gin, D.L., Synthesis and performance of polymerizable room-temperature ionic liquids as gas separation membranes, *Ind. Eng. Chem. Res.* **46**, 5397–5404 (2007).

33  Bara, J.E., Hatakeyama, E.S., Gabriel, C.J., Zeng, X., Lessman, S., Gin, D.L., and Noble, R.D., Synthesis and light gas separations in cross-linked gemini room temperature ionic liquid polymer membranes, *J. Membrane Sci.* **316**, 186–191 (2010).

34  Zhao, D., Fei, Z., Scopelliti, R., and Dyson, P.J., Synthesis and characterization of ionic liquids incorporating the nitrile functionality, *Inorg. Chem.* **43**, 2197 (2004).

35  Hudiono, Y.C., Carlisle, T.K., Bara, J.E., Zhang, Y., Gin, D.L., and Noble, R.D., A three-component mixed-matrix membrane with enhanced CO$_2$ separation properties based on zeolites and ionic liquid materials, *J. Membrane Sci.* **350**, 117–123 (2010).

36  Hudiono, Y.C., Carlisle, T.K., LaFrate, A.L., Gin, D.L., and Noble, R.D., Novel mixed matrix membranes based on polymerizable room-temperature ionic liquids and SAPO-34 particles to improve CO$_2$ separation, *J. Membrane Sci.*, **370**, 141–148 (2011).

37  Voss, B.A., Bara, J.E., Gin, D.L., and Noble, R.D., Physically gelled ionic liquids: Solid membrane materials with liquid-like CO$_2$ gas transport, *Chem. Mater.* **21**, 3027–3029 (2009).

38  Schäfer, T., Di Paolo, R.E., Franco, R., and Crespo, J.G., Elucidating interactions of ionic liquids with polymer films using confocal raman spectroscopy, *Chem. Commun.* **20**, 2594–2596 (2005).

39  Fortunato, R., Branco, L.C., Afonso, C.A.M., Benavente, J., and Crespo, J.G., Electrical impedance spectroscopy characterisation of supported ionic liquid membranes, *J. Membr. Sci.* **270** (1–2), 42–49 (2006).

40  De los Ríos, A.P., Hernández-Fernández, F.J., Tomás-Alonso, F., Palacios, J.M., Gómez, D., Rubio, M., and Víllora, G., A SEM–EDX study of highly stable

supported liquid membranes based on ionic liquids, *J. Membr. Sci.* **300**, 88–94 (2007).

41 Neves, L.A., Benavente, J., Coelhoso, I.M., and Crespo, J.G., Design and characterisation of nafion membranes with incorporated ionic liquids cations, *J. Membr. Sci.* **347**,42–52 (2010).

42 Izak, P., Hovorka, S., Bartovsky, T., Bartovsky, L., and Crespo, J.G., Swelling of polymeric membranes in room temperature ionic liquids, *J. Membr. Sci.* **296**, 131–138 (2007).

43 Neves, L., Sebastião, P., Coelhoso, I.M., and Crespo, J.G., Proton, NMR relaxometry study of nafion membranes modified with ionic liquid cations, *J. Phys. Chem. B* **115**, 8713–8723 (2011).

44 Bara, J.E., Kaminski, A.K., Noble, R.D., and Gin, D.L., Influence of nanostructure on light gas separations in cross-linked lyotropic liquid crystal membranes, *J. Membr. Sci.* **288**, 13–19 (2007).

45 Robeson, L.M., Correlation of separation factor versus permeability for polymeric membranes, *J. Membr. Sci.* **62**, 165–185 (1991).

46 Freeman, B.D., Basis of permeability/selectivity tradeoff relations in polymeric gas separation membranes, *Macromolecules* **32**, 375–380 (1999).

47 Robeson, L.M., The upper bound revisited, *J. Membr. Sci.* **320**, 390–400 (2008).

48 Baker, R.W., and Lokhandwala, K., Natural gas processing with membranes: An overview, *Ind. Eng. Chem. Res.* **47**, 2109–2121 (2008).

49 Baker, R.W., Future directions of membrane gas separation technology, *Ind. Eng. Chem. Res.* **41**, 1393–1411 (2002).

50 Favre, E., Carbon dioxide recovery from post-combustion processes: Can gas permeation membranes compete with absorption?, *J. Membr. Sci.* **294**, 50–59 (2007).

51 Camper, D., Bara, J.E., Gin, D.L., and Noble, R.D., Room-temperature ionic liquid–amine solutions: Tunable solvents for efficient and reversible capture of $CO_2$, *Ind. Eng. Chem. Res.* **47**, 8496–8498 (2008).

52 LaFrate, A.L., Gin, D.L., and Noble, R.D., High water vapor flux membranes based on novel diol-imidazolium polymers, *Ind. Eng. Chem. Res.* **49**, 11914–11919 (2010).

53 LaFrate, A.L., Bara, J.E., Gin, D.L., and Noble, R.D., Synthesis of diol-functionalized imidazolium-based room-temperature ionic liquids with bis(trifluoromethane sulfonimide) anions that exhibit switchable water miscibility, *Ind. Eng. Chem. Res.* **48** (18) 8757–8759 (2009).

54 Matsumoto, M., Inomoto, Y., and Kondo, K., Selective separation of aromatic hydrocarbons through supported liquid membranes based on ionic liquids, *J. Membrane Sci.* **246**, 77–81 (2005).

55 Hernández-Fernández, F.J., De los Ríos, A.P., Rubio, M., Tomas-Alonso, F., Gomez, D., and Víllora, G., A novel application of supported liquid membranes based on ionic liquids to the selective simultaneous separation of the substrates and products of a transesterification reaction, *J. Membrane Sci.* **293**, 73–80 (2007).

56 Matsumoto, M., Inomoto, Y., and Kondo, K., Comparison of solvent extraction and supported liquid membrane permeation using an ionic liquid for concentrating penicillin G, *J. Membrane Sci.* **289**, 92–96 (2007).

57  Marták, J., Schlosser, Š., and Vlčková, S., Pertraction of lactic acid through supported liquid membranes containing phosphonium ionic liquid, *J. Membrane Sci.* **318**, 298–310 (2008).

58  De los Ríos, A.P., Hernandez-Fernandez, F.J., Presa, H., Gomez, D., and Villora, G., Tailoring supported ionic liquid membranes for the selective separation of transesterification reaction compounds, *J. Membrane Sci.* **328**, 81–85 (2009).

59  Han, S., Wong, H.T., and Livingston, A.G., Application of organic solvent nanofiltration to separation of ionic liquids and products from ionic liquid mediated reactions, *Chem. Eng. Design* **83**, 309–316 (2005).

60  Wong, H.T., See-Toh, Y.H., Ferreira, F.C., Crook, R., and Livingston, A.G., Organic solvent nanofiltration in asymmetric hydrogenation: Enhancement of enantioselectivity and catalyst stability by ionic liquids, *Chem. Commun.* **19**, 2063–2065 (2006).

61  Ferreira, F.C., Branco, L.C., Kaushal, K.V.A., Crespo, J.G., and Afonso, C.A.M., Application of nanofiltration to re-use the sharpless asymmetric dihydroxylation catalytic system, *Tetrahedron-Asymmetr.* **18**, 1637–1641 (2007).

62  Wong, H.T., Pink, C.J., Ferreira, F.C., and Livingston, A.G., Recovery and reuse of ionic liquids and palladium catalyst for Suzuki reactions using organic solvent nanofiltration, *Green Chem.* **8**, 373–379 (2006).

63  Fuller, J., Breda, A.C., and Carlin, R.T., Ionic liquid-polymer gel electrolytes, *J. Electrochem. Soc.* **144**, L67-L70 (1997).

64  Navarra, M.A., Panero, S., and Scrosati, B., Novel, ionic-liquid-based, gel-type proton membranes, *Electrochem. Solid St.* **8**, A324–A327 (2005).

65  Susan, M.A., Kaneko, T., Noda, A., and Watanabe, M., Ion gels prepared by *in situ* radical polymerization of vinyl monomers in an ionic liquid and their characterization as polymer electrolytes, *J. Am. Chem. Soc.* **127**, 4976–4983 (2005).

66  Armand, M., Endres, F., MacFarlane, D.R., Ohno, H., and Scrosati, B., Ionic-liquid materials for the electrochemical challenges of the future, *Nature Mater.* **8**, 621–629 (2009).

67  Miyako, E., Maruyama, T., Kamiya, N., and Goto, M., Enzyme-facilitated enantioselective transport of (S)-ibuprofen through a supported liquid membrane based on ionic liquids, *Chem. Commun.* **23**, 2926–2927 (2003).

68  Moniruzzaman, M., Kamiya, N., and Goto, M., Activation and stabilization of enzymes in ionic liquids, *Org. Biomol. Chem.* **8**, 2887–2899 (2010).

69  Mori, M., Garcia, R.G., Belleville, M.P., Paolucci-Jeanjean, D., Sanchez, J., Lozano, P., Vaultier, M., and Rios, G., A new way to conduct enzymatic synthesis in an active membrane using ionic liquids as catalyst support, *Catal. Today* **104**, 313–317 (2005).

70  Fujita, K., MacFarlane, D.R., Forsyth, M., Yoshizawa-Fujita, M., Murata, K., Nakamura, N., and Ohno, H., Solubility and stability of cytochrome c in hydrated ionic liquids: Effect of oxo acid residues and kosmotropicity, *Biomacromolecules* **8**, 2080–2086 (2007).

71  Fujita, K., Forsyth, M., MacFarlane, D.R., Reid, R.W., and EIliott, G.D., Unexpected improvement in stability and utility of cytochrome c by solution in biocompatible ionic liquids, *Biotechnol. Bioeng.* **94**, 1209–1213 (2006).

72  Howarth, J., Hanlon, K., Fayne, D., and McCormac, P., Moisture stable dialkylimid-azolium salts as heterogeneous and homogeneous Lewis acids in the Diels-Alder reaction, *Tetrahedron Lett.* **38**, 3097–3100 (1997).

73  Bica, K., and Gaertner, P., Applications of chiral ionic liquids, *Eur. J. Org. Chem.* **19**, 3235–3250 (2008).

74  Gausepohl, R., Buskens, P., Kleinen, J., Bruckmann, A., Lehmann, C.W., Klanker-mayer, J., and Leitner, W., Highly enantioselective aza-baylis-hillman reaction in a chiral reaction medium, *Angew. Chem.-Int. Edit.* **45**, 3689–3692 (2006).

75  Afonso, C.A.M., and Crespo, J.G., Recent advances in chiral resolution using membrane-based approaches, *Angew.Chem.Int.Edit.* **43**,5293–5295(2004).

76  Ding, J., Welton, T., and Armstrong, D.W., Chiral ionic liquids as stationary phases in gas chromatography, *Anal. Chem.* **76**, 6819–6822 (2004).

77  Okuno, M., Hamaguchi, H.O., and Hayashi, S., Magnetic manipulation of materials in a magnetic ionic liquid, *Appl. Phys. Lett.* **89**, 132506 (2006).

78  Zhao, Y., Li, M., and Lu, Q.H., Tunable wettability of polyimide films based on electrostatic self-assembly of ionic liquids, *Langmuir* **24**, 3937–3943 (2008).

# 5 Engineering Simulations

DAVID ROONEY and NORFAIZAH AB MANAN

QUILL Research Centre, School of Chemistry and Chemical Engineering, Belfast, UK

## ABSTRACT

Over recent years, ionic liquids have emerged as a class of novel fluids that have inspired the development of a number of new products and processes. The ability to design these materials with specific functionalities and properties means that they are highly relevant to the growing philosophy of chemical-product design. This is particularly appropriate in the context of a chemical industry that is becoming increasingly focussed on small-volume, high-value added products with relatively short times to market. To support such product and process development, a number of tools can be utilised. A key requirement is that the tool can predict the physical properties and activity coefficients of multi-component mixtures and, if required, model the process in which the materials will be used.

Multi-scale simulations that span density functional theory (DFT) to process-engineering computations can address the relevant time and length scales and have increased in usage with the availability of cheap and powerful computers. Herein we will discuss the area of engineering calculations relating to the design of ionic liquid processes, that is, the computational tools that bridge this gap and allow for process simulation tools to utilise and assist in the design of ionic liquids.

It will be shown that, at present, it is possible to use available tools to estimate many important properties of ionic liquids and mixtures containing them with a sufficient level of accuracy for preliminary design and selection.

*Ionic Liquids Further UnCOILed: Critical Expert Overviews*, First Edition.
Edited by Natalia V. Plechkova and Kenneth R. Seddon.
© 2014 John Wiley & Sons, Inc. Published 2014 by John Wiley & Sons, Inc.

## 5.1 INTRODUCTION

Over the last decade or so, ionic liquids have emerged as a class of novel fluids that have inspired researchers to develop new products and processes. The synthesis of these materials leads naturally to approaches being adopted which consider desired functionalities and properties at the outset, and thus ionic liquids are ideally placed to benefit from modern chemical-product design philosophies. Diverse applications spanning sensors, thermal fluids, energetic materials, ion propulsion, embalming fluids, and many more have all led to specific products in recent years. At the larger scale, their utilisation within full and pilot scale industrial processes has also been realised. Here again, success has relied on either the phase behaviour or catalytic properties of the ionic liquids generated or used, thus demonstrating the importance of specific and desired physical or chemical properties.

The ability to design ionic liquids with specific functionalities is one of the greatest attractions, particularly when combined with other well-publicised properties such as significantly lower vapour pressures when compared to molecular counterparts. While such designer functionalities are not limited to the field of ionic liquids, these materials have arguably exemplified the modern approach to chemical-product design to a greater extent than any other recent technology. Sustained growth in ionic liquids research is therefore likely to continue for some time, particularly given that this area has coincided with a marked shift within a chemical industry that is now increasingly focussing on small-volume, high-value added products as opposed to bulk low-value products.

To support the industrial design process, "conceptual" or "process synthesis" tools are increasingly used. Such tools offer users a method of evaluating large numbers of potential routes to the manufacture of a specific product using a combination of experimental and mathematical analyses. Typically, these tools have been applied to the design of continuous processes by combining knowledge of the necessary design parameters, that is, kinetics for reaction, activity coefficients for distillation, and so on, with accurate data on the physical properties of the streams involved and how these change with temperature, pressure, and composition. These models can subsequently be used to theoretically test different designs and maximise efficiency of a single unit or the whole plant. Increasingly important is the use of such tools for the design of energy recycle and efficiency systems, and for the consideration of environmental impact. Here, the steady-state behaviour inherent to continuous processes is beneficial to developing the mathematical tools needed to both select and then optimise a process design or flowsheet. Flowsheet simulations themselves have their origins in a number of published papers relating to unit operation models in the 1950s [1]. These individual models (i.e., distillation, mixing) were then combined together into full flowsheet programs in the 1960s. This was obviously assisted by the evolution of computer technology and the increasing speed and capacity of these machines for handling relatively large and integrated computations. Over time, these advances in terms of computational power, together with increased availability and enhanced program–user interface interaction, have allowed such tools to be used more than ever in process evaluation, and as such they have become part of typical undergraduate chemical engineering programmes.

In one of our early publications, which considered the engineering of ionic liquids, we highlighted that efficient process design necessitated a complete understanding of the behaviour of these materials under operating conditions [2]. At the time, important physical properties such as viscosity, density, heat capacity, and surface tension, which are necessary for inclusion into design equations for pumping systems, heat exchangers, and so on, were relatively scarce. The commonly available tools used to predict missing physical properties had of course been developed for molecular compounds, but not for ionic liquids. These models thus tended to require information such as the critical temperature and pressure, that is, properties that do not apply to ionic liquids but that nonetheless were subsequently calculated for ionic liquids [3–5]. Similarly, methods for predicting physical properties like surface tension, density, heat capacity, and so on using group contribution methods were simply not available due to a lack of published data, and again standard approaches developed for molecular fluids failed to account for organic salts and therefore could not be used with any degree of confidence.

Over the past decade, databases have become available, such as IUPAC's ILThermo [6]. Since becoming operational, these databases have collected a significant amount of data including common physical properties and their measurement methods. A particular advantage of such repositories is that they have facilitated research into the development of predictive tools to generate group contribution parameters, and other similar structure–property relationships, for these novel fluids. The recent growth in this area has also been recognised within this book, as a number of chapters relate to physical properties of ionic liquids.

In addition to their potential use as solvents or catalysts in large processes, it is important to remember that ionic liquids are increasingly finding application as products in their own right. This area commonly referred to as "product design" is rapidly growing within chemical engineering as a result of the increased importance of developing new and novel products. Here, important objectives are to match the market needs to identified chemicals and/or their mixtures and to quickly evaluate key economic, market, and process design issues in order to support decisions in the early stages of development. As opposed to process design, major objectives are often time-to-market and performance of the required functionality. Further complications arise from the fact that the processes used in their manufacture are likely to be batch and, here, obtaining all the required design information is time-consuming (relative to time-to-market) and costly, and thus uneconomical, particularly where the modelling is further complicated by the inherent nature of the non-steady-state processes, thus leading to challenging design problems. While "hardware" tools such as modular engineering and micro-fabricated devices are increasingly being used to help to bridge the gap between batch and continuous processes, thus maintaining the main advantages of both, "software" tools have also advanced considerably. Therefore, in addition to device manufacture, there has been a significant growth in advanced computational tools used to aid the development of products and processes for the specialty chemical, pharmaceutical, and microstructured product sectors.

It is clear from the foregoing discussion that the engineering and design tools useful for the various development stages cover a very broad spectrum. Hence an important question here is "What in fact are engineering computations for ionic liquids?" Given the broad range of computational tools used in modern approaches to the engineering of products and processes, there would appear to be no one specific answer to this question. Indeed, Grossmann et al. [7] argued that process design and product design formally belong to the wide body of knowledge known as process systems engineering (PSE). As the computational tools have developed over recent years, so too has the scope of PSE, which has now expanded from the microscopic (molecules) to the macroscopic (process plant). A common way of representing such a multi-scale engineering approach is given in Figure 5.1. Traditional viewpoints would suggest that ionic liquids engineering calculations should specifically focus on problems relating to the design of larger plant with length scales in the order of kilometres and timescales in the order of years. However multi-scale approaches including ionic liquids, such as that shown in Figure 5.1, are now beginning to appear in the literature. One recent example is that by Tian et al. [8], where a multi-scale simulation method was proposed to enable screening of ionic liquids as

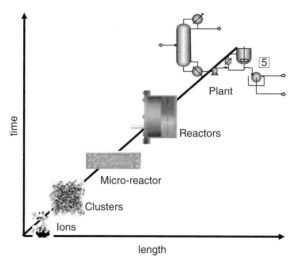

**Figure 5.1**   A multi-scale modelling approach to engineering computations using ionic liquids.

entrainers for the extractive distillation step in 1,3-butadiene production. Here, the specific term "multi-scale" was adopted, whereas others such as Brennecke et al., who had previously combined molecular simulation, experimental data, and process simulation in an effort to evaluate ionic liquids for gas separations, applied these principles but did not specifically refer to them [9, 10].

For product design applications, there is an obvious attraction towards the molecular level, which has promoted the creation of the area known as computer-aided design modelling (CADM). While this again is not specific to ionic liquids, it is a rapidly growing area and one which should significantly enhance ionic liquid design in the coming years. At the other end of this scale are modern process simulation packages, as discussed previously.

Herein we will focus on the engineering calculations relating to the design of industrial processes, that is, the computational tools that bridge this gap, and allow for process simulation tools to utilise ionic liquids. As will be shown, this can and does include aspects of CADM, as already demonstrated by Tian and Brennecke earlier.

## 5.2 ENGINEERING COMPUTATIONS FOR PROCESS DESIGN USING IONIC LIQUIDS

Process design itself is often described in the form of an "onion" diagram, as shown in Figure 5.2. Here the reactor forms the core of the process, with the separation and recycle systems being designed around the requirements of the reactor. Then, the heat integration recovery and waste treatment processes are included until the whole process is described. A similar approach can be used

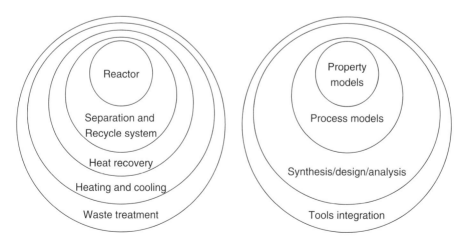

**Figure 5.2** Onion diagrams for process design in terms of unit operations (left) and calculation (right).

to describe the engineering computations used to support the design process. In this case, property models represent the core. Using these, the various unit operations can be designed based on their individual mathematical descriptions. As discussed earlier, many commercial flowsheeting software packages, such as ProSim, Pro/II, CHEMCAD, gProms, and Aspen Plus, include unit operation models of varying complexity and rigour, and give users the ability to link them together to form a complete process description.

It should be obvious that the validity of the simulation depends on the accuracy of the property correlation in describing the phase behaviour and properties of the mixture under the given process conditions, that is, temperature and pressure. Within a process, these conditions can vary widely and range from cryogenic to hundreds of degrees centigrade, and from vacuum to hundreds of bars, respectively. In terms of the general structure of algorithms for solving the equipment models and thermophysical properties correlation, several techniques, such as sequential-based modular, simultaneous-based modular, or equation-based techniques, have been developed to optimise and reduce the time required to solve the whole flowsheet of equipment. A review of these techniques has been published previously by Perkins [11].

Property models, which are at the heart of process design calculations and are specific to a discussion on ionic liquids, can be categorised into two areas: (1) those that can be used to correlate physical properties (i.e., heat capacity, thermal conductivity, viscosity, etc.) with temperature and others, and (2) thermodynamic models used for determining the state of matter under processing conditions (equations of state [EoS]) and the interactions between molecules (i.e., activity). As stated earlier, thermophysical properties have been discussed and reviewed elsewhere, and so the focus here will be on thermodynamic models, particularly those related to separation processes involving ionic liquids.

## 5.3   THERMODYNAMIC MODELS FOR IONIC LIQUIDS

In order to address the needs of engineers when undertaking product and process design of systems based on ionic liquids, investigation and measurement of phase equilibria and related thermophysical properties are essential. However, as previously discussed, experimental data collection for the desired range of temperature, pressure, and composition are both expensive and time-consuming. Therefore, thermodynamic modelling is a necessary step in determining the optimum operating conditions, process performance, and equipment sizing for a given unit operation. While considerable advancement has been achieved in estimating thermodynamic behaviour from simple molecular descriptors, this has not yet reached a sufficiently high accuracy to be used in isolation. Therefore, the application of any thermodynamic model to a particular system will require some experimental data in order to enhance and validate the applicability of the chosen model.

Studies of such models for ionic liquid-based systems began approximately a decade ago during investigation of aromatic extraction from mixtures of paraffin and aromatics using 1-ethyl-3-methylimidazolium triiodide, [C$_2$mim] [I$_3$], or 1-butyl-3-methylimidazolium triiodide, [C$_4$mim][I$_3$]. The non-random two liquid (NRTL) activity coefficient model [12] was chosen, as other Gibbs excess ($G^E$) models for liquid–liquid equilibria (LLE) such as van Laar and Wilson were unsuitable due to the non-ideality of this particular system. Popular alternatives, such as UNIQUAC and UNIFAC, required molecular parameters which were not available at that time. For similar reasons, EoS methods could not be evaluated due to a lack of critical properties, including acentric factors, dipole moment, polarisability, and so on. Furthermore, quantum chemical calculations, which are now increasingly common, were not sufficiently advanced in order to predict such behaviour for ionic liquid systems. However, as previously discussed, a significant increase in experimental data has occurred over the last decade, allowing for the correlation, modelling, and subsequent prediction of ionic liquid-based systems using a variety of the aforementioned models. Thus, the barriers derived from a lack of experimental data needed to determine group parameters, for example, and to verify such models, have diminished accordingly. Herein we will attempt to review recent progress in this area, and to aid this discussion the various models will be categorised into one of the following groups:

- Gibbs excess models
- EoS
- quantum chemical calculations

### 5.3.1   Gibbs Excess $G^E$ Models

After the initial work of Selvan et al. [12], Domanska and coworkers [13, 14] measured and correlated binary solid–liquid equilibrium (SLE) systems containing ionic liquids with alcohols using three different $G^E$ models (i.e., Wilson, modified NRTL [15], and UNIQUAC ASM [16]). The resulting correlation yielded standard mean deviations ranging from 0.7 to 4.77 when compared to the experimental data. The same group subsequently modelled the binary SLE and LLE systems of [C$_2$mim][PF$_6$] and [C$_4$mim][PF$_6$] with aromatic hydrocarbons using both UNIQUAC and NRTL with a much improved deviation of around 0.4 [17]. Since then measurements and correlations utilising the Wilson, NRTL, and UNIQUAC methods have been performed; however, these focus mainly on binary solid–liquid mixtures containing ionic liquids with another compound [18–47]. Similar work was carried out by Shiflett et al. for binary gas–ionic liquid systems [48–51] and LLE [52–57], respectively, and the results were again all correlated using the NRTL method. Ternary systems (LLE) have also been modelled using this method by Arce et al. [58–75], Pereiro et al. [76–84], Gonzalez et al. [85–91], and Letcher et al. [92–94], where prediction of the ternary phase diagram tie lines and binodal curve have been demonstrated. These correlated phase equilibria included azeotropic mixtures, in which azeotropic (alkane + alcohol), (ketone + alcohol), (alkane + ester), (alcohol + ester) mixtures are separated using an ionic liquid extraction solvent [76, 78–81]. Overall, the predictions were adequate, although sometimes poor correlations were obtained when compared to experimental results depending on the systems studied. This poor performance of the NRTL equation was found in cases that contained high concentrations of the solute, such as thiophene, as reported by Alonso et al. [62].

Modification of the NRTL method to model systems containing electrolytes was proposed by Chen et al. [95] in 1982. This model was initially developed for describing systems containing water and free ions, that is, electrolyte solutions, and was described as the electrolyte NRTL (e-NRTL) model. In essence, this model attempts to take into account long-range interaction between ions (Columbic interaction) as well as the short-range interactions between ion–molecule (electrostatic forces) and between molecule–molecule (dipole-dipole, induced dipole, van der Waals forces). From this description, it would appear that the e-NRTL model should be more representative of ionic liquids, and because of this it was applied to mixtures of ionic liquid and water by Belvèze et al. [96] and Chapeaux et al. [97], respectively. In Belvèze et al. [96], binary mixtures of quaternary ammonium ionic liquid using various anions at low concentrations with water were modelled for activity coefficient at 25°C. In this model, and due to the low ionic liquid concentration in the water, the ionic liquid was assumed to be completely dissociated within the water phase. From the resulting correlation, it was shown that the e-NRTL works very well for modelling the activity coefficients of this ionic liquid in water. However, limitations of this simple model appear at higher concentrations and for highly branched compounds, where the assumptions of complete dissociation no longer hold true. Hence, this model is unable to describe the effect of incomplete dissociation as well as micelle formation. Other works on e-NRTL modelling with ionic liquids to investigate salting out effects of the ionic liquid in a molecular solvent as well as its effect on azeotropic breaking have been reported by Orchillés et al. [98–104] for vapour–liquid equilibrium (VLE) and LLE systems, respectively.

Accepting the limitations of the e-NRTL model, another approach for predicting LLE of ionic liquids and mixed solvent systems was developed by Simoni et al. [105, 106]. Known as the asymmetric framework method, it is designed such that it calculates the ionic liquid-rich phase and high-average–dielectric-constant solvent-rich phase compositions separately using NRTL and e-NRTL methods, respectively. The theoretical assumption behind this work is that an ionic liquid associates and dissociates to different degrees depending on the phase in which it is dissolved; thus, different models are applied to each phase. One phase is characterised by a high concentration of ionic liquid, and when in a mixed solvent with a low dielectric constant it is assumed to have completely associated ions, that is, ion pairs (molecular-like), whereas the other phase that contains a low concentration of ionic liquid and in a mixed solvent of high dielectric constant is assumed to have complete dissociation of the ions. Although not extensively applied to date, the asymmetric framework appears to only work well in predicting LLE of ternary systems involving an aqueous phase that is dilute in ionic liquid, which is of course the intended application of this method.

More recently, another type of NRTL modification was proposed by Chen and Song [107] by introducing four different "conceptual" segments that are broadly characterised by various surface interaction characteristics of the molecules which measures the hydrophobicity, hydrophilicity, polarity, and solvation strength of the molecular surface areas known as molecular descriptors. This segmented calculation of the residual term of the activity coefficients is termed the NRTL segmented activity coefficients (NRTL-SACs) model. In a subsequent paper, Chen et al. [108] applied the NRTL-SAC method to ionic liquid systems by firstly correlating the infinite-dilution activity coefficients for organic solvents in ionic liquids and then using the resulting molecular descriptors to predict their phase behaviour in various mixtures. This method was found to predict VLE behaviour with comparable results to that obtained from Wilson, NRTL, and UNIQUAC [109], although it was less successful in predicting LLE phase behaviour. At present, the segments proposed in the NRTL-SAC method appear to overpredict the activity coefficients of alkylbenzenes with ionic liquids as well as alcohol, in this case 2-propanol, in ionic liquids with higher alkyl chain lengths. One particular advantage of using the NRTL-SAC method is that the predictions are based on pure component ionic liquid molecular descriptors rather than the binary interaction parameters needed for other $G^E$ models such as Wilson, NRTL, and UNIQUAC. This is a significant advantage over these other models as it reduces the amount of experimental data needed to describe a given ionic liquid system. However, NRTL-SAC is more numerically intensive and complex when compared to these alternative methods and it is unlikely that the descriptors are sufficiently broad enough at present to correctly describe ionic liquid behaviour in all situations.

Several other published works have used the UNIQUAC method to correlate binary and ternary systems containing ionic liquids [110–112]. Typically, the values $R_k$ (van der Waals volume) and $Q_k$ (van der Waals surface area) are estimated from the Bondi method [113] as described by Kato et al. [114]. Banerjee et al. [115], however, used the polarisable continuum model (PCM) to generate $R_k$ and $Q_k$, giving a 40% improvement in the root mean square deviation (RMSD) over NRTL. Santiago et al. [111, 112] meanwhile utilised quantum chemical calculations performed by the Gaussian 03 and GAMESS 7.1 packages, which included optimisation of a component's structure via DFT. The generated $R_k$ and $Q_k$ parameters were used in the UNIQUAC correlation

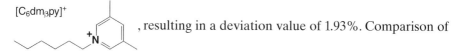

$[C_6dm_3py]^+$ , resulting in a deviation value of 1.93%. Comparison of

NRTL and UNIQUAC methods in correlating similar measured systems has been performed by a number of groups such as Pereiro et al. [116–118], Arce et al. [119–123], and Garcia et al. [124, 125]. In general, UNIQUAC was found to give a better correlation when compared to NRTL. For example, in Alonso et al. [119], the RMSD of the residual function was found to be 8.2 and 1.4 for NRTL and UNIQUAC, respectively, when correlating ternary systems containing ([$C_8$mim][NTf$_2$] + thiophene + toluene). Similar studies by Arce et al. [122], using [$C_6$dm$_3$py][NTf$_2$] to extract thiophene from mixtures containing toluene or heptane or trimethylpentane, respectively, also found that the use of the UNIQUAC model resulted in lower deviations than that given by NRTL. Studies by Pereiro et al. [116] on binary systems containing [$C_n$mim][PF$_6$] with 1-propanol, 1-butanol, or 1-pentanol at various temperatures indicated that both NRTL and UNIQUAC equations show a satisfactory fit to the experimental data. Further comparison of the Wilson, NRTL, and UNIQUAC methods have been reported by Domańska et al. [40] and Doker et al. [109]. The latter paper used these methods to correlate the binary VLE of [$C_2$mim][NTf$_2$] and [$C_4$mim][NTf$_2$] with propanone, 2-propanol, or water, respectively, giving average deviations of 3.92%, 1.45%, and 1.53%. As expected, the poor correlation of the Wilson equation for the binary system is due to its inability to describe the miscibility gap of the water–ionic liquid system, whereas good agreement was found for both NRTL and UNIQUAC. Using the same $G^E$ model parameters, predictions were made and compared with experimental data sets with reported errors when using Wilson, NRTL, and UNIQUAC of 5.61%, 7.22%, and 5.02%, respectively, which shows that the $G^E$ models could provide acceptable results for binary and ternary phase equilibria at typically low pressures and mild temperatures.

Predictive versions of this $G^E$ model such as UNIFAC have long been used to estimate unmeasured systems through prior determination of the binary interaction parameters of the respective functional groups. Until recently, functional group parameters for ionic liquids were not available in published databases; however, as experimental data have increased, the calculation of these parameters has since been made possible. Here both Wang et al. [126] and Lei et al. [127] have shown the applicability of the UNIFAC model to predict thermodynamic properties and phase behaviour of systems containing ionic liquid. In the paper by Wang et al. [126], the ionic liquids are segmented into several neutral functional groups as suggested by Kim et al. [128], in which the bulk cation and anion of the ionic liquid are treated as one main group. However, closely adjacent atoms or groups which can be subjected to electrophilic effects by strong electronegative atoms or electrophilic groups such as nitrogen, oxygen, hydroxyl, and carbonyl are most likely not separable. This thus leads to two alternative approaches which either include or exclude the adjacent groups connected to the electronegative atom of the cation as part of the main group of the ionic liquid. In their work, the resulting binary parameters were correlated from measured data of infinite-dilution activity coefficient of several ionic liquids with molecular solutes such as alkanes, alkenes, aromatics, alcohols, ketones, and water and used to predict the vapour pressure of binary mixtures containing ionic liquid at various temperatures with an RMSD of less than 4.2%. Similarly, results reported by Lei et al. [127] also showed that after further extending the group parameters for the UNIFAC model to include other ionic liquid main groups beyond the reported imidazolium-based ionic liquids with the $[NTf_2]^-$ anion, prediction of VLE for a system with ionic liquid could be performed with reasonable accuracy.

However, a limitation was found for predicting non-polar solute–ionic liquid systems due to the UNIFAC model slightly overpredicting the homogeneity of the liquid phase when liquid–liquid demixing should have taken place. Work by Alevizou et al. [129] was based on treating the imidazolium and hexafluorophosphate group as separate entities each having its individual value of $R_k$, $Q_k$, and group binary parameters determined simultaneously by fitting binary LLE and infinite-dilution activity coefficient experimental data. This was then further tested to predict ternary LLE of systems containing $[C_nmim][PF_6]$ with alkanes, alcohols, and water, respectively. In this work, the UNIFAC model successfully predicted the phase behaviour of the ternary systems but with considerable underestimation for the size of the biphasic region for the longer alkyl chain length of the imidazolium cation (i.e., $[C_6mim]$ $[PF_6]$ and $[C_8mim][PF_6]$). It is also worth noting that the original UNIFAC model was adopted in all of the works mentioned earlier [126, 127, 129]. An augmented UNIFAC (Do) model was proposed in 1987 by Gmehling and coworkers [130–134], which introduced modifications to the combinatorial part by including temperature-dependent group interaction parameters simultaneously calculated for VLE, LLE, enthalpy of mixing ($H^E$), and gamma infinity ($\gamma^\infty$) data. Naturally, this requires a massive databank of phase behaviour measurements over a wide temperature range. Despite this requirement, application of the modified UNIFAC (Do) model to systems containing ionic liquid has been reported [135–143], in which, for the most part, the modified UNIFAC results in better agreement with experimental data especially for the calculation of $H^E$ and $\gamma^\infty$ due to the temperature-dependent parameters. In one of the group's earliest papers relating to ionic liquids, Kato and Gmehling [135] described the main imidazolium cation and bis{(trifluoromethyl)sulfonyl}amide anion as separate individual groups, where the relative van der Waals volumes ($R_k$) and surface areas ($Q_k$) were obtained using the Bondi method [113]. As before, Santiago et al. [144] proposed using quantum chemical calculation through molecular geometry optimisation via DFT, and later using the PCM model for the area and volume calculation.

A comparison of several different $G^E$ models for modelling the extraction of alcohols from water using $[C_6mim][NTf_2]$ through the development of ternary phase diagrams has been reported by Chapeaux et al. [145]. In this work, four different $G^E$ models (consisting of NRTL, e-NRTL, UNIQUAC, and UNIFAC) were used to predict the behaviour of ternary LLE systems of types I and II, respectively, based solely on parameters determined from binary data measurements. UNIQUAC was found to be the best model for predicting the type I system (although NRTL provided the most accurate tie lines in this case), and e-NRTL and UNIQUAC were the better models for predicting the type II system. Similar comparison work was also reported by Simoni et al. [146] for predicting ternary systems of (ionic liquid + water + alcohol), (ionic liquid + aromatic + aliphatic), (ionic liquid + ether + alcohol), and (ionic liquid + ether + water) systems, respectively.

### 5.3.2   EoS Models

In 2003, Shariati and Peters [147] reported one of the first EoS models used for correlating the VLE of systems containing $[C_2mim][PF_6]$ and trifluoro-methane ($CHF_3$). This work used the Peng–Robinson (PR) EoS at temperatures and pressures ranging from 309.3 to 367.5 K and 1.6 to 51.6 MPa, respectively. Similar work by Liu et al. [148] reported the VLE modelling of a ternary system consisting of $CO_2$, $[C_4mim][PF_6]$, and methanol mixtures at elevated pressures. The corresponding solubility and viscosity of the liquid phases at equilibrium conditions were also modelled using the same EoS method. Carvalho and coworkers [149–151], in their series of published works on high-pressure carbon dioxide phase behaviour in ionic liquids, also used the PR EoS coupled with the Wong–Sandler mixing rules. Here the UNIQUAC model was used for calculating the activity coefficients and to describe Henry's constants of $CO_2$ with imidazolium-based and phosphonium-based ionic liquids, respectively, at temperatures of up to 363 K and pressure of up to 74 MPa. Good agreement was found with the experimental data. Several other authors have also reported the application of the PR EoS to describe binary VLE for systems containing ionic liquids with refrigerant gasses [152], compressed $CO_2$ [153], supercritical fluids, and hydrocarbons [154], respectively. Trindade et al. [155] further studied the applicability of PR EoS combined with the Mathias–Klotz–Prausnitz (MKP) mixing rule to correlate the binary LLE of imidazolium-based ionic liquids $[C_2mim][NTf_2]$, $[C_{10}mim][NTf_2]$, and $[C_2mim][OTf]$ with 1-propanol, 1,2-propanediol, 1,3-propanediol, and glycerol (1,2,3-propanetriol), respectively. In this work, satisfactory correlation was achieved with most of the systems studied, except in the case of 1,3-propanediol.

Recently, Arce et al. [156] used a modified version of the PR EoS called PR + Stryjek–Vera (PRSV) EoS model to describe the phase equilibria of 17 binary gas and ionic liquid systems at high pressures, which achieved satisfactory results. In this work, comparison of two different mixing rules based on phi–phi (van der Waals) and gamma–phi (Wong–Sandler + Van Laar) approaches showed comparable deviations to the experimental results, with the phi–phi approach requiring less computational time in comparison with gamma–phi. Meanwhile, Ren et al. [157] studied the effect of cation alkyl chain length of imidazolium-based ionic liquids with the $[NTf_2]^-$ anion on $CO_2$ solubility at several temperatures and pressures of up to 25 MPa, and modelled these using the PR EoS in combination with the van der Waals two-parameter mixing rule. In this work, and at temperatures below the $CO_2$ critical point, that is, 304 K, the various multiphase equilibria, viz. VLE, LLE, and VLLE were satisfactorily modelled by the PR EoS within the experimental pressure range. However, above the correlated pressure, that is, 25 MPa, extrapolation of the model predicts physically unrealistic dew points, ionic liquid solubility in the liquid $CO_2$, and mixture critical point.

Shiflett and Yokozeki measured considerable amounts of solubility data for various gases in a variety of ionic liquids over a wide range of pressures, which were then modelled using a modified Redlich–Kwong (RK) EoS [158–169].* In their series of published works, and from the measured $P$-$T$-$x$ data of gas solubility in ionic liquid, excess properties such as enthalpy ($H^E$), entropy ($S^E$), and Gibbs energy ($G^E$) of the mixtures could be calculated using the modified RK EoS correlation. This can be used to imply the strength of attractive interaction between the gas and ionic liquid through the positive or negative values of the excess properties. For example, in one paper [162], very strong attractive intermolecular interactions such as hydrogen bonding have been suggested to exist in the studied binary mixture of ammonia and ionic liquid which is deduced from the highly negative calculated values of $H^E$ and $G^E$. This is in accordance with experimental observations of high ammonia solubility in ionic liquids as well as the homogenous nature of the mixture. In all of these works which involved classical cubic EoS, for example, PR-EoS or RK-EoS, ionic liquids are considered neutral clusters of ions. In terms of calculation, the vapour phase was assumed to contain a negligible amount of ionic liquid, allowing the fugacity coefficient of the components in the vapour phase to be calculated. This then utilises the $K$-value (distribution coefficient) to determine the fugacity coefficients in the liquid phase and eventually the binary interaction parameters of the ionic liquid with the individual component at the respective temperature. The chosen EoS methods were satisfactory in terms of describing the experimental bubble point, but were only able to qualitatively predict the solubility of ionic liquid in the supercritical solvent-rich phase due to the very low mole fraction, particularly near the critical region.

In order to better describe the solubility of gases such as $CO_2$ in ionic liquids at high pressure, molecular-based EoSs were applied to characterise the interaction of these gases with respective ionic liquids through several theoretical models for the mixture. Here these EoSs are based solely on statistical mechanics that allow physical interpretation of the effect of molecular structure on thermodynamic properties to be classified and quantified. Using a truncated perturbed chain polar statistical associating fluid theory (tPC-PSAFT EoS), Kroon et al. [170] reported the correlation and prediction of $CO_2$ phase behaviour in various imidazolium-based ionic liquids with different cation alkyl chain lengths and different anions at pressures up to 100 MPa. This work is based on a theory expressed through the residual Helmholtz energy where

---

* See also Shiflett, M.B. and Yokozeki, A., Phase behaviour of gases in ionic liquids, in *Ionic Liquids UnCOILed: Critical Expert Overviews*, eds. N.V. Plechkova and K.R. Seddon (Wiley, Hoboken, New Jersey, 2012), pp. 349–398.

molecular dipolar interactions are described between ionic liquid molecules, which are themselves assumed to be neutral ion pairs. Between $CO_2$ molecules quadrupolar interactions are used, and interactions between the ionic liquid and $CO_2$ molecules are described by Lewis acid–base-type associations. Overall, good agreement with VLE experimental data was obtained for all cases. Karakatsani et al. [171, 172] later on re-estimated the parameters of tPC-PSAFT for imidazolium-based ionic liquid by fitting available density data over a wide range of temperature, which resulted in a much lower and improved values of binary interaction parameters for the gas and ionic liquid [170].

Andreu and Vega [173] reported an application of another type of SAFT EoS, called soft-SAFT, to model $CO_2$ solubility in $[C_n mim][BF_4]$ and $[C_n mim]$ $[PF_6]$ at various temperatures and pressures of up to 90 MPa. In this work, interactions are categorised into four classifications: a reference term that was based on Lennard-Jones spherical fluids, chain formation, the association, and the polar interaction of molecules. Molecular parameters and association parameters of the pure components that are required prior to applying the EoS to a given mixture were obtained by initially fitting the parameters to experimental density data. The soft-SAFT EoS qualitatively described the solubility of $CO_2$ in ionic liquids although successful parameterisation requires accurate models of pure ionic liquids based on molecular simulation. Although the equation is simple, the mathematical complexity which requires high numbers of different parameters in solving could be a significant hindrance for wider implementation. Andreu and Vega [174] later applied the soft-SAFT EoS to model solubility of $CO_2$ as well as other gases such as $H_2$ and xenon in $[C_n mim][NTf_2]$ at several different temperatures and pressure of up to 100 MPa. More recently, Vega et al. [175] continued to extend the soft-SAFT EoS for modelling the solubility of $BF_3$ in $[C_n mim][BF_4]$ at various temperatures and pressures. However, while previous soft-SAFT models of $CO_2$ solubility in ionic liquids did not require cross-association interaction [173, 174], this is not the case with $BF_3$ due to the 1:1 weak interaction between the gas and ionic liquid "molecule." Hence, explicit cross-association interactions between the two components are required, resulting in a higher number of parameters for the soft-SAFT model.

Other groups such as Wang et al. [176] used the square-well chain-fluid EoS (SWCF-EoS) to model the $P$-$V$-$T$ behaviour of pure ionic liquids and their respective phase behaviour with gases by representing ionic liquids as "diblock compounds" with the cation alkyl chain length group as one block and the imidazolium ring–anion pair as the other. Correlation of VLE for 22 binary systems containing imidazolium-based ionic liquids and solvents provided satisfactory results with an average deviation of 5.38%. Li et al. [177] further refined this EoS by introducing variable interaction ranges for the disperse contribution of the Helmholtz function as well as treating the chain formation contribution separately into respective hard spheres and square-well potentials. From this treatment, the molecular parameters were obtained from the correlation of the experimental density data of pure ionic liquids, which were then coupled with the fluid mixture rule and temperature-independent binary interaction parameters obtained from experimental solubility data, thus enabling calculation of the properties of ionic liquid mixtures. Comparison between the calculated results and the experimental data showed satisfactory results for data at low pressures. Other forms of the EoS based on chemical association theory using lattice-fluid model (LF-EoS) were extended to model the ionic liquid systems. Initially proposed by Yang et al. [178], these combine computer simulation results with statistical mechanics to obtain expression for the Helmholtz energy of mixing. In this work, infinite-dilution activity coefficients, binary VLE, and LLE of systems containing ionic liquids were correlated satisfactorily. However, due to the close-packed characteristics of the proposed model in this work, it was unable to calculate the effect of pressure on phase behaviour. In order to overcome this limitation and enable the calculation of $P$-$V$-$T$ properties and phase equilibria for systems containing ionic liquid, a modification was proposed by Xu et al. [179], which introduced holes into the lattice structure. In addition, a new parameter determined from Monte Carlo simulation data was introduced to characterise the long-range correlations beyond the close contact pairs. Using this model, the $P$-$V$-$T$ properties, VLE, and LLE of binary systems containing ionic liquids were satisfactorily described. Here correlation of binodal curves for ternary LLE were obtained through determination of component-specific binary interaction parameters.

Approaches for using group contribution EoS (GC-EoS) in predicting phase behaviour of systems containing ionic liquids have been developed and proposed by several groups in an attempt to predict solubility of gases in ionic liquids at high pressure. Kim et al. [128] used a GC non-random lattice-fluid EoS (GCNRLF-EoS) to predict solubilities of $CO_2$ in six methylimidazolium-based ionic liquids at pressures of up to 1 MPa. Each ionic liquid was segmented into functional groups where binary interaction parameters between the groups were determined from fitting experimental data. In general, the predicted values of $CO_2$ solubility in ionic liquids were in good agreement with the experimental data; however, no prediction for higher pressures (>1 MPa) is reported. Breure et al. [180] applied a GC-EoS, which was originally proposed by Skjold-Jørgensen [181, 182] in the 1980s to calculate the VLE of non-ideal mixtures for pressures of up to 30 MPa. This GC-EoS is based on a generalised van der Waals function in combination with a local composition principle, and was expressed in terms of residual Helmholtz energy, which took into account the attractive term and free volume term, respectively. From the determined pure group constants, pure group attractive energy parameters, as well as group–group binary interaction parameters, were found. At pressures of up to 20 MPa, the predicted results were in very good agreement with experimental data, with absolute average deviations ranging from 1% to 4%, and for pressures as high as 100 MPa, predictions were qualitatively in good agreement with the experimental data. Martin et al. [183] applied a similar approach to correlate the phase behaviour of several gases ($CO_2$, $O_2$, or $SO_2$)

in ionic liquids with the $[NTf_2]^-$ anion and cations of the families 1-alkyl-2,3-dimethylimidazolium, 1-alkyl-1-methyl-pyrrolidinium, and 1-alkyl-3-methyl-pyridinium. The newly obtained parameters were later used to predict the VLE of other ionic liquids belonging to the same families with the gases mentioned earlier. This GC EoS was able to correctly predict the trend of phase behaviour with respect to the experimental results; however, deviation was higher for data at low pressures and temperatures. Bermejo et al. [184] later extended the GC EoS to predict the binary system of $CO_2$ with imidazolium-based ionic liquids employing the nitrate anion for pressures of up to 12 MPa with an average deviation of 4.75%. However, adjustment of the ionic liquid critical diameters was necessary in order to achieve a good correlation.

### 5.3.3   Quantum Chemical Calculations

A more *a priori* effort was made by Diedenhofen et al. [185] through the application of **CO**nductor-like**S**creening**MO**del of Real Solvents (COSMO-RS) in predicting the infinite-dilution activity coefficients of 38 compounds with three different ionic liquids. In this work, quantum chemical calculations were initially performed by parameterisation of the compound structure through an *ab initio* calculation, which generates a surface charge density ($\sigma$) profile of the individual compound. This was then followed by a COSMO-RS thermodynamic model which utilised these sigma profiles to describe the interaction of the compounds within the mixture and, hence, the respective chemical potential. This departure of the component's chemical potential in the mixture from its pure compound chemical potential then allows thermodynamic properties such as activity coefficient to be calculated. This *a priori* thermodynamic calculation of properties for ionic liquids obtained simply from the chemical structures was able to provide a predictive tool with comparable accuracy to that observed for normal organic solvents. Jork et al. [186] further discussed the use of COSMO-RS as a tool for pre-selection of ionic liquids as entrainers through identification of favourable structural variations of the ionic liquids. Through the activity coefficient calculation, variations in the predicted results were identified due to the different conformations of the volatile components and ionic liquids, respectively. In the case of VLE systems containing ionic liquids, the effect of solvent conformations increased with the polarity of the solvents, whereas the ionic liquid conformer effect on VLE was secondary; hence, modelling the VLE system of ionic liquid and polar solvent requires the consideration of all solvent conformers within the homogenous liquid phase. Conversely, prediction of LLE for systems containing ionic liquids using COSMO-RS is significantly dependent on the ionic liquid conformers, especially for the composition of the ionic liquid-rich phase. Overall, qualitative agreement of the predicted solubility was obtained with experimental results with increased deviation as the concentration of solvent increases in the ionic liquid-rich phase. This is due to the inability of COSMO-RS to accurately describe the long-range interactions (electrostatic contribution).

Banerjee and coworkers used COSMO-RS to predict binary VLE of systems containing five different imidazolium-based ionic liquids with various solvents, in which the RMSD for pressure prediction was 6% in comparison with the Wilson, NRTL, and UNIQUAC models, which resulted in RMSDs of 4%, 1.45%, and 3.13%, respectively [187]. An extension of similar work for LLE was performed by the same group for ternary systems containing ionic liquids, ethanol, and alkenes [188]. In this work, complete dissociation of cation and anion was assumed and conformer correction of the sigma profile of the alkenes was adopted prior to the calculation of tie lines in the ternary system. The effect of this treatment results in the average RMSD improving from 22.6% to 10.6%. Using this dissociating assumption, COSMO-RS was able to qualitatively predict the correct concentration of ethanol in ionic liquid-rich phase. However, this also resulted in the ethanol concentration in alkene-rich phase being poorly predicted especially for the case of high ethanol concentrations in the alkene-rich phase. Preliminary screening of suitable ionic liquids using COSMO-RS for specific applications of ionic liquids in extraction process, such as thiophene from hydrocarbon [189], phenol from aqueous solution [190], and alcohol or tetrahydrofuran in aqueous systems [191], have also been studied by the same group. Freire et al. [192–194], in a series of publications, evaluated the use of COSMO-RS in predicting mutual solubilities of water in ionic liquids, VLE, and LLE of binary mixtures of imidazolium- and pyridinium-based ionic liquids and alcohols, as well as VLE and LLE of binary systems containing ionic liquids and water, respectively. In general, COSMO-RS was able to give a good qualitative prediction of the VLE system in comparison with the experimental data. However, LLE prediction of the system became poorer as the alkyl chain length of the ionic liquid or solvent (i.e., alcohol) increases, or for very small molecules such as methanol [193]. An increase in the deviation for LLE prediction was also observed as the anion hydrophilicity increases [194].

Further comparison of COSMO-RS in predicting VLE and infinite dilution of binary systems containing ionic liquids with the original and modified UNIFAC method have been reported by Kato et al. [135] and Lei et al. [195], respectively. For LLE systems, Sahandzhieva et al. reported a comparison of a COSMO-RS prediction and UNIQUAC correlation for $[C_4mim][PF_6]$ with three different alcohols (ethanol, 1-propanol, and 1-butanol) at several temperatures. Although a better representation of the experimental data was achieved using UNIFAC and UNIQUAC methods, due to the nature of the COSMO-RS method, in that no prior experimental data is necessary, it can be, and has been, used widely as a tool for qualitative prediction and screening of ionic liquid properties and its phase behaviour with other solvents [196].

However, it is worth noting that the COSMO-RS model alone is not sufficient to provide the relationships of fugacity or chemical potential of a component in a phase with the physicochemical properties of that phase, such as caloric properties (e.g., heat capacity) or volumetric properties (e.g., density), which EoS and $G^E$ thermodynamic models can perform.

## 5.4   CONCLUSIONS

The sections in this chapter have attempted to discuss the various methods used to describe the thermodynamic properties of ionic liquids. As can be seen, no one method has had universal acceptance. This is, of course, unsurprising; no one method has ever been applied to non-ionic liquid systems, and hence process simulators such as Aspen Plus continue to offer a number of different physical property options. In addition to the normal problems with identifying a suitable physical property model, it is worth remembering that ionic liquids do pose some additional complications in that they have properties common to both salts and normal molecular solvents. These materials can dissociate, cluster, exchange with other ions in the system, or interact with other molecules using a variety of methods, and therefore, depending on the system being studied, it may be better to consider the ionic liquid as being composed of individual ions, ion pairs, or ionic liquid clusters. To date, the NRTL group of methods appear to be by far the most applied models for describing ionic liquid VLE and LLE. This is likely due to the fact that this method is relatively robust and easy to use, and has shown success for a wide range of systems. Extensions to this including e-NRTL have been used in an attempt to better describe the dissociation of the ionic liquid in aqueous systems, again with some success particularly for dilute systems in aqueous media. Apart from its wide applicability, the fact that it does not require information on the relative van der Waals volume ($R_k$) and surface parameters ($Q_k$), required for other local composition methods such as UNIQUAC and UNIFAC, has also likely contributed to its popularity. More recently, the asymetric framework method has been used to try and bridge the gap between both these methods, again showing success for those systems to which it is most relevant. In addition, recent attempts to simplify the NRTL method, in terms of the overall number of interaction parameters needed, by using an SAC approach also show promise. This will be especially true if segments can be included which better describe the charged nature of the ionic liquid. EoS models have also been investigated in recent years, spanning well-known correlations such as PR to more elaborate methods such as the GCNRLF-EoS. Again, these have shown some success particularly for gas phase systems. Furthermore, as computational power has increased in recent years, so too has the ability for molecular dynamics and quantum chemical calculations to assist in prediction of phase equilibria. Of the methods discussed within the literature, COSMO-RS has been used by a number of groups with varying success. However, again, an understanding of the actual structure in terms of conformers and so on appears important for correctly defining the relevant phases. This highlights that understanding the physical nature of the ionic liquid and associated molecules is crucially important.

Overall, the area of engineering computations has increased significantly in recent years and currently it is possible to use available tools to estimate the properties of ionic liquids and mixtures containing them with sufficient accuracy for preliminary design and selection. Such tools vary in complexity, with most requiring significantly less time and effort than experimentally determining them. Over the coming decade, it is anticipated that multi-scale engineering approaches will be routinely used for the design of ionic liquid systems.

## REFERENCES

1   Rosen, E. M., and Pauls, A. C., Computer aided chemical process design: the FLOWTRAN system, *Comput. Chem. Eng.* **1**, 11–21 (1977).

2   Seddon, K. R., and Rooney, D. W., Ionic liquid, in *Handbook of Solvents*, ed. G. Wypych (ChemTec Publishing, Ontario, 2001), p. 1675.

3   Valderrama, J. O., and Robles, P. A., Critical properties, normal boiling temperatures, and acentric factors of fifty ionic liquids, *Ind. Eng. Chem. Res.* **46**, 1338–1344 (2007).

4   Valderrama, J. O., Sanga, W. W., and Lazzus, J. A., Critical properties, normal boiling temperature, and acentric factor of another 200 ionic liquids, *Ind. Eng. Chem. Res.* **47**, 1318–1330 (2008).

5   Valderrama, J. O., and Rojas, R. E., Critical properties of ionic liquids. Revisited, *Ind. Eng. Chem. Res.* **48**, 6890–6900 (2009).

6   Dong, Q., Muzny, C. D., Kazakov, A., Diky, V., Magee, J. W., Widegren, J. A., Chirico, R. D., Marsh, K. N., and Frenkel, M., ILThermo: A free-access web database for thermodynamic properties of ionic liquids, *J. Chem. Eng. Data* **52**, 1151–1159 (2007).

7   Biegler, L. T., Grossmann, I. E., and Westerberg, A. W., Issues and trends in the teaching of process and product design, *AIChE J.* **56**, 1120–1125 (2010).

8   Tian, X. A., Zhang, X. P., Wei, L., Zeng, S. J., Huang, L., and Zhang, S. J., Multi-scale simulation of the 1,3-butadiene extraction separation process with an ionic liquid additive, *Green Chem.* **12**, 1263–1273 (2010).

9   Brennecke, J. F., Kestel, D. E., Lopez-Castillo, Z. K., Anderson, J. L., and de la Fuente J. C., Carbon dioxide capture using ionic liquids, (American Chemical Society, 2008), pp. IEC-012.

10  Goodrich, B. F., Brennecke, J. F., and Ficke, L. E., Ionic liquids for post-combustion $CO_2$ capture: Tunable capacities and enthalpies, (American Chemical Society, 2010), pp. FUEL-325.

11  Perkins, J. D., Equation-oriented flowsheeting, in *Second International Conference on Foundations of Computer-Aided Design*, eds. A. W. Westerberg and H. H. Chien (CACHE Publications, Ann Arbor, Michigan, 1984).

12  Selvan, M. S., McKinley, M. D., Dubois, R. H., and Atwood, J. L., Liquid-liquid equilibria for toluene plus heptane+1-ethyl-3-methylimidazolium triiodide and toluene plus heptane+1-butyl-3-methylimidazolium triiodide, *J. Chem. Eng. Data* **45**, 841–845 (2000).

13  Domańska, U., and Bogel-Łukasik, E., Measurements and correlation of the (solid plus liquid) equilibria of [1-decyl-3-methylimidazolium chloride plus Alcohols (C-2-C-12)], *Ind. Eng. Chem. Res.* **42**, 6986–6992 (2003).

14 Domańska, U., Bogel-Łukasik, E., and Bogel-Łukasik, R., Solubility of 1-dodecyl-3-methyl-imidazolium chloride in alcohols (C-2-C-12), *J. Phys. Chem. B* **107**, 1858–1863 (2003).

15 Nagata, I., Nakamiya, Y., Katoh, K., and Koyabu, J., Ternary liquid-liquid equilibria and their representation by modified NRTL equations, *Thermochim. Acta* **45**, 153–165 (1981).

16 Nagata, I., On the thermodynamics of alcohol-solutions. Phase equilibria of binary and ternary mixtures containing any number of alcohols, *Fluid Phase Equilibr.* **19**, 153–174 (1985).

17 Domańska, U., and Marciniak, A., Solubility of 1-alkyl-3-methylimidazolium hexafluorophosphate in hydrocarbons, *J. Chem. Eng. Data* **48**, 451–456 (2003).

18 Domańska, U., and Bogel-Łukasik, E., Solid–liquid equilibria for systems containing 1-butyl-3-methylimidazolium chloride, *Fluid Phase Equilibr.* **218**, 123–129 (2004).

19 Domańska, U., and Marciniak, A., Solubility of ionic liquid [emim][PF₆] in alcohols, *J. Phys. Chem. B* **108**, 2376–2382 (2004).

20 Domańska, U., and Mazurowska, L., Solubility of 1,3-dialkylimidazolium chloride or hexafluoro-phosphate or methylsulfonate in organic solvents: Effect of the anions on solubility, *Fluid Phase Equilibr.* **221**, 73–82 (2004).

21 Domańska, U., Solubilities and thermophysical properties of ionic liquids, *Pure Appl. Chem.* **77**, 543–557 (2005).

22 Domańska, U., and Bogel-Łukasik, R., Solubility of ethyl-(2-hydroxyethyl)-dimethylammonium bromide in alcohols (C-2-C-12), *Fluid Phase Equilibr.* **233**, 220–227 (2005).

23 Domańska, U., and Bogel-Łukasik, R., Physicochemical properties and solubility of alkyl-(2-hydroxyethyl)-dimethylammonium bromide, *J. Phys. Chem. B* **109**, 12124–12132 (2005).

24 Domańska, U., and Marciniak, A., Experimental liquid-liquid equilibria of 1-methylimidazole with hydrocarbons and ethers, *Fluid Phase Equilibr.* **238**, 137–141 (2005).

25 Domańska, U., Thermophysical properties and thermodynamic phase behavior of ionic liquids, *Thermochim. Acta* **448**, 19–30 (2006).

26 Domańska, U., Pobudkowska, A., and Eckert, F., Liquid-liquid equilibria in the binary systems (1,3-dimethylimidazolium, or 1-butyl-3-methylimidazolium methylsulfate plus hydrocarbons), *Green Chem.* **8**, 268–276 (2006).

27 Domańska, U., Pobudkowska, A., and Eckert, F., (Liquid plus liquid) phase equilibria of 1-alkyl-3-methylimidazolium methylsulfate with alcohols, or ethers, or ketones, *J. Chem. Thermodyn.* **38**, 685–695 (2006).

28 Domańska, U., and Casas, L. M., Solubility of phosphonium ionic liquid in alcohols, benzene, and alkylbenzenes, *J. Phys. Chem. B* **111**, 4109–4115 (2007).

29 Domańska, U., Ługowska, K., and Pernak, J., Phase equilibria of didecyldimethylammonium nitrate ionic liquid with water and organic solvents, *J. Chem. Thermodyn.* **39**, 729–736 (2007).

30 Domańska, U., and Marciniak, A., Liquid phase behaviour of 1-butyl-3-methylimidazolium 2-(2-methoxyethoxy)-ethylsulfate with organic solvents and water, *Green Chem.* **9**, 262–266 (2007).

31 Domańska, U., and Marciniak, A., Phase behaviour of 1-hexyloxymethyl-3-methyl-imidazolium and 1,3-dihexyloxymethyl-imidazolium based ionic liquids with alcohols, water, ketones and hydrocarbons: The effect of cation and anion on solubility, *Fluid Phase Equilibr.* **260**, 9–18 (2007).

32 Domańska, U., and Morawski, P., Influence of high pressure on solubility of ionic liquids: experimental data and correlation, *Green Chem.* **9**, 361–368 (2007).

33 Domańska, U., Pobudkowska, A., and Królikowski, M., Separation of aromatic hydrocarbons from alkanes using ammonium ionic liquid $C_2NTf_2$ at T=298.15 K, *Fluid Phase Equilibr.* **259**, 173–179 (2007).

34 Domańska, U., Pobudkowska, A., and Żołek-Tryznowska, Z., Effect of an ionic liquid (IL) cation on the ternary system (IL plus p-xylene plus hexane) at T=298.15 K, *J. Chem. Eng. Data* **52**, 2345–2349 (2007).

35 Domańska, U., Laskowska, M., and Marciniak, A., Phase equilibria of (1-ethyl-3-methylimidazolium ethylsulfate plus hydrocarbon, plus ketone, and plus ether) binary systems, *J. Chem. Eng. Data* **53**, 498–502 (2008).

36 Domańska, U., Marciniak, A., and Królikowski, M., Phase equilibria and modeling of ammonium ionic liquid, $C_2NTf_2$, solutions, *J. Phys. Chem. B* **112**, 1218–1225 (2008).

37 Domańska, U., and Paduszyński, K., Phase equilibria study in binary systems (tetra-*n*-butylphosphonium tosylate ionic liquid + 1-alcohol, or benzene, or *n*-alkylbenzene), *J. Phys. Chem. B* **112**, 11054–11059 (2008).

38 Domańska, U., Rękawek, A., and Marciniak, A., Solubility of 1-alkyl-3-ethylimidazolium-based ionic liquids in water and 1-octanol, *J. Chem. Eng. Data* **53**, 1126–1132 (2008).

39 Domańska, U., and Królikowski, M., Phase equilibria study of the binary systems (1-butyl-3-methylimidazolium tosylate ionic liquid plus water, or organic solvent), *J. Chem. Thermodyn.* **42**, 355–362 (2009).

40 Domańska, U., Królikowski, M., and Paduszyński, K., Phase equilibria study of the binary systems (N-butyl-3-methylpyridinium tosylate ionic liquid plus an alcohol), *J. Chem. Thermodyn.* **41**, 932–938 (2009).

41 Domańska, U., Królikowski, M., Pobudkowska, A., and Letcher, T. M., Phase equilibria study of the binary systems (N-butyl-4-methylpyridinium tosylate ionic liquid plus organic solvent, or water), *J. Chem. Eng. Data* **54**, 1435–1441 (2009).

42 Domańska, U., Królikowski, M., and Ślesińska, K., Phase equilibria study of the binary systems (ionic liquid plus thiophene): Desulphurization process, *J. Chem. Thermodyn.* **41**, 1303–1311 (2009).

43 Domańska, U., Laskowska, M., and Pobudkowska, A., Phase equilibria study of the binary systems (1-butyl-3-methylimidazolium thiocyanate ionic liquid plus organic solvent or water), *J. Phys. Chem. B* **113**, 6397–6404 (2009).

44 Domańska, U., and Paduszyński, K., (Solid plus liquid) and (liquid plus liquid) phase equilibria measurements and correlation of the binary systems {tri-iso-butyl(methyl)phosphonium tosylate plus alcohol, or plus hydrocarbon}, *Fluid Phase Equilibr.* **278**, 90–96 (2009).

45 Domańska, U., Redhi, G. G., and Marciniak, A., Activity coefficients at infinite dilution measurements for organic solutes and water in the ionic liquid 1-butyl-

1-methylpyrrolidinium trifluoromethanesulfonate using GLC, *Fluid Phase Equilibr.* **278**, 97–102 (2009).

46  Domańska, U., Żołek-Tryznowska, Z., and Pobudkowska, A., Separation of hexane/ethanol mixtures. LLE of ternary systems (ionic liquid or hyperbranched polymer plus ethanol plus hexane) at T=298.15 K, *J. Chem. Eng. Data* **54**, 972–976 (2009).

47  Domańska, U., Królikowska, M., and Królikowski, M., Phase behaviour and physico-chemical properties of the binary systems {1-ethyl-3-methylimidazolium thiocyanate, or 1-ethyl-3-methylimidazolium tosylate plus water, or plus an alcohol}, *Fluid Phase Equilibr.* **294**, 72–83 (2010).

48  Shiflett, M. B., Harmer, M. A., Junk, C. P., and Yokozeki, A., Solubility and diffusivity of difluoromethane in room-temperature ionic liquids, *J. Chem. Eng. Data* **51**, 483–495 (2006).

49  Shiflett, M. B., Harmer, M. A., Junk, C. R., and Yokozeki, A., Solubility and diffusivity of 1,1,1,2-tetrafluoroethane in room-temperature ionic liquids, *Fluid Phase Equilibr.* **242**, 220–232 (2006).

50  Shiflett, M. B., and Yokozeki, A., Gaseous absorption of fluoromethane, fluoroethane, and 1,1,2,2-tetrafluoroethane in 1-butyl-3-methylimidazolium hexafluorophosphate, *Ind. Eng. Chem. Res.* **45**, 6375–6382 (2006).

51  Shiflett, M. B., and Yokozeki, A., Solubility and diffusivity of hydrofluorocarbons in room-temperature ionic liquids, *AIChE J.* **52**, 1205–1219 (2006).

52  Shiflett, M. B., and Niehaus, A. M. S., Liquid-liquid equilibria in binary mixtures containing substituted benzenes with ionic liquid 1-ethyl-3-methylimidazolium bis(trifluoromethylsulfonyl)imide, *J. Chem. Eng. Data* **55**, 346–353 (2010).

53  Shiflett, M. B., Niehaus, A. M. S., and Yokozeki, A., Liquid-liquid equilibria in binary mixtures containing chlorobenzene, bromobenzene, and iodobenzene with ionic liquid 1-ethyl-3-methylimidazolim bis(trifluoromethylsulfonyl)imide, *J. Chem. Eng. Data* **54**, 2090–2094 (2009).

54  Shiflett, M. B., and Yokozeki, A., Liquid-liquid equilibria of hydrofluoroethers and ionic liquid 1-ethyl-3-methylimidazolium bis(trifluoromethylsulfonyl)imide, *J. Chem. Eng. Data* **52**, 2413–2418 (2007).

55  Shiflett, M. B., and Yokozeki, A., Hydrogen substitution effect on the solubility of perhalogenated compounds in ionic liquid [bmim][PF$_6$], *Fluid Phase Equilibr.* **259**, 210–217 (2007).

56  Shiflett, M. B., and Yokozeki, A., Liquid-liquid equilibria in binary mixtures of 1,3-propanediol plus ionic liquids [bmim][PF$_6$], [bmim][BF$_4$], and [emim][BF$_4$], *J. Chem. Eng. Data* **52**, 1302–1306 (2007).

57  Shiflett, M. B., and Yokozeki, A., Liquid-liquid equilibria in binary mixtures containing fluorinated benzenes and ionic liquid 1-ethyl-3-methylimidazolium bis(trifluoromethylsulfonyl)imide, *J. Chem. Eng. Data* **53**, 2683–2691 (2008).

58  Alonso, L., Arce, A., Francisco, M., Rodríguez, O., and Soto, A., Liquid-liquid equilibria for systems composed by 1-methyl-3-octylimidazolium tetrafluoroborate ionic liquid, thiophene, and *n*-hexane or cyclohexane, *J. Chem. Eng. Data* **52**, 1729–1732 (2007).

59  Alonso, L., Arce, A., Francisco, M., Rodríguez, O., and Soto, A., Gasoline desulfurization using extraction with [C$_8$mim][BF$_4$] ionic liquid, *AIChE J.* **53**, 3108–3115 (2007).

60 Alonso, L., Arce, A., Francisco, M., and Soto, A., Measurement and correlation of liquid-liquid equilibria of two imidazolium ionic liquids with thiophene and methylcyclohexane, *J. Chem. Eng. Data* **52**, 2409–2412 (2007).

61 Alonso, L., Arce, A., Francisco, M., and Soto, A., Phase behaviour of 1-methyl-3-octylimidazolium bis[trifluoromethylsulfonyl]imide with thiophene and aliphatic hydrocarbons: The influence of *n*-alkane chain length, *Fluid Phase Equilibr.* **263**, 176–181 (2008).

62 Alonso, L., Arce, A., Francisco, M., and Soto, A., (Liquid plus liquid) equilibria of [C$_8$mim][NTf$_2$] ionic liquid with a sulfur-component and hydrocarbons, *J. Chem. Thermodyn.* **40**, 265–270 (2008).

63 Alonso, L., Arce, A., Francisco, M., and Soto, A., Solvent extraction of thiophene from *n*-alkanes (C-7, C-12, and C-16) using the ionic liquid [C$_8$mim][BF$_4$], *J. Chem. Thermodyn.* **40**, 966–972 (2008).

64 Arce, A., Earle, M. J., Katdare, S. P., Rodríguez, H., and Seddon, K. R., Phase equilibria of mixtures of mutually immiscible ionic liquids, *Fluid Phase Equilibr.* **261**, 427–433 (2007).

65 Arce, A., Earle, M. J., Rodríguez, H., Seddon, K. R., and Soto, A., 1-ethyl-3-methylimidazolium bis{(trifluoromethyl)sulfonyl}amide as solvent for the separation of aromatic and aliphatic hydrocarbons by liquid extraction—extension to C-7- and C-8-fractions, *Green Chem.* **10**, 1294–1300 (2008).

66 Arce, A., Earle, M. J., Rodríguez, H., Seddon, K. R., and Soto, A., Bis{(trifluoromethyl)sulfonyl} amide ionic liquids as solvents for the extraction of aromatic hydrocarbons from their mixtures with alkanes: Effect of the nature of the cation, *Green Chem.* **11**, 365–372 (2009).

67 Arce, A., Earle, M. J., Rodríguez, H., Seddon, K. R., and Soto, A., Isomer effect in the separation of octane and xylenes using the ionic liquid 1-ethyl-3-methylimidazolium bis{(trifluoromethyl)sulfonyl}amide, *Fluid Phase Equilibr.* **294**, 180–186 (2010).

68 Arce, A., Marchiaro, A., Rodríguez, O., and Soto, A., Essential oil terpenless by extraction using organic solvents or ionic liquids, *AlChE J.* **52**, 2089–2097 (2006).

69 Arce, A., Pobudkowska, A., Rodríguez, O., and Soto, A., Citrus essential oil terpenless by extraction using 1-ethyl-3-methylimidazolium ethylsulfate ionic liquid: Effect of the temperature, *Chem. Eng. J.* **133**, 213–218 (2007).

70 Arce, A., Rodríguez, H., and Soto, A., Purification of ethyl tert-butyl ether from its mixtures with ethanol by using an ionic liquid, *Chem. Eng. J.* **115**, 219–223 (2006).

71 Arce, A., Rodríguez, H., and Soto, A., Effect of anion fluorination in 1-ethyl-3-methylimidazolium as solvent for the liquid extraction of ethanol from ethyl tert-butyl ether, *Fluid Phase Equilibr.* **242**, 164–168 (2006).

72 Arce, A., Rodríguez, H., and Soto, A., Use of a green and cheap ionic liquid to purify gasoline octane boosters, *Green Chem.* **9**, 247–253 (2007).

73 Arce, A., Rodríguez, O., and Soto, A., Experimental determination of liquid-liquid equilibrium using ionic liquids: Tert-amyl ethyl ether plus ethanol plus 1-octyl-3-methylimidazolium chloride system at 298.15 K, *J. Chem. Eng. Data* **49**, 514–517 (2004).

74   Arce, A., Rodríguez, O., and Soto, A., Tert-amyl ethyl ether separation from its mixtures with ethanol using the 1-butyl-3-methylimidazolium trifluoromethanesulfonate ionic liquid: Liquid-liquid equilibrium, *Ind. Eng. Chem. Res.* **43**, 8323–8327 (2004).

75   Soto, A., Arce, A., and Khoshkbarchi, M. K., Partitioning of antibiotics in a two-liquid phase system formed by water and a room temperature ionic liquid, *Sep. Purif. Technol.* **44**, 242–246 (2005).

76   Pereiro, A. B., Tojo, E., Rodríguez, A., Canosa, J., and Tojo, J., HMImPF$_6$ ionic liquid that separates the azeotropic mixture ethanol plus heptane, *Green Chem.* **8**, 307–310 (2006).

77   Pereiro, A. B., Canosa, J., and Rodríguez, A., Liquid-liquid equilibria of 1,3-dimethylimidazolium methyl sulfate with ketones, dialkyl carbonates and acetates, *Fluid Phase Equilibr.* **254**, 150–157 (2007).

78   Pereiro, A. B., and Rodríguez, A., Ternary (liquid plus liquid) equilibria of the azeotrope (ethyl acetate+2-propanol) with different ionic liquids at T=298.15 K, *J. Chem. Thermodyn.* **39**, 1608–1613 (2007).

79   Pereiro, A. B., and Rodríguez, A., Ternary liquid-liquid equilibria ethanol+2-butanone+1-butyl-3-methylimidazolium hexafluorophosphate, 2-propanol+2-butanone+1-butyl-3-methylimidazolium hexafluorophosphate, and 2-butanone+2-propanol+1,3-dimethylimidazolium methyl sulfate at 298.15 K, *J. Chem. Eng. Data* **52**, 2138–2142 (2007).

80   Pereiro, A. B., and Rodríguez, A., Azeotrope-breaking using [BMIM][MeSO$_4$] ionic liquid in an extraction column, *Sep. Purif. Technol.* **62**, 733–738 (2008).

81   Pereiro, A. B., and Rodríguez, A., Phase equilibria of the azeotropic mixture hexane plus ethyl acetate with ionic liquids at 298.15 K, *J. Chem. Eng. Data* **53**, 1360–1366 (2008).

82   Pereiro, A. B., Deive, F. J., Esperança, J., and Rodríguez, A., Alkylsulfate-based ionic liquids to separate azeotropic mixtures, *Fluid Phase Equilibr.* **291**, 13–17 (2009).

83   Pereiro, A. B., and Rodríguez, A., An ionic liquid proposed as solvent in aromatic hydrocarbon separation by liquid extraction, *AlChE J.* **56**, 381–386 (2009).

84   Pereiro, A. B., and Rodríguez, A., Application of the ionic liquid Ammoeng 102 for aromatic/aliphatic hydrocarbon separation, *J. Chem. Thermodyn.* **41**, 951–956 (2009).

85   González, E. J., Calvar, N., Canosa, J., and Domínguez, A., Effect of the chain length on the aromatic ring in the separation of aromatic compounds from methylcyclohexane using the ionic liquid 1-ethyl-3-methylpyridinium ethylsulfate, *J. Chem. Eng. Data* **55**, 2289–2293.

86   González, E. J., Calvar, N., Gomez, E., and Domínguez, A., Separation of benzene from alkanes using 1-ethyl-3-methylpyridinium ethylsulfate ionic liquid at several temperatures and atmospheric pressure: Effect of the size of the aliphatic hydrocarbons, *J. Chem. Thermodyn.* **42**, 104–109 (2009).

87   González, E. J., Calvar, N., González, B., and Domínguez, A., Liquid-liquid equilibrium for ternary mixtures of hexane plus aromatic compounds plus [EMpy][ESO$_4$] at T=298.15 K, *J. Chem. Eng. Data* **55**, 633–638 (2010).

88   González, E. J., Calvar, N., González, B., and Domínguez, A., Measurement and correlation of liquid-liquid equilibria for ternary systems {cyclooctane plus aromatic hydrocarbon+1-ethyl-3-methylpyridinium ethylsulfate} at T=298.15 K and atmospheric pressure, *Fluid Phase Equilibr.* **291**, 59–65 (2009).

89   González, E. J., Calvar, N., González, B., and Domínguez, A., (Liquid plus liquid) equilibria for ternary mixtures of (alkane plus benzene plus [EMpy][ESO₄]) at several temperatures and atmospheric pressure, *J. Chem. Thermodyn.* **41**, 1215–1221 (2009).

90   González, E. J., Calvar, N., González, B., and Domínguez, A., Measurement and correlation of liquid-liquid equilibria for ternary systems {cyclooctane plus aromatic hydrocarbon+1-ethyl-3-methylpyridinium ethylsulfate} at T=298.15 K and atmospheric pressure, *Fluid Phase Equilibr.* **291**, 59–65 (2010).

91   González, E. J., Calvar, N., González, B., and Domínguez, A., Separation of toluene from alkanes using 1-ethyl-3-methylpyridinium ethylsulfate ionic liquid at T=298.15 K and atmospheric pressure, *J. Chem. Thermodyn.* **42**, 752–757 (2010).

92   Letcher, T. M., and Deenadayalu, N., Ternary liquid-liquid equilibria for mixtures of 1-methyl-3-octyl-imidazolium chloride plus benzene plus an alkane at *T*=298.2 K and 1 atm, *J. Chem. Thermodyn.* **35**, 67–76 (2003).

93   Letcher, T. M., Deenadayalu, N., Soko, B., Ramjugernath, D., and Naicker, P. K., Ternary liquid-liquid equilibria for mixtures of 1-methyl-3-octylimidazolium chloride plus an alkanol plus an alkane at 298.2 K and 1 bar, *J. Chem. Eng. Data* **48**, 904–907 (2003).

94   Letcher, T. M., and Reddy, P., Ternary liquid-liquid equilibria for mixtures of 1-hexyl-3-methylimidozolium (tetrafluoroborate or hexafluorophosphate) plus ethanol plus an alkene at T=298.2 K, *Fluid Phase Equilib.* **219**, 107–112 (2004).

95   Chen, C. C., Britt, H. I., Boston, J. F., and Evans, L. B., Local composition model for excess gibbs energy of electrolyte systems. Part 1. Single solvent, single completely dissociated electrolyte systems, *AIChE J.* **28**, 588–596 (1982).

96   Belvèze, L. S., Brennecke, J. F., and Stadtherr, M. A., Modeling of activity coefficients of aqueous solutions of quaternary ammonium salts with the electrolyte-NRTL equation, *Ind. Eng. Chem. Res.* **43**, 815–825 (2004).

97   Chapeaux, A., Simoni, L. D., Stadtherr, M. A., and Brennecke, J. F., Liquid phase behavior of ionic liquids with water and 1-octanol and modeling of 1-octanol/water partition coefficients, *J. Chem. Eng. Data* **52**, 2462–2467 (2007).

98   Orchillés, A. V., Miguel, P. J., Vercher, E., and Martínez-Andreu, A., Isobaric vapor-liquid equilibria for ethyl acetate plus ethanol+1-ethyl-3-methylimidazolium trifluoromethanesulfonate at 100 kPa, *J. Chem. Eng. Data* **52**, 2325–2330 (2007).

99   Orchillés, A. V., Miguel, P. J., Vercher, E., and Martínez-Andreu, A., Isobaric vapor-liquid equilibria for methyl acetate plus methanol plus 1-ethyl-3-methylimidazolium trifluoromethanesulfonate at 100 kPa, *J. Chem. Eng. Data* **52**, 915–920 (2007).

100  Orchillés, A. V., Miguel, P. J., Vercher, E., and Martínez-Andreu, A., Ionic liquids as entrainers in extractive distillation: Isobaric vapor-liquid equilibria for acetone plus methanol plus 1-ethyl-3-methylimidazolium trifluoromethanesulfonate, *J. Chem. Eng. Data* **52**, 141–147 (2007).

101 Orchillés, A. V., Miguel, P. J., Vercher, E., and Martínez-Andreu, A., Isobaric vapor-liquid and liquid-liquid equilibria for chloroform plus ethanol+1-ethyl-3-methylimidazolium trifluoromethanesulfonate at 100 kPa, *J. Chem. Eng. Data* **53**, 2642–2648 (2008).

102 Orchillés, A. V., Miguel, P. J., Vercher, E., and Martínez-Andreu, A., Isobaric vapor-liquid equilibria for 1-propanol plus water plus 1-ethyl-3-methylimidazolium trifluoromethanesulfonate at 100 kPa, *J. Chem. Eng. Data* **53**, 2426–2431 (2008).

103 Orchillés, A. V., Miguel, P. J., Vercher, E., and Martínez-Andreu, A., Using 1-ethyl-3-methylimidazolium trifluoromethanesulfonate as an entrainer for the extractive distillation of ethanol plus water mixtures, *J. Chem. Eng. Data* **55**, 1669–1674 (2010).

104 Orchillés, A. V., Miguel, P. J., Vercher, E., and Martínez-Andreu, A., Isobaric vapor-liquid and liquid-liquid equilibria for chloroform plus methanol+1-ethyl-3-methylimidazolium trifluoromethanesulfonate at 100 kPa, *J. Chem. Eng. Data* **55**, 1209–1214 (2010).

105 Simoni, L. D., Brennecke, J. F., and Stadtherr, M. A., Asymmetric framework for predicting liquid-liquid equilibrium of ionic liquid-mixed-solvent systems. 1. theory, phase stability analysis, and parameter estimation, *Ind. Eng. Chem. Res.* **48**, 7246–7256 (2009).

106 Simoni, L. D., Chapeaux, A., Brennecke, J. F., and Stadtherr, M. A., Asymmetric framework for predicting liquid-liquid equilibrium of ionic liquid-mixed-solvent systems. 2. prediction of ternary systems, *Ind. Eng. Chem. Res.* **48**, 7257–7265 (2009).

107 Chen, C. C., and Song, Y. H., Solubility modeling with a nonrandom two-liquid segment activity coefficient model, *Ind. Eng. Chem. Res.* **43**, 8354–8362 (2004).

108 Chen, C. C., Simoni, L. D., Brennecke, J. F., and Stadtherr, M. A., Correlation and prediction of phase behavior of organic compounds in ionic liquids using the nonrandom two-liquid segment activity coefficient model, *Ind. Eng. Chem. Res.* **47**, 7081–7093 (2008).

109 Doker, M., and Gmehling, J., Measurement and prediction of vapor-liquid equilibria of ternary systems containing ionic liquids, *Fluid Phase Equilibr.* **227**, 255–266 (2005).

110 Naydenov, D., and Bart, H. J., Ternary liquid-liquid equilibria for six systems containing ethylacetate plus ethanol or acetic acid plus an imidazolium-based ionic liquid with a hydrogen sulfate anion at 313.2 K, *J. Chem. Eng. Data* **52**, 2375–2381 (2007).

111 Santiago, R. S., Santos, G. R., and Aznar, M., UNIQUAC correlation of liquid-liquid equilibrium in systems involving ionic liquids: The DFT-PCM approach. Part II, *Fluid Phase Equilibr.* **293**, 66–72 (2009).

112 Santiago, R. S., Santos, G. R., and Aznar, M., UNIQUAC correlation of liquid-liquid equilibrium in systems involving ionic liquids: The DFT-PCM approach, *Fluid Phase Equilibr.* **278**, 54–61 (2009).

113 Bondi, A., *Physical Properties of Molecular Crystals, Liquids and Glasses* (Wiley, New York, 1968).

114 Kato, R., Krummen, M., and Gmehling, J., Measurement and correlation of vapor-liquid equilibria and excess enthalpies of binary systems containing ionic liquids and hydrocarbons, *Fluid Phase Equilibr.* **224**, 47–54 (2004).

115 Banerjee, T., Singh, M. K., Sahoo, R. K., and Khanna, A., Volume, surface and UNIQUAC interaction parameters for imidazolium based ionic liquids via Polarizable Continuum Model, *Fluid Phase Equilibr.* **234**, 64–76 (2005).

116 Pereiro, A. B., and Rodríguez, A., Experimental liquid-liquid equilibria of 1-alkyl-3-methylimidazolium hexafluorophosphate with 1-alcohols, *J. Chem. Eng. Data* **52**, 1408–1412 (2007).

117 Pereiro, A. B., and Rodríguez, A., Study on the phase behaviour and thermodynamic properties of ionic liquids containing imidazolium cation with ethanol at several temperatures, *J. Chem. Thermodyn.* **39**, 978–989 (2007).

118 Pereiro, A. B., and Rodríguez, A., Binary mixtures containing OMIM PF$_6$: density, speed of sound, refractive index and LLE with hexane, heptane and 2-propanol at several temperatures, *Phys. Chem. Liq.* **46**, 172–184 (2008).

119 Alonso, L., Arce, A., Francisco, M., and Soto, A., Liquid-liquid equilibria for [C$_8$mim][NTf$_2$] +thiophene+2,2,4-trimethylpentane or plus toluene, *J. Chem. Eng. Data* **53**, 1750–1755 (2008).

120 Alonso, L., Arce, A., Francisco, M., and Soto, A., Liquid-liquid equilibria of ([C$_2$mim][EtSO$_4$] plus thiophene plus 2,2,4-trimethylpentane) and ([C$_2$mim][EtSO$_4$] plus thiophene plus toluene): Experimental data and correlation, *J. Solution Chem.* **37**, 1355–1363 (2008).

121 Alonso, L., Arce, A., Francisco, M., and Soto, A., Thiophene separation from aliphatic hydrocarbons using the 1-ethyl-3-methylimidazolium ethylsulfate ionic liquid, *Fluid Phase Equilibr.* **270**, 97–102 (2008).

122 Arce, A., Francisco, M., and Soto, A., Evaluation of the polysubstituted pyridinium ionic liquid [hmmpy][Ntf2)] as a suitable solvent for desulfurization: Phase equilibria, *J. Chem. Thermodyn.* **42**, 712–718 (2010).

123 Francisco, M., Arce, A., and Soto, A., Ionic liquids on desulfurization of fuel oils, *Fluid Phase Equilibr.* **294**, 39–48 (2010).

124 García, J., Fernández, A., Torrecilla, J. S., Oliet, M., and Rodríguez, F., Ternary liquid-liquid equilibria measurement for hexane and benzene with the ionic liquid 1-butyl-3-methylimidazolium methylsulfate at T = (298.2, 313.2, and 328.2) K, *J. Chem. Eng. Data* **55**, 258–261.

125 García, J., Fernández, A., Torrecilla, J. S., Oliet, M., and Rodríguez, F., Liquid-liquid equilibria for {hexane plus benzene+1-ethyl-3-methylimidazolium ethylsulfate} at (298.2, 313.2 and 328.2) K, *Fluid Phase Equilibr.* **282**, 117–120 (2009).

126 Wang, J. F., Sun, W., Li, C. X., and Wang, Z. H., Correlation of infinite dilution activity coefficient of solute in ionic liquid using UNIFAC model, *Fluid Phase Equilibr.* **264**, 235–241 (2008).

127 Lei, Z. G., Zhang, J. G., Li, Q. S., and Chen, B. H., UNIFAC model for ionic liquids, *Ind. Eng. Chem. Res.* **48**, 2697–2704 (2009).

128 Kim, Y. S., Choi, W. Y., Jang, J. H., Yoo, K. P., and Lee, C. S., Solubility measurement and prediction of carbon dioxide in ionic liquids, *Fluid Phase Equilibr.* **228**, 439–445 (2005).

129 Alevizou, E. I., Pappa, G. D., and Voutsas, E. C., Prediction of phase equilibrium in mixtures containing ionic liquids using UNIFAC, *Fluid Phase Equilibr.* **284**, 99–105 (2009).

130 Weidlich, U., and Gmehling, J., A modified UNIFAC model. 1. Prediction of VLE, hE, and gamma-Infinity, *Ind. Eng. Chem. Res.* **26**, 1372–1381 (1987).

131 Gmehling, J., Li, J.D., and Schiller, M., A modified UNIFAC model. 2. Present parameter matrix and results for different thermodynamic properties, *Ind. Eng. Chem. Res.* **32**, 178–193 (1993).

132 Gmehling, J., Lohmann, J., Jakob, A., Li, J.D., and Joh, R., A modified UNIFAC (Dortmund) model. 3. Revision and extension, *Ind. Eng. Chem. Res.* **37**, 4876–4882 (1998).

133 Gmehling, J., Wittig, R., Lohmann, J., and Joh, R., A modified UNIFAC (Dortmund) model. 4. Revision and extension, *Ind. Eng. Chem. Res.* **41**, 1678–1688 (2002).

134 Jakob, A., Grensemann, H., Lohmann, J., and Gmehling, J., Further development of modified UNIFAC (Dortmund): Revision and extension 5, *Ind. Eng. Chem. Res.* **45**, 7924–7933 (2006).

135 Kato, R., and Gmehling, J., Systems with ionic liquids: Measurement of VLE and gamma (infinity) data and prediction of their thermodynamic behavior using original UNIFAC, mod. UNIFAC(Do) and COSMO-RS(O1), *J. Chem. Thermodyn.* **37**, 603–619 (2005).

136 Nebig, S., Bolts, R., and Gmehling, J., Measurement of vapor-liquid equilibria (VLE) and excess enthalpies (H-F) of binary systems with 1-alkyl-3-methylimidazolium bis(trifluoromethylsulfonyl) imide and prediction of these properties and gamma(infinity) using modified UNIFAC (Dortmund), *Fluid Phase Equilibr.* **258**, 168–178 (2007).

137 Liebert, V., Nebig, S., and Gmehling, J., Experimental and predicted phase equilibria and excess properties for systems with ionic liquids, *Fluid Phase Equilibr.* **268**, 14–20 (2008).

138 Mu, T. C., and Gmehling, J., Conductor-like screening model for real solvents (COSMO-RS), *Prog. Chem,* **20**, 1487–1494 (2008).

139 Nebig, S., Liebert, V., and Gmehling, J., Measurement and prediction of activity coefficients at infinite dilution (gamma(infinity)), vapor-liquid equilibria (VLE) and excess enthalpies (H-E) of binary systems with 1,1-dialkyl-pyrrolidinium bis(trifluoromethylsulfonyl)imide using mod. UNIFAC (Dortmund), *Fluid Phase Equilibr.* **277**, 61–67 (2009).

140 Westerholt, A., Liebert, V., and Gmehling, J., Influence of ionic liquids on the separation factor of three standard separation problems, *Fluid Phase Equilibr.* **280**, 56–60 (2009).

141 Cehreli, S., and Gmehling, J., Phase equilibria for benzene-cyclohexene and activity coefficients at infinite dilution for the ternary systems with ionic liquids, *Fluid Phase Equilibr.* **295**, 125–129 (2010).

142 Mokhtarani, B., and Gmehling, J., (Vapour plus liquid) equilibria of ternary systems with ionic liquids using headspace gas chromatography, *J. Chem. Thermodyn.* **42**, 1036–1038 (2010).

143   Nebig, S., and Gmehling, J., Measurements of different thermodynamic properties of systems containing ionic liquids and correlation of these properties using modified UNIFAC (Dortmund), *Fluid Phase Equilibr.* **294**, 206–212 (2010).

144   Santiago, R. S., Santos, G. R., and Aznar, M., Liquid-liquid equilibrium in ternary ionic liquid systems by UNIFAC: New volume, surface area and interaction parameters. Part I, *Fluid Phase Equilibr.* **295**, 93–97 (2009).

145   Chapeaux, A., Simoni, L. D., Ronan, T. S., Stadtherr, M. A., and Brennecke, J. F., Extraction of alcohols from water with 1-hexyl-3-methylimidazolium bis(trifluoromethylsulfonyl)imide, *Green Chem.* **10**, 1301–1306 (2008).

146   Simoni, L. D., Lin, Y., Brennecke, J. F., and Stadtherr, M. A., Modeling liquid-liquid equilibrium of ionic liquid systems with NRTL, electrolyte-NRTL, and UNIQUAC, *Ind. Eng. Chem. Res.* **47**, 256–272 (2008).

147   Shariati, A., and Peters, C. J., High-pressure phase behavior of systems with ionic liquids: Measurements and modeling of the binary system fluoroform+1-ethyl-3-methylimidazolium hexafluorophosphate, *J. Supercrit. Fluids* **25**, 109–117 (2003).

148   Liu, Z. M., Wu, W. Z., Han, B. X., Dong, Z. X., Zhao, G. Y., Wang, J. Q., Jiang, T., and Yang, G. Y., Study on the phase behaviors, viscosities, and thermodynamic properties of $CO_2/[C_4mim][PF_6]$/methanol system at elevated pressures, *Chem. Eur. J.* **9**, 3897–3903 (2003).

149   Carvalho, P. J., Álvarez, V. H., Machado, J. J. B., Pauly, J., Daridon, J. L., Marrucho, I. M., Aznar, M., and Coutinho, J. A. P., High pressure phase behavior of carbon dioxide in 1-alkyl-3-methyl-imidazolium bis(trifluoromethylsulfonyl)imide ionic liquids, *J. Supercrit. Fluids* **48**, 99–107 (2009).

150   Carvalho, P. J., Álvarez, V. H., Marrucho, I. M., Aznar, M., and Coutinho, J. A. P., High pressure phase behavior of carbon dioxide in 1-butyl-3-methylimidazolium bis(trifluoromethylsulfonyl)imide and 1-butyl-3-methylimidazolium dicyanamide ionic liquids, *J. Supercrit. Fluids* **50**, 105–111 (2009).

151   Carvalho, P. J., Álvarez, V. H., Marrucho, I. M., Aznar, M., and Coutinho, J. A. P., High carbon dioxide solubilities in trihexyltetradecylphosphonium-based ionic liquids, *J. Supercrit. Fluids* **52**, 258–265 (2010).

152   Ren, W., and Scurto, A. M., Phase equilibria of imidazolium ionic liquids and the refrigerant gas, 1,1,1,2-tetrafluoroethane (R-134a), *Fluid Phase Equilibr.* **286**, 1–7 (2009).

153   Nwosu, S. O., Schleicher, J. C., and Scurto, A. M., High-pressure phase equilibria for the synthesis of ionic liquids in compressed $CO_2$ for 1-hexyl-3-methylimidazolium bromide with 1-bromohexane and 1-methylimidazole, *J. Supercrit. Fluids* **51**, 1–9 (2009).

154   Álvarez, V. H., and Aznar, M., Thermodynamic modeling of vapor-liquid equilibrium of binary systems ionic liquid plus supercritical {$CO_2$ or $CHF_3$} and ionic liquid plus hydrocarbons using Peng-Robinson equation of state, *J. Chin. Inst. Chem. Eng,* **39**, 353–360 (2008).

155   Trindade, C. A. S., Visak, Z. P., Bogel-Łukasik, R., Bogel-Łukasik, E., and da Ponte, M. N., Liquid-liquid equilibrium of mixtures of imidazolium-based ionic liquids with propanediols or glycerol, *Ind. Eng. Chem. Res.* **49**, 4850–4857 (2010).

156 Arce, P. F., Robles, P. A., Graber, T. A., and Aznar, M., Modeling of high-pressure vapor-liquid equilibrium in ionic liquids + gas systems using the PRSV equation of state, *Fluid Phase Equilibr.* **295**, 9–16 (2010).

157 Ren, W., Sensenich, B., and Scurto, A. M., High-pressure phase equilibria of {carbon dioxide ($CO_2$) + *n*-alkyl-imidazolium bis(trifluoromethylsulfonyl)amide} ionic liquids, *J. Chem. Thermodyn.* **42**, 305–311 (2010).

158 Shiflett, M. B., and Yokozeki, A., Solubilities and diffusivities of carbon dioxide in ionic liquids: [bmim][$PF_6$] and [bmim][$BF_4$], *Ind. Eng. Chem. Res.* **44**, 4453–4464 (2005).

159 Shiflett, M. B., and Yokozeki, A., Vapor-liquid-liquid equilibria of hydrofluoro-carbons+1-butyl-3-methylimidazolium hexafluorophosphate, *J. Chem. Eng. Data* **51**, 1931–1939 (2006).

160 Yokozeki, A., and Shiflett, M.B., Global phase behaviors of trifluoromethane in ionic liquid [bmim][$PF_6$], *AIChE J.* **52**, 3952–3957 (2006).

161 Shiflett, M. B., and Yokozeki, A., Solubility of $CO_2$ in room temperature ionic liquid [hmim][$Tf_2N$], *J. Phys. Chem. B* **111**, 2070–2074 (2007).

162 Yokozeki, A., and Shiflett, M. B., Vapor-liquid equilibria of ammonia plus ionic liquid mixtures, *Appl. Energy* **84**, 1258–1273 (2007).

163 Yokozeki, A., and Shiflett, M. B., Ammonia solubilities in room-temperature ionic liquids, *Ind. Eng. Chem. Res.* **46**, 1605–1610 (2007).

164 Shiflett, M. B., and Yokozeki, A., Binary vapor-liquid and vapor-liquid-liquid equilibria of hydrofluorocarbons (HFC-125 and HFC-143a) and hydrofluoro-ethers (HFE-125 and HFE-143a) with ionic liquid [emim][$Tf_2N$], *J. Chem. Eng. Data* **53**, 492–497 (2008).

165 Yokozeki, A., and Shiflett, M.B., Binary and ternary phase diagrams of benzene, hexafluorobenzene, and ionic liquid [emim][$Tf_2N$] using equations of state, *Ind. Eng. Chem. Res.* **47**, 8389–8395 (2008).

166 Shiflett, M. B., and Yokozeki, A., Phase behavior of carbon dioxide in ionic liquids: [emim][acetate], [emim][trifluoroacetate], and [emim][acetate] plus [emim][trifluoroacetate] mixtures, *J. Chem. Eng. Data* **54**, 108–114 (2009).

167 Yokozeki, A., and Shiflett, M. B., Separation of carbon dioxide and sulfur dioxide gases using room-temperature ionic liquid [hmim][$Tf_2N$], *Energy Fuels* **23**, 4701–4708 (2009).

168 Shiflett, M. B., and Yokozeki, A., Separation of $CO_2$ and $H_2S$ using room-temper-ature ionic liquid [bmim][$PF_6$], *Fluid Phase Equilibr.* **294**, 105–113 (2010).

169 Shiflett, M. B., and Yokozeki, A., Separation of carbon dioxide and sulfur dioxide using room-temperature ionic liquid [bmim][$MeSO_4$], *Energy Fuels* **24**, 1001–1008 (2010).

170 Kroon, M. C., Karakatsani, E. K., Economou, I. G., Witkamp, G. J., and Peters, C. J., Modeling of the carbon dioxide solubility in imidazolium-based ionic liquids with the tPC-PSAFT equation of state, *J. Phys. Chem. B* **110**, 9262–9269 (2006).

171 Karakatsani, E. K., Economou, I. G., Kroon, M. C., Peters, C. J., and Witkamp, G. J., TPC-PSAFT modeling of gas solubility in imidazolium-based ionic liquids, *J. Phys. Chem. C* **111**, 15487–15492 (2007).

172  Karakatsani, E. K., Economou, I. G., Kroon, M. C., Bermejo, M. D., Peters, C. J., and Witkamp, G. J., Equation of state modeling of the phase equilibria of ionic liquid mixtures at low and high pressure, *PCCP* **10**, 6160–6168 (2008).

173  Andreu, J. S., and Vega, L. F., Capturing the solubility behavior of $CO_2$ in ionic liquids by a simple model, *J. Phys. Chem. C* **111**, 16028–16034 (2007).

174  Andreu, J. S., and Vega, L. F., Modeling the solubility behavior of $CO_2$, $H_2$, and Xe in [$C_n$-mim][$Tf_2N$] ionic liquids, *J. Phys. Chem. B* **112**, 15398–15406 (2008).

175  Vega, L. F., Vilaseca, O., Llovell, F., and Andreu, J. S., Modeling ionic liquids and the solubility of gases in them: Recent advances and perspectives, *Fluid Phase Equilibr.* **294**, 15–30 (2010).

176  Wang, T. F., Peng, C. J., Liu, H. L., Hu, Y., and Jiang, J. W., Equation of state for the vapor-liquid equilibria of binary systems containing imidazolium-based ionic liquids, *Ind. Eng. Chem. Res.* **46**, 4323–4329 (2007).

177  Li, J. L., He, Q., He, C. C., Peng, C. J., and Liu, H. L., Representation of phase behavior of ionic liquids using the equation of state for square-well chain fluids with variable range, *Chin. J. Chem. Eng.* **17**, 983–989 (2009).

178  Yang, J. Y., Peng, C. J., Liu, H. L., and Hu, Y., Calculation of vapor-liquid and liquid-liquid phase equilibria for systems containing ionic liquids using a lattice model, *Ind. Eng. Chem. Res.* **45**, 6811–6817 (2006).

179  Xu, X. C., Peng, C. J., Liu, H. L., and Hu, Y., Modeling pVT properties and phase equilibria for systems containing ionic liquids using a new lattice-fluid equation of state, *Ind. Eng. Chem. Res.* **48**, 11189–11201 (2009).

180  Breure, B., Bottini, S. B., Witkamp, G. J., and Peters, C. J., Thermodynamic modeling of the phase behavior of binary systems of ionic liquids and carbon dioxide with the group contribution equation of state, *J. Phys. Chem. B* **111**, 14265–14270 (2007).

181  Skjold-Jørgensen, S., Gas solubility calculations. 2. Application of a new group-contribution equation of state, *Fluid Phase Equilibr.* **16**, 317–351 (1984).

182  Skjold-Jørgensen, S., Group contribution equation of state (GC-EOS)—a predictive method for phase-equilibrium computations over wide ranges of temperature and pressures up to 30 MPa, *Ind. Eng. Chem. Res.* **27**, 110–118 (1988).

183  Martin, A., Mendez, D., and Bermejo, M. D., Application of a group contribution equation of state for the thermodynamic modeling of binary systems (gas plus ionic liquids) with bis[(trifluoromethyl) sulfonyl]imide anion, *J. Chem. Thermodyn.* **42**, 524–529 (2009).

184  Bermejo, M. D., Martin, A., Foco, G., Cocero, M. J., Bottini, S. B., and Peters, C. J., Application of a group contribution equation of state for the thermodynamic modeling of the binary systems $CO_2$-1-butyl-3-methyl imidazolium nitrate and $CO_2$-1-hydroxy-1-propyl-3-methyl imidazolium nitrate, *J. Supercrit. Fluids* **50**, 112–117 (2009).

185  Diedenhofen, M., Eckert, F., and Klamt, A., Prediction of infinite dilution activity coefficients of organic compounds in ionic liquids using COSMO-RS, *J. Chem. Eng. Data* **48**, 475–479 (2003).

186  Jork, C., Kristen, C., Pieraccini, D., Stark, A., Chiappe, C., Beste, Y. A., and Arlt, W., Tailor-made ionic liquids, *J. Chem. Thermodyn.* **37**, 537–558 (2005).

187 Banerjee, T., Singh, M. K., and Khanna, A., Prediction of binary VLE for imidazolium based ionic liquid systems using COSMO-RS, *Ind. Eng. Chem. Res.* **45**, 3207–3219 (2006).

188 Banerjee, T., Verma, K. K., and Khanna, A., Liquid-liquid equilibrium for ionic liquid systems using COSMO-RS: Effect of cation and anion dissociation, *AIChE J.* **54**, 1874–1885 (2008).

189 Kumar, A. A. P., and Banerjee, T., Thiophene separation with ionic liquids for desulphurization: A quantum chemical approach, *Fluid Phase Equilibr.* **278**, 1–8 (2009).

190 Mohanty, S., Banerjee, T., and Mohanty, K., Quantum chemical based screening of ionic liquids for the extraction of phenol from aqueous solution, *Ind. Eng. Chem. Res.* **49**, 2916–2925 (2010).

191 Verma, V. K., and Banerjee, T., Ionic liquids as entrainers for water plus ethanol, water+2-propanol, and water plus THF systems: A quantum chemical approach, *J. Chem. Thermodyn.* **42**, 909–919 (2010).

192 Freire, M. G., Neves, C., Carvalho, P. J., Gardas, R. L., Fernandes, A. M., Marrucho, I. M., Santos, L., and Coutinho, J. A. P., Mutual solubilities of water and hydrophobic ionic liquids, *J. Phys. Chem. B* **111**, 13082–13089 (2007).

193 Freire, M. G., Santos, L., Marrucho, I. M., and Coutinho, J. A. P., Evaluation of COSMO-RS for the prediction of LLE and VLE of alcohols plus ionic liquids, *Fluid Phase Equilibr.* **255**, 167–178 (2007).

194 Freire, M. G., Ventura, S. P. M., Santos, L., Marrucho, I. M., and Coutinho, J. A. P., Evaluation of COSMO-RS for the prediction of LLE and VLE of water and ionic liquids binary systems, *Fluid Phase Equilibr.* **268**, 74–84 (2008).

195 Lei, Z. G., Arlt, W., and Wasserscheid, P., Selection of entrainers in the 1-hexene/n-hexane system with a limited solubility, *Fluid Phase Equilibr.* **260**, 29–35 (2007).

196 Diedenhofen, M., and Klamt, A., COSMO-RS as a tool for property prediction of IL mixtures-A review, *Fluid Phase Equilibr.* **294**, 31–38 (2010).

# 6 Molecular Simulation of Ionic Liquids: Where We Are and the Path Forward

JINDAL K. SHAH

Department of Chemical and Biomolecular Engineering, University of Notre Dame, Notre Dame, Indiana, USA
The Center for Research Computing, University of Notre Dame, Notre Dame, Indiana, USA

EDWARD J. MAGINN

Department of Chemical and Biomolecular Engineering, University of Notre Dame, Notre Dame, Indiana, USA

## ABSTRACT

Molecular-based simulation methods are an essential tool for the development of new ionic liquids. Simulations have led to the discovery of new ionic liquid phenomena. They enable predictions of physical properties to be made for systems that have not even been synthesised yet, and are thus an important partner with experimental studies in ionic liquid research. In this chapter, we provide a brief introduction to the topic, and then discuss the kinds of phenomena that can be probed by a molecular simulation. These include calculations of thermodynamic and transport properties, as well as insight into the behaviour of these systems at the molecular level. We highlight the kinds of properties and phenomena that have been computed with some success in recent years, and then give our views on the areas where additional work is needed. These include vapour–liquid (VLE), liquid–liquid (LLE), and solid–liquid phase equilibria (SLE), and the development of a database of force fields and simulation results. We also discuss topics that have run their course and areas where more simulation research is not needed.

*Ionic Liquids Further UnCOILed: Critical Expert Overviews*, First Edition.
Edited by Natalia V. Plechkova and Kenneth R. Seddon.

## 6.1 INTRODUCTION

The term molecular simulation can be defined broadly as the use of computational methods to describe the behaviour of matter at the atomistic or molecular level. There is a clear distinction between this and continuum-based modelling, in which atomic-level phenomena are neglected. Molecular simulation also differs from techniques such as equation of state modelling, quantitative structure–property relationship modelling, and related approaches, in that these other techniques rely upon empirical parameters regressed against experimental data to develop correlations for these properties. Molecular simulations, on the other hand, attempt to capture the underlying physics of the system and in so doing are much more predictive than these other techniques. Molecular simulations also enable one to probe a wide range of properties and develop physical insights into why a particular material behaves as it does. These capabilities are especially important for the young field of ionic liquids because there are many gaps in the experimental data set and the dimensionality of possible "compound space" is enormous, as has been pointed out on many occasions [1].

According to the definition above, molecular simulations encompass both computational quantum chemistry methods and classical atomistic simulation methods such as molecular dynamics (MD) and Monte Carlo (MC). In this chapter, we will only focus on classical atomistic simulations, but quantum chemical methods are an equally important tool for discovery in this field and in fact play a large role in the development of the potential functions used in classical atomistic simulations. Thus, when we speak of "molecular simulations," we are only focussing on classical MD and MC.

Before a molecular simulation may be carried out, a classical potential function ("force field") must be developed which can capture both intramolecular and intermolecular interactions. A typical functional form is given in Equation (6.1):

$$V(\mathbf{r}) = \sum_{\text{bonds}} k_b (r - r_0)^2 + \sum_{\text{angles}} k_\theta (\theta - \theta_0)^2 + \sum_{\text{dihedrals}} k_\chi [1 + \cos(n_0 \chi - \delta_0)]$$
$$+ \sum_{i=1}^{N-1} \sum_{j=i+1}^{N} \left( 4\varepsilon_{ij} \left[ \left( \frac{\sigma_{ij}}{r_{ij}} \right)^{12} - \left( \frac{\sigma_{ij}}{r_{ij}} \right)^{6} \right] + \frac{q_i q_j}{r_{ij}} \right), \tag{6.1}$$

where $V(\mathbf{r})$ is the total potential energy, which is made up of bond length, bond angle, dihedral angle, and improper angle terms, as well as non-bonded van der Waals and electrostatic terms. Equation (6.1) shows typical functional forms, although others can be used.

The parameters in Equation (6.1) typically are developed using a combination of quantum chemical calculations and analogies with other compounds for which accurate force fields have been developed. Given such a force field for describing the interactions in the model system, all the properties of the model are set. All that is left to do is conduct a simulation to discover what those properties are. Because of this, the development of accurate force fields is one of the essential research needs of this field. More will be said about this later in Section 6.7.4.

Once the force field has been determined, a numerical procedure must be used to generate configurations of a collection of molecules at the desired state point. From these configurations, one can compute the thermodynamic and (if time-dependent configurations were obtained) transport properties of the model system by applying well-known statistical mechanical methods. By far, MD is the most frequently used method for doing this in the ionic liquids research field. In an MD simulation, a set of initial momenta and positions are assigned to each atom of the system, and variants of the Newtonian or Hamiltonian equations of motion are used to generate time-dependent positions and momenta. The natural thermodynamic constraints of the standard equations of motion are number of particles $(N)$, volume $(V)$, and total energy $(E = V + K)$, where $K$ is the kinetic energy. This set of constraints is consistent with the microcanonical statistical mechanical ensemble. Often one wishes to compute properties such as density at a fixed temperature and pressure; various "extended Hamiltonian" methods can be used to conduct MD in other ensembles such as the isothermal–isobaric ensemble.

Molecular dynamics is an attractive method to use for several reasons. First, one can obtain thermodynamic information on the system from knowledge of the positions of the atoms in the system *and* dynamical information from knowledge of the time dependence of the positions and velocities of the atoms. As will be discussed later, MC methods do not give time-dependent information. Second, there are many general, fast, and well-documented MD codes available for use. Setting up and conducting a simulation of an ionic liquid can be done relatively quickly, if a force field is available. Third, most MD codes are highly parallelised, so one can take advantage of newer multi-core processors to speed the calculations. It is no longer necessary to have dedicated supercomputer resources to obtain meaningful results from an MD simulation.

However, MD has some drawbacks that one must be aware of when applying this technique to ionic liquids. First, because MD methods are deterministic and time dependent, the more sluggish the dynamics of the system are, the longer the simulation will take. This is an especially insidious problem with many ionic liquids, where there are often multiple, slow dynamical relaxation processes, even far from the melting point or glass transition temperature. This means that assuring an MD simulation is sampling an equilibrium probability distribution is a tricky business. At best, it can take a very long time; at worst, one can be fooled into thinking the system under study is at equilibrium when in fact the simulations are simply sampling local motions around a (arbitrary) non-equilibrium state. Second, typical MD simulations are limited to timescales on the order of tens of nanoseconds. There are heroic calculations in which microsecond (and even millisecond) MD simulations have been conducted, but for most situations an MD simulation of 10 ns is considered quite long. This means that one simply cannot study, in a direct way, any dynamical process that occurs on a timescale longer than 10 ns. For example, assuming that diffusive behaviour occurs on length scales several times larger than a molecular diameter, the time limitation of MD implies that self-diffusivities smaller than $1 \times 10^{-11} \, \mathrm{m^2 s^{-1}}$ are inaccessible to standard MD simulations. This is about the value of ambient temperature self-diffusivities for many common ionic liquids [2]. The literature contains many examples where simulations *much shorter* than 10 ns were used to predict self-diffusivities of ionic liquids; the self-diffusivities determined in these papers must be viewed with a sceptical eye, even if they happen to obtain results similar to published experimental data. Of course, if higher temperatures are studied and/or the ionic liquid is far above its melting point or glass transition temperature, self-diffusivities will be higher, and shorter simulations are sufficient for obtained self-diffusivities.

The large separation of timescales was convincingly demonstrated by Urahata and Ribeiro [3], who used MD to compute various single particle time correlation functions for [C$_4$mim]Cl, and showed that there is a vast separation of timescales in this system. The correlation functions associated with the ring centre of mass and the alkyl chain dihedral angles decorrelate relatively fast. However, reorientational motion of the ring takes place on timescales that are orders of magnitude slower. Not surprisingly, the mean square displacement (MSD) over these timescales (which is directly related to the self-diffusivity) also shows distinct regions; simulations much longer than 1 ns are required to observe diffusive motion. The point is that one must be very careful when computing self-diffusivities for ionic liquids.

The second problem with standard MD is that it is difficult to apply it to compute properties related to free energies. MD is well suited for averaging over "mechanical" properties such as density or pressure, but it is much more difficult to use MD to compute "statistical" properties. For example, if one wants to compute the partitioning of an ionic liquid between two phases (say a liquid and a porous solid), or the solubility of a vapour in an ionic liquid, one needs to ensure that the chemical potentials of the different species are the same in both phases before number averages can be taken. An MD simulation would require one to track molecules as they diffuse from one phase to another and, more importantly, would have to use a system that is large enough to ensure that surface effects are minimised. For most situations, the timescale and length scales involved in interfacial mass transfer are orders of magnitude longer and larger than anything accessible to MD. There are ways around this, such as the use of "thermodynamic integration" procedures [4], but in general MD is not the best choice for simulating phase behaviour.

MC methods have been used much less frequently than MD in the ionic liquids community, but offer many advantages. We were the first to use MC to study ionic liquids [5] and later developed MC methods to compute gas solubility in ionic liquids [6, 7]. There are two major advantages of MC. First, molecular configurations are not generated in a deterministic, time-dependent manner, but instead are the result of stochastic moves. Bold rearrangements of configuration space can be made using advanced biasing procedures [8], meaning that equilibration of ionic liquid systems can be done, at least in principle, even when the natural dynamical processes of the system are slower than those that can be tracked with MD. Second, MC methods are ideally suited for computing free energies and for simulating phase behaviour. Over the last 20 years or so, powerful MC methods have been developed that enable computation of phase equilibria (gas–liquid, solid–liquid) in a more robust manner than is possible with MD.

Despite these advantages, the ionic liquids community has been slow to adopt MC. Why is that? We believe there are two main reasons. MC does not yield dynamical information like MD, so if this is important, researchers will always turn to MD. Even if only thermodynamic properties are of interest, however, MD is still used more often than MC because there are few general-purpose and easy-to-use MC codes capable of simulating ionic liquids. This latter problem will go away with the emergence of new, powerful open-source MC codes, a development we expect will occur soon. We therefore anticipate more users turning to MC as a tool for computing phase behaviour and other thermodynamic properties of ionic liquids.

## 6.2   GOALS OF A MOLECULAR SIMULATION

What is the goal of an ionic liquid molecular simulation study? As mentioned earlier, molecular simulations yield atomic configurations consistent with a given set of thermodynamic constraints. If MD is used, momenta are also available, meaning dynamical properties can also be determined. As such, not only can any macroscopic thermodynamic or transport property be obtained from a simulation, but also one can "see" what is going on at the atomic level. Thus, simulations have a twofold purpose: to compute properties of ionic liquids and to give molecular-level insight into the behaviour of ionic liquids.

### 6.2.1   Properties

Many studies have been carried out in which properties of ionic liquids have been computed from a molecular simulation. Of course, these physical properties are actually the properties of the model system, but to the extent that the force field captures the physics of the ionic liquid, these properties should conform to those of the real material. Thus, one important goal of a molecular simulation is *to predict the thermodynamic and transport properties of an ionic liquid*. This can be done for known ionic liquids as well as for "hypothetical" ionic liquids that may only exist as structures on the computer. The simplest and most widely computed property is the liquid density as a function of temperature and pressure. Heat capacities have also been computed with some regularity, but other thermodynamic properties of pure ionic liquids such as phase transition temperatures (solid–solid, solid–liquid, liquid–vapour), refractive indices, surface tensions, speeds of sound, critical points, enthalpies of vapourisation, and vapour pressures are rarely, if ever, computed. Mixture thermodynamic properties such as solubilities, enthalpies, and entropies of mixing and LLE are even less widely computed. Why is this the case? One reason is that molecular simulations are typically used in a "post-predictive" manner, meaning that comparisons with experimental data are made and the level of agreement between simulations and experiment is assessed. Many of the properties mentioned earlier have simply not been measured experimentally, so simulators have not tried to compute these properties. We believe this argues all the more strongly for why they should be computed! While post-predictive simulations are an essential exercise for validation purposes, the real value of molecular simulations is the ability to make predictions that can lead to the discovery of new ionic liquids with desired properties. Moreover, simulations can, and should, be used as a surrogate for experiments when the experiments are difficult or expensive to carry out. Simulations are not a replacement for experiments, but they certainly can complement experimental property measurement. Journal reviewers are often loathe to allow publication of molecular simulation papers that predict properties with no experimental verification for the prediction; this needs to change.

In addition to thermodynamic properties, transport properties such as self-diffusivities, viscosities, thermal conductivities, and ionic conductivities may be determined from MD simulations. Self-diffusivities are easiest to obtain from a simulation because they are based on individual molecules. The other transport properties are *collective* properties, and are an order of magnitude more difficult to obtain from a simulation [9]. As discussed in Section 6.3.3, even "simple" self-diffusivities can require extraordinarily long simulations, and thus transport properties are quite difficult to obtain. Nevertheless, there have been many examples of self-diffusivities and viscosities computed from a molecular simulation, as we will discuss later. Other transport properties are rarely computed.

A second reason many thermodynamic properties are not computed in a molecular simulation is simply because they are difficult to compute. Phase transition properties are better suited for calculation with MC, but most ionic liquid researchers use MD. Mixture properties are often not computed because force fields have still not been validated on pure component properties. We believe that as robust validated methods and force fields become more available, more thermodynamic properties will be determined.

Often one finds that the properties computed in a simulation do not agree perfectly with experimental data. Usually this means the simulations are "wrong," but not always. For example, Marsh and coworkers showed, using benchmark round-robin tests, that experimental properties such as heat capacities can vary up to 7% as a result of experimental uncertainty and variations in sample quality [10]. For properties such as the solubility of $CO_2$, variations among experimental measurements of 20% or more were common. Conversely, this means that if a simulation "agrees" with a single experimental measurement, it is not necessarily a confirmation that the force field or simulation method is "right." Simulations do provide an independent means of assessing the reasonableness of experimental property measurements, however, and should be used more often in this way, especially when unusual results are reported. For example, Hert and coworkers [11] reported that the presence of $CO_2$ significantly enhances the solubility of $O_2$ in $[C_6mim][NTf_2]$. Molecular simulations were performed on this system to test the experiments [7], and no such enhancement was found. Based on these simulations, the experimental data were re-analysed and it was found that the original experimental results were in error [12]. Used in this way, molecular simulations can serve as an important partner to experimental property measurement, each helping the other achieve greater accuracy.

### 6.2.2  Trends and Insight

Many times, it is not necessary to obtain quantitative accuracy in property predictions from a molecular simulation. More often *trends* in properties are more useful to the discovery effort and are more easily obtained from a simulation. Instead of having to obtain viscosities for a particular ionic liquid to 1% accuracy, it can be more useful to predict the relative viscosity of a range of ionic liquids. Which ones are expected to be least viscous? How will the viscosity change with water content and temperature? What is the underlying mechanism responsible for a high viscosity, and can this be overcome? There are many examples of property trends being predicted from molecular simulations; some examples are given later in this chapter.

Perhaps the quantity most widely predicted in molecular simulations is liquid structure in the form of a radial distribution function. Simulations have been shown to be accurate at capturing the detailed liquid structure of ionic liquids via comparison with neutron diffraction experiments [13]. It was through molecular simulations that the first indications of the now-famous nanoscale segregation of ionic liquids into polar and non-polar domains were predicted [14–16]. Canongia Lopes and Padua [16] showed that as the length of cation alkyl chains increased, the non-polar regions became larger, eventually percolating the entire simulation box. Subsequent experimental work [17] confirmed the basic picture presented in the simulations. There are many other examples of molecular simulations being used to make predictions of the behaviour of ionic liquids, only to have experimental work confirm the predictions of the simulations. For example, it was molecular simulations that first predicted the enthalpy of vapourisation of an ionic liquid [5, 18], although these liquids were commonly referred to at the time as "non-volatile" liquids. Many subsequent predictions of the enthalpy of vapourisation have been made with molecular simulations [19–21], and these have turned out to be accurate, especially in terms of predicting trends [22–24]. Molecular simulations have also elucidated the nature of the ions in the gas phase [25, 26]. These predictions have largely been confirmed experimentally [27–29].

So molecular simulations can provide not only access to physical properties but also detailed information on the nature of ionic liquids, trends in properties, and ultimately a better understanding of why they behave the way they do. This is a powerful tool that must be integrated along with experimentation in any serious effort aimed at developing new ionic liquids or at understanding the nature of these fascinating liquids. In the next sections, we provide a more detailed look at what can be explored with molecular simulations and the technological focus areas where simulations are making a difference. We close this review with a discussion of the advances that are needed to further the role of molecular simulations in ionic liquid research.

## 6.3  PROPERTY PREDICTIONS

### 6.3.1  Setting the Force Field

Before an MC or MD simulation of ionic liquids may be carried out, the parameters in Equation (6.1) must be specified. A common strategy in the ionic liquid simulation community is to assign parameters based on a combination of quantum calculations and those available in the literature. For example, routinely the van der Waals interaction parameters are assigned from the well-established force field databases such as OPLS [30], AMBER [31], and CHARMM [32]. Often times, the force constants that describe intramolecular degrees of freedom are directly adopted from these force fields, or quantum calculations are conducted to determine the coefficients. The partial charges are usually obtained from gas phase quantum calculations performed on isolated ions or an ion pair. Later we will provide a sampling of force fields available in the literature that cover a broad range of cations and anions. It is not clear, however, once a set of parameters is chosen, how to systematically modify force field parameters and improve property prediction accuracy.

Hanke et al. [33] carried out the first atomistic simulations of the cations $[C_1mim]^+$ and $[C_2mim]^+$ paired with $Cl^-$ and $[PF_6]^-$ anions, compounds that can be classified as ionic liquids under the current definition. The partial charges on the atoms were obtained from a quantum mechanical calculation based on the second-order Møller–Plesset (MP2) correlated charge density of each ion, while the van der Waals interaction parameters were taken from the literature [34–36]. Shah et al. [5] followed a similar approach for the united-atom force field parameterisation of a room temperature ionic liquid, $[C_4mim][PF_6]$, and carried out the first MC simulations at a series of temperatures and pressures. Several other force fields have been developed and refined since the early days of the molecular simulations of ionic liquids. For example, our group has developed force fields for an all-atom model of $[C_4mim][PF_6]$ [18], several alkylpyridinium cations paired with $[NTf_2]^-$ [37], and a range of triazolium ionic liquids (such as 1,2,4-triazolium, 1,2,3-triazolium, 4-amino-1,2,4-triazolium, and 1-methyl-4-amino-1,2,4-triazolium) [38] in combination with either nitrate $[NO_3]^-$ or perchlorate $[ClO_4]^-$ anions. Additionally, our group has also proposed force fields for the ionic liquid, 1-(3-amino)propyl-3-methylimidazolium bistriflamide [39], a task-specific ionic liquid for $CO_2$ capture, and its reaction products with $CO_2$.

Pádua and coworkers have also developed a large database of force field parameters for ionic liquids based on the 1,3-dialkylimidazolium, 1-alkylpyridinium, and tetraalkylphosphonium families of cations, and anions such as $[NO_3]^-$, $[PF_6]^-$, and $Cl^-$ [40–42]. Recently, the same group published force fields for the cation families 1,2,3-trialkylimidazolium and alkoxycarbonylimidazolium [43], trihydroxymethylimidazolium, dimethoxy-2-methylimidazolium, and fluoroalkylimidazolium [44], and anions such as alkylsulfonate, alkylsulfate [43], bis(fluorosulfonyl)amide, perfluoroalkanesulfonylamide, and fluoroalkylfluorophosphate [44]. This collection of force fields is used widely by many other research groups. Liu et al. [45] proposed a "refined" force field for imidazolium-based ionic liquids, which they claim does a better job matching experimental properties. Zhou and coworkers [46] developed a force field for ionic liquids containing the $[P_{4444}]^+$ cation combined with a series of amino acid-based anions. Liu et al. [47] developed force field parameters for $N,N,N',N'$-tetramethylguanidinium ionic liquids paired with methanoate, lactate, perchlorate, trifluoroethanoate, and trifluoromethanesulfonate anions. In addition, they also proposed a force field for 11 different cyclic guanidinium cations paired with the nitrate anion. Working with Smit and Bell, Liu also developed an improved united-atom force field for 1-alkyl-3-methylimidazolium chloride [48, 49]. Bedrov and coworkers [50] have developed polarisable force fields for a number of ionic liquids within their "APPLE&P" framework. They have shown that dynamic properties, as well as enthalpies of vapourisation, are captured well with this force field. The force fields discussed in this section do not comprise an exhaustive list of ionic liquid force fields; there are many other examples in the literature. It is not that we have too few force fields, but perhaps we have too few fully validated force fields. Later in the chapter we discuss needs in this vital area.

### 6.3.2   Thermodynamic Properties

*6.3.2.1   Density.*   Density is one of the easiest thermodynamic quantities to compute. It is determined by conducting simulations in the isothermal–isobaric ensemble and calculating the average volume. Also, this is the property for which probably the largest experimental database exists, enabling an initial test of a proposed force field. For several united-atom and all-atom models developed for a range of ionic liquids, the computed densities have been generally found to agree within 1–5% of the corresponding experimental values [5, 18, 20, 40, 41, 43–45, 47, 51–53]. Good agreement between the predicted densities and experiments, however, does not necessarily indicate that the force field can accurately predict other properties. Large deviations, however, do suggest potential problems with force field parameterisation and can lead to inaccuracies in the prediction of other properties [54].

Directly related to the density predictions are the calculations of derivative quantities such as volume expansivity and isothermal compressibility. Two approaches are commonly used in a molecular simulation to calculate these quantities: simulations are carried out as a function of temperature or pressure and the resulting densities are fit as a linear function of temperature to determine volume expansivity, or to a Tait equation [55] for the isothermal compressibility. An alternate approach is to monitor fluctuations in volume or the product of enthalpy and volume to estimate volume expansivity and isothermal compressibility, respectively. We find that the former procedure is more effective from a practical standpoint [5]. These derivative quantities are much more sensitive to the quality of the force field than single point densities and are therefore a more stringent test of the predictive power of a force field. In most cases, it is difficult to obtain agreement within 30% between computed and experimental compressibilities and expansivities [5, 18, 51]. Although the deviations are larger than those expected for density predictions, molecular simulations are able to make predictions of qualitative trends. For example, simulations [56] predicted that a new class of "biomimetic" ionic liquids [57] containing chain lengths of 18 carbon atoms should be among the most compressible ionic liquids yet made, comparable with the tetraalkylphosphonium ionic liquids studied by Rebelo and coworkers [58]. Experimental confirmation (or refutation) awaits.

#### 6.3.2.2 *Cohesive Energy Density/Enthalpy of Vapourisation.* Recent experimental investigations have shown that ionic liquids possess finite vapour pressure and can be distilled to some extent [24]. Molecular simulations have suggested this possibility since 2002, based on the fact that the calculated enthalpy of vapourisation of ionic liquids, although considerably larger than that of common substances, is not infinite. Our group has computed the enthalpy of vapourisation of 1,3-dialkylimidazolium ionic liquids such as [$C_4$mim][$PF_6$], [$C_4$mim][$NO_3$], three different 1-alkylpyridinium ionic liquids paired with the [$NTf_2$] anion, and seven different triazolium-based ionic liquids. Under the assumption that the vapour phase of ionic liquids is composed of neutral ion pairs, the calculated enthalpy of vapourisation for this broad range of ionic liquids was found to vary between 148 and 238 kJ mol$^{-1}$. The enthalpies of vapourisation generally increase with increasing alkyl chain length. 1,3-Dialkylimidazolium cations had lower enthalpies of vapourisation than either related pyridinium or triazolium cations. The [$NTf_2$]$^-$ anion serves to lower the enthalpy of vapourisation relative to others investigated. It was also observed that the enthalpy of vapourisation decreases with increasing temperature. The increase in enthalpy of vapourisation of ionic liquids with alkyl chain length has also been reported by Ludwig and coworkers [20] using a refined force field for imidazolium-based ionic liquids paired with the [$NTf_2$]$^-$

anion. The calculations by Liu et al. [45] also suggest that for a number of 1,3-dialkylimidazolium cations in combination with the $[PF_6]^-$, $[BF_4]^-$, and $Cl^-$ anions, the enthalpy of vapourisation is in the range specified earlier. Simulations show that the smaller the anion, the larger is the enthalpy of vapourisation. Molecular simulations show that the enthalpy of vapourisation of guanidinium-based ionic liquids tends to be in the range of 175–210 kJ mol$^{-1}$ [59]. In addition to the calculation of the enthalpy of vapourisation, these studies suggest that the high enthalpy of vapourisation of ionic liquids is due to the strong intermolecular electrostatic interactions in the condensed phase.

***6.3.2.3  Heat Capacity.***    The constant pressure heat capacity can be calculated either from the fluctuation formula or its definition by finite difference approach, Equation (6.2):

$$C_P(T, P) = \left(\frac{\partial H}{\partial T}\right)_P \approx \left(\frac{\Delta \langle H \rangle}{\Delta T}\right)_P, \tag{6.2}$$

where $\langle H \rangle$ is the ensemble averaged enthalpy, a quantity readily available from a simulation. In practice, direct application of Equation (6.2) with classical force fields of ionic liquids typically does not yield accurate heat capacities. This is due to the fact that most force fields utilise harmonic potentials for bond stretching and angle bending. These approximations tend to overestimate the energy storage capacity of molecules and thus result in heat capacities that are too high. We therefore suggest that only the intermolecular "excess" portion of the heat capacity be determined from a simulation, and that the remaining intramolecular "ideal" portion be determined from quantum calculations. In most molecular simulation studies, gas phase quantum calculations are carried out nonetheless for the assignment of partial charges and can provide an accurate estimate of the ideal heat capacity. A detailed discussion of this approach is provided in the published work of Cadena et al. [37] and Gutowski and Maginn [39]. Molecular simulation results of the heat capacities are generally found to be in good agreement with experimental measurements. In line with experiments, simulations have yielded a range of heat capacities varying between 1.2 and 2.6 J g$^{-1}$ K$^{-1}$, with imidazolium- and pyridinium-based ionic liquids at the lower end of the spectrum [37], while phosphonium-based ionic liquids are found to possess large heat capacities [46]. Triazolium-based ionic liquids occupy an intermediate range of the heat capacity window [38]. As the ionic liquid ions contain a large number of intramolecular degrees of freedom, it is not uncommon to find the ideal gas heat capacity contributions up to 70% of the total heat capacity.

### 6.3.3  Transport Properties

***6.3.3.1  Self-Diffusivity.***    The self-diffusivity is a single molecule property and is the easiest transport property to compute from a simulation. It is determined by computing the rate at which MSD of the molecules increases as a function of time over a sufficiently long period of time. Unlike the thermodynamic quantities described earlier, transport properties are accessible only from MD simulations. One of the challenges with reliable computation of ionic liquid self-diffusivity is that the sluggish dynamics of ionic liquids preclude adequate translational sampling to establish diffusive behaviour (i.e., loss of correlations with initial positions or velocities of molecules). The small self-diffusivities of ionic liquids (typically on the order of $10^{-11} \, m^2 s^{-1}$ at room temperature) imply that ions undergo an MSD of only $1 \, Å^2$ over $1 \, ns$. Since these displacements are much smaller than the size of the ions themselves, it is clear that only local dynamics is probed over these small timescales. To overcome this difficulty, simulations on the order of $10 \, ns$ or longer must be carried out to observe diffusive behaviour at room temperature. Alternatively, shorter simulations may be performed at higher temperatures to take advantage of faster dynamics and the results can be extrapolated to a lower temperature.

To test for the diffusive behaviour, scaling of the MSD ($\Delta r^2$) with time may be monitored by calculating the exponent $\beta$ [60] defined by Equation (6.3):

$$\beta = \frac{d \log(\Delta r^2)}{d \log(t)}. \tag{6.3}$$

As $\beta$ approaches unity, the system should exhibit diffusive motion. Another useful indicator of the diffusive behaviour is the non-Gaussian parameter $\alpha_2$ [61], Equation (6.4):

$$\alpha_2 = \frac{3}{5} \frac{\left\langle |\Delta r(t)|^4 \right\rangle}{\left\langle |\Delta r(t)|^2 \right\rangle^2} - 1. \tag{6.4}$$

As $\alpha_2$ decays to zero, random walk dynamics are recovered and the system should therefore exhibit diffusive motion. It has been observed by Del Pópolo and Voth that, for ionic liquids, the approach of the $\beta$ parameter to unity is generally faster than the decay of the non-Gaussian parameter to zero, indicating that ionic liquids exhibit dynamic heterogeneity [62].

When self-diffusivities are calculated from very short MD simulations (for example, less than 0.5–1 ns), the apparent self-diffusivities have been found to be an order of magnitude lower than the experimental value, suggestive of the sub-diffusive motion of the ions [37]. However, very early simulation studies [63] fortuitously obtained the self-diffusivity from rather short simulations (tens of picoseconds) and found the same order of magnitude as determined experimentally using nuclear magnetic resonance (NMR) spectroscopy [64]. Simulations correctly predict that the self-diffusivity of ionic liquids increases with temperature and this dependence is of Arrhenius type with activation energies between 35 and $50 \, kJ \, mol^{-1}$ for a range of ionic liquids [37, 59, 65].

Based on the calculation of self-diffusivities of ionic liquids, it has been concluded that the ions do not exhibit free diffusion, but rather exist as pairs or clusters [37]. It has been observed that bulky imidazolium cations diffuse faster than the smaller cations [18, 45, 56, 66, 67], consistent with many experiments. This behaviour is not universal, however [37], and a discussion of such a dynamic heterogeneity effect has been given by Urahata and Rebeiro [3].

Although simulations do not always yield self-diffusivities in quantitative agreement with experiment, they are usually effective in predicting relative trends. For example, simulations have shown that the dynamics of task specific ionic liquids drop dramatically upon complexation with $CO_2$, a result consistent with experiment [39]. Moreover, the simulations provided a mechanistic explanation for this viscosity increase, and have enabled other task specific ionic liquids to be developed which do not exhibit an increase in viscosity [68]. Most simulations that utilise a fixed-charge model tend to underestimate the dynamics of the ionic liquid by anywhere from a factor of 2 to a factor of 10. Recent studies have shown that polarisable force fields tend to give faster dynamics than fixed-charge models [50, 69]. The temperature dependence of the self-diffusivity has also been observed to be different when polarisable force fields are employed [50]. This argues in favour of the use of polarisable force fields when one is interested in capturing absolute values of the dynamics, although (as will be discussed later) this does not necessarily have to be the case.

**6.3.3.2  *Viscosity.***  Unlike self-diffusivity, viscosity is a collective property and is thus much more difficult to obtain from simulations. From equilibrium MD (EMD) simulations, the shear viscosity $\eta$ is determined from a Green–Kubo expression involving the pressure tensor, Equation (6.5):

$$\eta = \frac{V}{k_B T} \int_0^\infty \langle P_{ij}(0) P_{ij}(t) \rangle \, dt, \tag{6.5}$$

where $k_B$ is the Boltzmann's constant, $V$ is the volume, $T$ is the temperature, and $P_{ij}$ is the $ij$ component of the pressure tensor, $i \neq j$.

There are several problems associated with the application of Equation (6.5). The pressure tensor is a widely fluctuating quantity in a simulation, resulting in poor convergence of the ensemble average. Second, the rapid decay of the integral in Equation (6.5) and oscillations about zero for a long time make it difficult to accurately compute the "long time tail" contribution to the integral. Finally, as already pointed out, the sluggish dynamics of ionic liquids require that very long simulations be used to obtain reliable viscosities. As with self-diffusivities, fixed-charge potential models tend to give slower dynamics (i.e., higher viscosities) than the corresponding polarisable models [50, 69].

An indirect approach for calculating the viscosity of ionic liquids is to compute the (easy) self-diffusivity and estimate the (hard) viscosity using the

Stokes–Einstein relationship. In some cases, computed viscosities have been found to be in good agreement with experimental values. However, this methodology has been called in question by Ludwig and coworkers [70] on the basis that the ionic liquids display dynamic heterogeneity. Kelkar and coworkers [71] have also pointed out that, as the cations often exhibit greater self-diffusivities than the anions, application of the Stokes–Einstein relationship to imidazolium-based ionic liquids leads to the wrong conclusion that the cation is smaller than the anion. It has also been suggested that a fractional Stokes–Einstein relationship such that $D/T \propto (1/\eta)^t$ may be more appropriate for the calculation of viscosity from the calculated or measured diffusivity [72]. In our opinion, one should be wary of using the Stokes–Einstein relationship because the correlated motion of ions present in the ionic liquid is not accounted for. We believe there is an incomplete understanding of the relationship between self-diffusivity and viscosity for ionic liquids. Simulations are an ideal means for studying this relationship.

Non-equilibrium MD (NEMD) and reverse-NEMD (RNEMD) methods provide another avenue for the calculation of viscosity. In such approaches, the equilibrium of the system is perturbed by imposing either a shear stress or a shear strain. The response of the system is then monitored and an extrapolation procedure is used to obtain the Newtonian viscosity. Kelkar and Maginn [73] applied this methodology to compute the viscosity of $[C_2mim][NTf_2]$ over a range of temperatures. They observed that the simulation results were in good agreement with experimental measurements and a non-Arrhenius dependence on temperature was qualitatively captured. They also noted that the RNEMD method was highly effective in obtaining the transport properties of the system. The same authors published results on the viscosity calculations of the ionic liquid $[C_2mim][C_2H_5SO_4]$ and its mixtures with water [71]. Zhao et al. [74] demonstrated the applicability of such an approach for calculating the viscosity of a highly viscous ionic liquid $[C_4mim][PF_6]$ ($>100\,cP$).

Although well suited for the calculation of viscosity, such non-equilibrium approaches require careful data analysis to obtain reliable viscosities. A wide spectrum of shear rates corresponding to the Newtonian and shear-thinning regimes must be sampled. To extract the zero shear viscosity from the viscosities obtained at finite shear rates, a model is fitted to the simulation results and an extrapolation to zero shear rate is performed. It has been pointed out that for ionic liquids, the crossover from the shear-thinning regime to the Newtonian plateau may occur at shear rates much lower than that indicated by the largest relaxation time in the system [75]. This is important since many extrapolation procedures used for simple fluids assume that the crossover frequency corresponds to the inverse of the longest relaxation time. Borodin et al. [76] pointed out that different functional forms of the extrapolation function can lead to different estimates of the viscosity of ionic liquids. Despite these problems, we believe non-equilibrium methods are an effective choice for calculating viscosities, and they remain the *only* method that gives shear rate-dependent viscosities.

***6.3.3.3  Ionic Conductivity.***    Ionic liquids are prime candidates for electrolytes in electrochemical devices, where knowledge of ionic conductivity is extremely important. Molecular simulations have made some progress in this direction, and researchers have computed this transport property from MD simulations using the Nernst–Einstein (NE) equation, Equation (6.6), connecting self-diffusivity ($D$) and ionic conductivity ($\sigma$):

$$\sigma = \frac{Ne^2 D}{Vk_B T},\tag{6.6}$$

where $N$ is the total number of ions in the simulation cell of volume $V$ and $e$ is the electronic charge. It can also be calculated from a Green–Kubo expression, Equation (6.7), as the time integral of the electric-current function:

$$\sigma = \frac{1}{3k_B TV} \int_0^\infty \langle j(t)j(0)\rangle dt,\tag{6.7}$$

where the electric current $j(t)$ is given by Equation (6.8):

$$j(t) = \sum_{i=1}^N q_i v_i^c(t),\tag{6.8}$$

where $q_i$ is the charge on ion $i$ with centre-of-mass velocity $v_i^c(t)$.

Lee et al. [67] applied Equation (6.6) to compute the ionic conductivity of the $[C_4mim]^+$ cation paired with a range of fluorine-containing anions. Their calculations showed that the simulations captured the correct trend of conductivities, but the results were consistently lower than experimental measurements. Bhargava and Balasubramanian [66] published results on the conductivity of $[C_1mim]Cl$ at 425 K and found that the conductivity obtained by Equation (6.6) (0.069 S cm$^{-1}$) is very much larger than that computed from the Green–Kubo relationship (0.0089 S cm$^{-1}$) (Equation 6.7). The authors suggested that this is likely due to the fact that the NE equation ignores cross-correlation between the ionic currents. Moreover, the NE equation contains an additional contribution due to the movement of ions as pairs. Such motion does not result in any measurable electrical conductivity. As in the works of Lee et al. [67], the reported conductivities were much lower than the experimental values. Picálek and Kolafa [77] have also reported a similar observation, that the conductivities calculated from Equation (6.6) can be as high as a factor of 2 as compared with those obtained from the Green–Kubo expression. The study by Kowsari et al. [78] also reached a similar conclusion for a range of imidazolium-based cations paired with Cl$^-$, $[PF_6]^-$, and $[NO_3]^-$ anions. In our opinion, calculation of ionic conductivity from either the NE or the Green–Kubo expression suffers from the same issues outlined for the self-diffusivity and viscosity calculations. It is important that the simulations are conducted for sufficiently long time to ensure that all relevant relaxation processes are sampled. It also highlights the importance of estimating the contribution due to ion-pair diffusion on the self-diffusivity of ions themselves. This remains a relatively unexplored topic for molecular simulations of ionic liquids, and a detailed investigation is warranted. Borodin and Smith [79] estimated that the degree of uncorrelated ion motion is in the range of 0.6–0.65 and is independent of temperature. However, there is a disagreement between the temperature dependence of this quantity, and in fact different classes of ionic liquids show different behaviour [80]. The work of Borodin and Smith suggested that a polarisable force field can quantitatively predict transport coefficients of ionic liquids when sufficiently long (up to 16 ns) simulation times are probed. For the Green–Kubo expression, it is not clear what the upper limit of the time integral should be in order to minimise the numerical errors associated with the oscillations of the autocorrelation function. There exists no simulation study that takes advantage of NEMD or RNEMD methods to compute the ionic conductivity of ionic liquids. We believe that the time is ripe for more simulation studies on the relationship between ionic conductivity, other transport properties, and ionic liquid structure and composition.

***6.3.3.4 Thermal Conductivity.*** To our knowledge, there has been only one molecular simulation study in which the thermal conductivity of an ionic liquid ($[C_2mim][C_2H_5SO_4]$) was computed [71]. In our view, thermal conductivity calculation is fundamentally no more challenging than the calculation of viscosity although, like the heat capacity, it is known that classical force fields tend to overestimate the thermal conductivity. Given the potential for using ionic liquids in heat transfer applications, this is a research area that needs further attention.

## 6.4 GAS–LIQUID, LIQUID–LIQUID, AND SOLID–LIQUID INTERFACES

Applications of ionic liquids in gas separations, extraction in biphasic systems, and electrolytes in batteries involve contact with gases, liquids, and solid electrode materials. The properties of such interfaces determine the efficiency of the physical and chemical processes occurring across the phase boundary. For example, the transport of gaseous species across an ionic liquid interface will be influenced by the structural arrangement of ions presented at the gas–liquid interface. Given that the interfacial regions are inhomogeneous and ions experience unbalanced forces, the structural, transport, and orientational properties are likely to be fundamentally different from those observed in the bulk. Atomistically detailed simulations have provided a wealth of information on the properties exhibited by ionic liquids when they are exposed to vacuum or a gas, another liquid, or solid surface.

### 6.4.1   Ionic Liquid–Gas Interface

The first MD study of a vacuum–liquid interface of an ionic liquid was for $[C_1mim]Cl$ at 400 and 500 K [81]. One of the methyl groups of the cation was found to preferentially orient towards the vacuum, while the other points inward towards the bulk region. The imidazolium ring was found to orient parallel to the interface, while a maximum in the cation number density was located just below the interface. A tendency of the anions to segregate at the interface was also noted. The study also discussed the effect of water concentration on the structural properties and deduced that the cations were progressively replaced at the interface by water molecules. Calculated surface tensions of the pure ionic liquid interface and its mixtures with water were also reported. A qualitative trend of reduction in surface tension with increase in water concentration was observed. Subsequent study with the ionic liquids $[C_4mim]^+$ paired with $[PF_6]^-$, $[BF_4]^-$, and $Cl^-$ also revealed similar orientational characteristics [82]. Bhargava and Balasubramanian [83] carried out interfacial simulations of the ionic liquid $[C_4mim][PF_6]$ with 256 and 512 ion pairs to ensure that the centre of the simulation cell represented the true bulk region. The authors reached very similar conclusions as previously reported by Lynden-Bell and Del Pópolo [82]. In addition, the simulations resolved the apparent discrepancy between the absence of oscillations in electron density in experiments and the oscillations in number density observed in simulations. Müller-Plathe and coworkers [84] reported that, in the interfacial region, the reorientational dynamics of the butyl chain in the ionic liquid $[C_4mim][PF_6]$ is much slower than that in the bulk, presumably due to the alignment of the chains in this region. However, accelerated translational motion of the cation was observed. The presence of the interface had a negligible influence on the dynamical properties of the anion [84].

The interface of the ionic liquid $[C_6mim][NTf_2]$ with vacuum was probed by Pádua and coworkers at 300, 350, and 423 K [85]. In addition to trends in the orientation of the cation and alkyl chains, the authors found that the hydrophobic $-CF_3$ groups on the anion also protrude towards vacuum. The surface tension calculations, based on a mechanical definition, showed that it was difficult to accurately determine this quantity at lower temperatures due to significant contributions arising from the bulk region. This led the authors to conclude that, at low temperatures, the system may not be in local equilibrium. At the highest temperature, the calculated surface tension was found to be in good agreement with the extrapolated value determined from experiments at lower temperatures.

Perez-Blanco and Maginn reported MD simulation results for the $CO_2$–$[C_4mim][NTf_2]$ interface [86]. The ionic liquid–vacuum interface results were similar to those previously reported, with alkyl tails and anionic fluorine groups preferring to stick out into vacuum. When in contact with high pressure $CO_2$, a dense adsorbed layer of $CO_2$ formed at the interface region, but at very low pressures no such layer was observed. Potential of mean force calculations showed that the free energy barrier for $CO_2$ to cross the interface is small, and indeed many gas–liquid and liquid–gas crossing events were observed during the course of the simulation.

As with the simulations of bulk ionic liquids, polarisable force fields have been evaluated for the study of interfaces involving ionic liquids. Voth and coworkers investigated the performance of polarisable and non-polarisable force fields for $[C_2mim][NO_3]$ at a vacuum interface and predicted similar orientational features of the ions in the interfacial region [87]. The inclusion of polarisability, however, resulted in a significant reduction in the surface tension of the ionic liquid by as much as $30 \, \text{mN} \, \text{m}^{-1}$. Moreover, the polarisable force field predicted the segregation of the cation at the interface, in direct contrast to that of anions observed with non-polarisable force field in this study and others cited earlier. In contrast, Chang and Dang [88] observed the segregation of anions in $[C_1mim]^+$ paired with $Cl^-$, $Br^-$, and $I^-$ when a polarisable force field was used. Consistent with previous results, surface tensions appeared to be lower when a polarisable force field was used; for $[C_1mim]Cl$, the polarisable force field of Chang and Dang yielded a value of $68 \, \text{mN} \, \text{m}^{-1}$, while the fixed-charge model of Lynden-Bell [81] resulted in a value of $100 \, \text{mN} \, \text{m}^{-1}$. Of course, part of this difference could be due to other differences in the models, but indications are that inclusion of polarisability will lower surface tensions.

### 6.4.2 Ionic Liquid–Liquid Interface

Lynden-Bell and coworkers [89] examined the liquid–liquid interface of the ionic liquid $[C_1mim]Cl$ with Lennard-Jones fluids differing in the strength of their well depth. The weak Lennard-Jones fluid wetted the ionic liquid interface and a density maximum was observed for the Lennard-Jones fluid near the interfacial region. Due to the interaction of the ionic liquid with the Lennard-Jones fluid, the tendency of the imidazolium ring to lie parallel in the outer region decreased. This effect was more pronounced when the ionic liquid was in contact with the weak Lennard-Jones fluid. The authors also reported that the interface of the ionic liquid with water was not stable.

The ionic liquid interface with a non-polar solvent such as 1-hexene has been investigated by Sieffert and Wipff [90]. The demixing simulations of $[C_4mim][PF_6]$, 1-hexene, and a number of ligands and reaction intermediates of hydroformylation of 1-hexene demonstrated that the separation of phases requires simulation times up to several hundred nanoseconds. The characteristic orientation of the imidazolium ring and the butyl chain in the interfacial region is similar to that observed with the vacuum interface; the butyl chains are perpendicular to the interface and project outwards in the hexene phase.

### 6.4.3 Ionic Liquid–Solid Interface

A detailed understanding of microstructure at the ionic liquid–solid substrate can be fruitfully exploited for optimisation of system performance in applications of ionic liquids in electrochemical devices, dye-sensitised solar cells, supercapacitors, and batteries. Experimental measurements at such interfaces rely on the information extracted from techniques such as sum-frequency generation spectroscopy.

Molecular dynamics simulations of a large number of ionic liquids in contact with surfaces such as graphite, rutile, quartz, and even electrified surfaces have appeared in the literature. These publications probe the interfacial population of ions, orientation of various cation groups, and the effect of the surfaces on transport properties such as surface diffusion of ions. In addition, the ordering of ions induced by a surface has also been investigated. For example, MD simulations of ionic liquids composed of $[C_4mim]^+$ with $Cl^-$, $[NTf_2]^-$, and $[PF_6]^-$ at a graphite surface showed that the mass density at the interface is almost twice that in the bulk, indicating strong adsorption of the ionic liquids at the surface. The density oscillations away from the surface suggested that there are three distinct adsorbed layers of ions [91]. With an increase in temperature up to $800\,K$, the ionic liquids with $[PF_6]^-$ and $[NTf_2]^-$ anions still retained all the layers. In the case of $Cl^-$, however, the third layer was found to be almost non-existent, suggesting a structural transition with temperature. The authors also reported a decrease in the surface diffusion coefficients of the ions due to strong adsorption. Wang et al. [92] have reported results for MD simulations of $[C_4mim]^+$ and $[C_8mim]^+$ with the $[PF_6]^-$ anion at a graphite interface. These authors also observed strong oscillations in the number densities of the cation and anion that extend up to $15\,Å$ in the bulk. The cations preferentially adsorb on the surface with the orientations of the imidazolium ring and the alkyl chains parallel to the interface, presumably to maximise the van der Waals interaction with the surface. The effect of the alkyl chain manifests itself in the more negative potential drop across the interface as the bulky $[C_8mim]^+$ results in a denser $[PF_6]^-$ packing at the interface. Wang et al. [93] also observed similar arrangement of ions of $[C_4mim][PF_6]$ on the rutile (110) interface, with the anion occupying the first layer. A similar segregation of the anion at the rutile surface was reported for the ionic liquid $[C_4mim][NO_3]$ [94]. In the MD simulation of a mixture of $[C_4mim][PF_6]$ and $CO_2$ at a rutile (110) surface, however, $CO_2$ molecules preferentially covered the surface, displacing the anion to the second layer [95].

Sieffert and Wipff [96] conducted a comprehensive study of several ionic liquids in contact with a quartz (001) surface with silanol and silane functionality representative of a hydrophilic and a hydrophobic surface, respectively. The authors considered the cation $[C_4mim]^+$ in combination with the anions $[PF_6]^-$, $[BF_4]^-$, $[NTf_2]^-$, and $Cl^-$. Additionally, the presence of a long alkyl chain $(C_8)$ in the cation was also investigated. The study reported that a mixture of cations and anions populated the interfacial region at the silanol surface. The orientation of the imidazolium ring and alkyl chains was found to be parallel to the surface, consistent with other surfaces as noted earlier. The anions interacted strongly with the hydroxylated surface through dynamic hydrogen bonding. The $[NTf_2]^-$ anion interacted in an amphiphilic manner at the surface, with $-SO_2$ groups pointing towards the surface and $-CF_3$ moieties directed towards the bulk. The interfacial structure at the hydrophobic surface was markedly different; the first layer was primarily occupied by the cations while the anions were repelled from a slightly negatively charged surface. Interestingly, the long alkyl chain led to the formation of apolar microdomains only when it was part of the anion $[C_8H_{17}SO_4]^-$. The presence of water did not affect the overall interfacial region, in direct contrast to the substantial effect of water on ionic liquid–vacuum interfaces and its effect on the bulk properties of ionic liquids.

Sha et al. [97] performed MD simulations to investigate the formation of a double layer of the ionic liquid $[C_4mim][PF_6]$ due to surface negative charge. Their work showed that, at moderate negative densities, two to three alternating layers of cations and anions were present. However, at high negative surface charge densities, the ionic liquid at the interface undergoes a structural transition to multiple double-layer stacking formation. The presence of such a transition was confirmed by an abrupt increase in the potential energy of the ionic liquid and a drop in the surface diffusion coefficients of the ions. Based on radial distribution function analysis, it was suggested that the aggregation of the alkyl domains appear as the surface charge densities become negative. This work clearly demonstrated that the structure of the ionic liquid at electrified interfaces is determined by a delicate balance of ion–ion, ion–surface, and short-ranged interactions. A similar conclusion can be drawn from the MD simulation study of the ionic liquid $[C_4mim][NO_3]$ confined between electrified surfaces [98]. It was shown that the behaviour of the anions was qualitatively different at the positive electrode in comparison with that of the cation at the negative electrode. Very modest accumulation of anions is observed at the negative electrode. However, significant adsorption of the cation occurs even at low surface charge densities. Moreover, the peak position in the number density of the anions moves towards the surface and the peak height increases with increasing positive surface charge densities. This observation implies that the charge–charge interactions play an important role at the positive electrode. On the other hand, the peak position of the cation remains essentially unaltered as the surface becomes more negative, indicating that the initial adsorption of the cation to the surface is driven by van der Waals interactions and additional charge–charge interaction drives an increase in the number density at the positive electrode. Confinement of ionic liquids in a solid matrix at molecular length scales can have a profound effect on phase behaviour and liquid state properties. It has been shown experimentally that the encapsulation of $[C_4mim][PF_6]$ in multi-walled carbon nanotubes (MWCNTs) leads to crystallisation of the ionic liquid at temperatures well above the bulk melting point [99]. Other researchers have observed that ionic liquids having $[C_2mim]^+$ cation combined with a range of anions have bulk-like mobilities in mesoporous silica [100]. The ionic liquid $[C_4mim][NTf_2]$ was reported to exhibit both solid and bulk-like characteristics depending on its proximity to the substrate [101] and the pore dimensions of the confining medium [102]. These studies demonstrate that rich sorption and diffusion behaviour can be expected for ionic liquids, and the combination of ionic liquid cation, anion, substrate chemistry, and confinement dimensions provides many degrees of freedom that can be adjusted to optimise the performance of ionic liquid systems.

In a recent MD simulation study [103] of solvation of CNTs by [$C_2$mim] [$BF_4$], the confinement dimension of single-walled nanotubes was found to have a pronounced effect on the structure of the ionic liquid inside the tube. As the diameter of the nanotube varied from 0.95 to 2.70 nm, several different structures such as single-file distribution, zigzag distributions with ion pairing, disordered, staggered pentagonal first solvation shell structures, and disordered octagonal configurations appeared. Dong et al. [104] reported a similar structural transition of [$C_4$mim][$PF_6$] when confined in (9,9) and (10,10) CNTs. In smaller nanotubes, single-file arrangement of the ions was observed while an alternating cation–anion pair structure was obtained with the large CNT. Based on free energy calculations, the authors concluded that the mechanism of insertion of ions into the CNT involves entry of the cation as it "drags" the anion from the bulk liquid phase. Singh et al. [105] simulated [$C_4$mim][$PF_6$] in MWCNTs with inner diameters of between 2.0 and 3.7 nm. The authors examined the effect of different pore loadings and pore diameters on the structural and dynamical properties of the ionic liquids. Simulation results predicted layers of cations and anions along the radial direction of the MWCNTs with local maximum in density located near the pore wall. With decreasing loading, the density at the centre of the wall approaches zero while significant accumulation of the ions occurs near the wall. The study suggested that the MSDs of the ions exhibit non-monotonic behaviour with respect to pore loading for a given pore size. The MSDs obtained in this study are lower than those obtained in the bulk system, indicating that the dynamics of the ionic liquid are affected by confinement.

Lynden-Bell and coworkers [106] examined the effect of inter-wall distance between two structureless walls on the structural and dynamical properties of [$C_1$mim]Cl using MD simulations. Significant accumulation of cations was observed at the wall and the ionic liquid density exhibited density maxima near the structureless wall. The imidazolium ring plane was found to be perpendicular to the surface normal. In contrast with other simulation studies of confinement of ionic liquids, the mobility of the ions was higher under confinement than in the bulk.

Sha et al. [107] used MD to study [$C_1$mim]Cl confined between graphite surfaces represented in atomic detail. They reported the phase transition of the ionic liquid as the confinement dimension fell below a critical value of 1.15 nm. The phase change was detected by monitoring the lateral diffusion coefficients of ions. Above the critical dimension, the ionic liquid exists as a liquid bilayer but transforms into a frozen state below 1.15 nm, as indicated by nearly zero lateral diffusion coefficients of the ions. In this study, the melting point of the frozen state was estimated to be in the range of 825–850 K, more than 400 K higher than the melting point of the bulk crystal (399 K).

## 6.5  MULTI-COMPONENT SYSTEMS

Molecular simulations of pure ionic liquids are crucial in testing the validity of a given force field, or improving an existing one, and enable prediction of thermodynamic and transport properties of yet unsynthesised ionic liquids. However, technological applications of ionic liquids invariably contain mixtures of ionic liquids with other substances. Given that a two or more component ionic liquid system may explore a range of structural regimes ranging from pure melt, clusters, micelles, neutral pairs, and isolated ions, properties of such mixtures are fundamentally different from those of pure ionic liquids. Molecular simulations of ionic liquid mixtures also indicate this possibility and have provided clues into the structures of these mixtures that lead to observed properties.

Molecular dynamics simulations of aqueous solutions of $[C_n mim]Br$ ($n = 2$, 4, 6, or 8) [108] revealed that the ions in $[C_2 mim]Br$ are distributed isotropically in their aqueous solutions, while weakly associated clusters form for $[C_4 mim]$ Br. $[C_6 mim]Br$ exist as small aggregates, while the solution of $[C_8 mim]Br$ exhibits aggregate formation and decreasing self-diffusivity of the ions. Feng and Voth [109] reported similar observations from MD simulations of aqueous solutions of $[C_4 mim][BF_4]$, $[C_8 mim][BF_4]$, and $[C_8 mim]Cl$. They observed that, at high water concentrations, the ionic network in $[C_4 mim][BF_4]$ breaks down while aggregation of $[C_8 mim]^+$ ions is observed, leading to micellar structures. Such aggregation behaviour has also been reported in MD simulations of $[C_4 mim]Cl$ and $[C_6 mim]Cl$ with propanol [110]. Molecular simulations of aqueous solution of the ionic liquid $[C_2 mim][C_2 H_5 SO_4]$ [71] showed that the mixtures display negative excess molar volumes and excess enthalpies over the entire range of composition. However, the excess properties were predicted to be more negative than experimental data [111] and this was attributed to a neglect of the polarisability of the water potential model.

Simulations have suggested that a structural transition from an ionic state at high ionic liquid concentrations to a molecular dipolar fluid at low ionic liquid concentrations may be responsible for the conductivity maximum per ion observed in a $[C_4 mim][PF_6]$–naphthalene mixture at the ionic liquid mole fractions of 0.15 [112]. Similarly, based on the structural arrangements deduced from molecular simulations of the ionic liquid $[C_2 mim][NTf_2]$ in benzene and 12 fluorinated benzene compounds, Pádua and coworkers have dissected the role of dipole and quadrupole moments on the solubilities of these compounds [113].

Molecular dynamics simulation studies [114] of mixtures containing $CO_2$ with imidazolium-based ionic liquids and the $[PF_6]^-$ anion at a $CO_2$ mol fraction of 10% indicated that the gas solubility is governed mainly by its interaction with the anion. At this concentration of $CO_2$, the underlying structure of the ionic liquid is not significantly perturbed. However, molecular simulations have calculated expansions of up to 40% in molar volumes when concentrations of $CO_2$ are much higher, for example, 70% [115]. At such high concentrations of $CO_2$, anion–anion interactions are altered, reducing the viscosity of the ionic liquid. In the case of $SO_2$, however, molecular simulations indicate that the melting of high temperature ionic liquid [$C_4$mim]Br is promoted by the influence of $SO_2$ on the long-range cation–anion order of the ionic liquid [116], while the short-range cation–anion interactions are enhanced. These examples demonstrate that molecular simulations are playing an important role in explaining experimental phenomena at molecular level.

## 6.6  SOLUBILITY IN IONIC LIQUIDS

Other applications involving mixtures can occur when ionic liquids are used as a solvent in a reaction or as a separation agent. The key thermodynamic quantity of interest is the solubility of a given species in the ionic liquid, and how this solubility varies with temperature, pressure, and the choice of ionic liquid. For gases, one is most often interested in the low solubility limit Henry's Law constant. For other solutes, activity coefficients over a range of concentrations or solubility limits are most useful. There has been a significant amount of experimental activity in this area, resulting in a wealth of data. The amount of molecular simulation work, in contrast, has been fairly minimal. Here, we highlight some of the work done in this area and discuss problems that must be overcome to enable simulations to be used more widely in this technical area.

To assess the solubility of a particular solute in an ionic liquid, one can compute the excess chemical potential of solutes. Such calculations are usually carried out with free energy schemes such as the Widom test particle insertion method [117], the expanded ensemble (EE) approach [118], or thermodynamic integration methods. The modelling community has applied these methodologies to study solvation of various species in ionic liquids. For example, Lynden-Bell and coworkers used thermodynamic integration to compute the excess chemical potentials of water, methanol, dimethyl ether, propanone, and propane in [$C_1$mim]Cl at 400 K. Based on the relative magnitude of excess chemical potentials, they determined relative rankings of the solubilities of these solutes. Our group applied the Widom test particle insertion method to calculate the Henry's law constant for $CO_2$ in [$C_4$mim][$PF_6$] over a range of temperatures. The calculations predicted the solubility of $CO_2$ to be two to three times higher than the experimental value [51]. Subsequently, we conducted MC simulations to compare Henry's Law constants obtained from the Widom test particle insertion scheme and an EE MC technique. The study focussed on water, carbon dioxide, methane, ethane, ethylene, and oxygen in [$C_4$mim][$PF_6$] [6]. The results from the EE method were found to be in qualitative agreement with experimental measurements, and in the case of water and carbon dioxide, quantitative agreement was obtained. This was an important finding, as it demonstrated that the method used to perform the simulation makes a difference in the results that are obtained. We now believe that single stage free energy perturbation schemes such as the Widom test particle insertion tend to perform poorly in highly dense and strongly interacting systems such as ionic liquids. Multistage free energy methods such as EE give better convergence because the solute is added to the system gradually. In addition, such methods allow for rearrangement of ionic liquids, leading to creation of cavities that accommodate solutes. It has been pointed out elsewhere [119] that care must be taken to understand the particular free energy method being used to ensure good results.

Deschamps and coworkers [120] determined the relative solubility of argon, methane, $O_2$, $N_2$, and carbon dioxide in [$C_4$mim][$PF_6$] and [$C_4$mim][$BF_4$] at 1 bar and a range of temperatures with thermodynamic integration. The simulations predicted correct relative rankings of solubilities, but the temperature dependence of non-polar gases such as dioxygen was opposite to that observed experimentally. Simulation results indicated that the solubility of non-polar gases decreases with an increase in temperature. However, for gases in which the enthalpic contributions dominate the free energy, a correct trend in solubility with temperature was obtained. The authors also showed, using quadrupolar models of $CO_2$ and $N_2$, that electrostatic interactions are important for accurately calculating the solubility. Finally, the authors carried out thermodynamic integration to calculate the solubility of water in [$C_4$mim][$PF_6$] and determined that the infinite-dilution activity coefficient of water in the ionic liquid is $4.7 \pm 3.6$, in accord with the experimental value of 5.36 [121].

In all the cases discussed earlier, excess chemical potentials/Henry's Law constants were computed using a single solute molecule. In this sense, they are mimics of the "infinite-dilution" solubility, although given system size limitations, even a single solute molecule can represent a relatively high concentration in a simulation. The first attempt to compute full gas absorption isotherms in an ionic liquid was carried out by Maurer and coworkers [122]. They calculated absorption isotherms of $CO_2$, CO, and $H_2$ in [$C_4$mim][$PF_6$] using the isothermal–isobaric Gibbs ensemble MC (GEMC) method [123]. Simulation temperatures ranged from 293 to 393 K, while pressures of up to 9 MPa were studied. The authors found that the simulation results agreed remarkably well with the available experimental data. The original publication contained a small conversion factor error, however, which when corrected yielded results that were not as close to experiments as suggested in the original study. Still, this was an encouraging development as it showed that simulations could be used to compute finite concentration solubilities.

In an effort to calculate gas absorption isotherms in ionic liquids reliably and efficiently, we developed an MC method that attempts to gradually create and destroy solute molecules rather than add or remove them in a single step as is commonly implemented in GEMC simulations. A successful application of the so-called continuous fractional component (CFC) MC method was reported by us for the calculation of $CO_2$ absorption isotherms in [$C_6$mim][$NTf_2$] [124]. The agreement between the molecular simulation results and experimental measurements was found to be quantitative for pressures up to

80 bar at 333 K. Subsequently, we employed the technique for prediction of absorption isotherms of $SO_2$, $O_2$, and $N_2$ [7] in the same ionic liquid. In addition, the mixed solubilities of $CO_2/SO_2$, $SO_2/N_2$, and $CO_2/O_2$ were also computed. The calculated Henry's Law constants of $SO_2$ at 298 and 333 K were found to be $0.9 \pm 0.3$ and $3.6 \pm 0.5$ bar, respectively, in very good agreement with the experimental Henry's constants of $1.64 \pm 0.01$ and $4.09 \pm 0.06$ bar at these temperatures [125]. The Henry's constants of $O_2$ and $N_2$ were also found to be in good agreement with experimental measurements. Given that the experimental determination of $O_2$ and $N_2$ absorption isotherms is challenging, due to very low solubilities of these gases, molecular simulations have played an important role in this area. Similarly, the simulation results of $SO_2/N_2$ mixed solubilities obtained in this study suggested that there is a slight decrease in the solubility of $SO_2$ in the presence of $N_2$ but nowhere close to the dramatic effect of reduction in solubility of $SO_2$ observed experimentally [126]. The discrepancy in the binary absorption isotherm results between molecular simulations and experiments is likely due to the fact that although single component absorption isotherm measurements are relatively straightforward, measurement of binary gas solubilities presents significant challenges. In a simulation, however, the calculations of a mixture absorption isotherm are not much more difficult to conduct than the corresponding pure gas absorption isotherms. This suggests a useful strategy in which the simulation results of the pure gas absorption are benchmarked against experimental findings. The validated models can then be used to predict mixed gas solubility from simulations, where experiments are much more difficult to carry out.

Absorption isotherms of ammonia were simulated in $[C_2mim][NTf_2]$ using the osmotic ensemble at 298, 322, and 348 K [127]. The calculated activity coefficients were found to be in the range of 0.5–0.8, indicating high solubility of ammonia and negative deviation from Raoult's law. The absolute average deviations between the computed and experimental isotherms at various temperatures ranged from 14% to 28%. The difference was ascribed to the inability of the ammonia potential model to capture saturation pressures accurately. Besides simply reproducing experimental isotherms, the calculations also provide important insight into the nature of the molecular-level interactions that are responsible for the observed solubility behaviour. For example, $CO_2$ and $SO_2$ absorb readily due to interactions with the anion. Decomposition of energetic contributions from various components suggests that the electrostatic interactions dominate in this case. For non-polar gases, such as $O_2$ and $N_2$, the dissolution process is controlled more by available free volume in the ionic liquid phase, while $NH_3$ tends to associate more strongly with cations via hydrogen-bonding interactions.

## 6.7   WHAT NEEDS TO BE DONE (AND WHAT DOES NOT)

It has now been just over 10 years since the first molecular simulation studies of ionic liquids. During that time, the number of papers reporting simulation results for ionic liquids has gone from a trickle to a torrent. So what still needs to be done? What areas and topics should be addressed in the next decade, and what barriers will need to be overcome? In contrast, are there topics and systems being studied within this torrent that probably no longer require investigation? In this section, we provide our opinions on these questions and, from our perspective, list the areas we think molecular simulation needs to focus on in the next 10 years.

### 6.7.1  VLE of Pure Ionic Liquids

One of the most intriguing aspects of ionic liquids is their extremely low vapour pressure. As noted earlier, molecular simulation researchers were the first to predict accurately the enthalpies of vapourisation of an ionic liquid— predictions that were subsequently confirmed experimentally several years later. Why were simulations the first to the punch? Quite simply, the extremely low vapour pressure of ionic liquids makes experimental investigation of ionic liquid VLE much harder to conduct than the simulations. Since this is the case, then it seems logical to expect that simulations should and will play an increasingly important role in this area.

A number of groups are now studying the vapour phase of ionic liquids, whether to measure enthalpies of vapourisation [23, 29] or to examine the nature of ions in the vapour phase [27, 28], or to investigate technological applications such as propulsion [128]. Enthalpies of vapourisation are a key thermodynamic quantity that can help explain the nature of the interactions that take place in the liquid phase. It is also an important data point that can be used to test the quality of intermolecular potentials. Vapour pressures are an equally important thermodynamic quantity that are routinely collected for most compounds and used in equation of state models. Finally, critical points are an essential thermodynamic quantity used in a host of correlations and equations of state [129]. While it is likely that the critical point of most ionic liquids is above the decomposition temperature, this has not stopped experimental groups [130] and group contribution modellers [131] from trying to estimate critical points. Hypothetical critical points are an important physical property and have already been used extensively in equation of state modelling of ionic liquids [132]. Unfortunately, obtaining an estimate of a critical point experimentally is quite difficult; critical points are expected to be above the decomposition temperature of the liquid and extrapolating properties to estimate the critical point (for example, using the temperature dependence of the surface tension) [130] introduces a great deal of uncertainty. We believe the time is ripe for molecular simulations to play a role in elucidating the VLE of ionic liquids. Understanding and predicting vapour–liquid coexistence curves, critical points, vapour pressures, and enthalpies of vapourisation are a key set of targets for molecular simulation. These are important thermodynamic quantities for any substance, and we simply have very little knowledge of them for ionic liquids. The reason is simple: experimentally determining them is difficult if not impossible. Computing these quantities with molecular simulation is also very difficult, but not impossible. How might this be done? GEMC [123] and related MC techniques are now well-established simulation methodologies that have been used for over 20 years to predict VLE of a wide range of organic compounds. Direct application of the GEMC to predict VLE of ionic liquids is not straightforward due to the complex structures of ionic liquids, high liquid densities, and strong Coulombic interactions. Conformational sampling of cations and anions is also extremely challenging, as they contain a large number of intramolecular degrees of freedom, branch points and cyclic groups. In addition, particle exchanges in the Gibbs ensemble require transfer of two or more ions to preserve charge neutrality. All of these factors make simulating ionic liquids with GEMC an order of magnitude (or more) difficult than conventional liquids. Over the next 10 years, however, as methodological advances continue and raw computing power increases, we believe these limitations will be overcome and VLE calculations of ionic liquids will become routine. If this stimulates additional experimental investigations in this area, it will have an added benefit to the molecular simulation community; the most common way of validating force fields for organic molecules is by comparing computed VLE against experimental data. Experiments and simulations mutually reinforce each other while at the same time greatly expanding our knowledge of this critical topic.

### 6.7.2  LLE

Now that the number of papers on ionic liquids has eclipsed 3000 per year, it is always interesting to find a topical area that is one-sided between experiments and simulations. When all the papers on a given topic are either experimental or simulation papers, it is a good clue that either (a) there are tremendous opportunities for the method that is not well represented in the area, but it is probably very hard to do the research, or (b) nobody really cares about the topic but it is easy for either the experimentalists or the simulators to generate results and publish papers. In the case of LLE, case (a) is clearly operative. LLE between an ionic liquid and one or more solutes are of extraordinary technological importance, given the potential of using ionic liquids in extraction-based separation processes. From a fundamental standpoint, the rich phase behaviour that has been observed with ionic liquids (upper critical solution behaviour, lower critical solution behaviour, and both) [133, 134] suggests that there is much we do not yet understand about LLE. While the experimental literature is rich with binary and even ternary LLE data sets [135, 136], to our knowledge there has been almost no molecular simulation studies of LLE. The only work we are aware of has been MD simulations of demixing/aggregation [137, 138], which has provided intriguing information about the interface between solute and ionic liquid phases. This method, however, is poorly suited for mapping out equilibrium coexistence compositions because of the timescales associated with interfacial mass transfer and the effect the interfaces themselves have on the systems. Once again, we believe that MC methods, which can enforce equality of chemical potentials and do not require an explicit interface, are the tools of choice for these types of studies.

For such calculations, all the methodological advances necessary for VLE calculations of ionic liquids will be needed, but the simulations are even more challenging. The difficulty arises for two reasons. First, unlike VLE calculations, LLE calculations will require transfer of ionic liquids and solutes between *two* dense phases instead of between a dense phase and a low density phase. Insertion and deletion of molecules in a dense phase is a very low probability event, and now that low probability is squared. Second, due to charge neutrality considerations, equilibrating the ionic liquid will require that at least two counter-ions need to be exchanged between the phases, a process that is considerably more difficult than the transfer of neutral molecules. These difficulties mean that LLE simulations of ionic liquids will remain uncommon not because they are unimportant, but because of the intrinsic difficulties of the simulations. Methodological advances are clearly called for to address these problems.

### 6.7.3 SLE

The melting point is one of the most important properties of an ionic liquid since its relatively low value distinguishes ionic liquids from conventional salts. Given that ionic liquids can be designed to have a wide range of melting points, and our inability to predict with confidence what the melting point will be until a sample is made and tested, it is clear that understanding the link between structure, chemical composition, and melting point is one of the grand challenges of ionic liquid research. There have been several efforts to develop empirical correlations [139] and quantitative structure–property relationship models [140] that can "predict" melting points of ionic liquids. All these methods require experimental data, often large amounts of it, to develop the correlations. They tend to not work outside the compound space for which they were parameterised, and they give almost no molecular-level indications of the factors that a synthetic researcher could control that would raise or lower a melting point. Once again, this is a great opportunity for molecular simulations to play a role.

The very first SLE calculation on an ionic liquid was performed by Alavi and Thompson [141]. They conducted MD simulations over a range of temperatures to compute the melting point of $[C_2mim][PF_6]$. Their simulations mimicked experiments, in that they gradually heated a crystal at constant pressure and looked for signatures of a first-order phase transition (in this case, an abrupt change in density, intermolecular energy, and Lindemann index). Since the free energy barrier for homogeneous nucleation of a liquid in a crystal is large and the length- and timescales of an MD simulation are small, it is virtually impossible to observe melting behaviour at the thermodynamic melting point directly in an MD simulation. Instead, "melting" is not observed until temperatures well beyond the thermodynamic melting point are reached. To overcome this superheating phenomenon, Alavi and Thompson introduced random voids in the crystal by removing ion pairs. As the concentration of voids increased, the apparent melting point decreased due to a reduction in the nucleation free energy barrier at the voids. Eventually, the introduction of additional voids does not cause much change in the melting point, and so the temperature at which this occurs was taken as the thermodynamic melting point. Using this approach, Alavi and Thompson determined the melting point of $[C_2mim][PF_6]$ to be $375\,K \pm 10\,K$, which is 10% higher than the experimental value of $333\,K$. The same authors applied this "void-induced" melting method to the calculation of the melting point of 1-propyl-4-amino-1,2,4-triazolium bromide (the paper title incorrectly details the ionic liquid as 1-butyl-4-amino-1,2,4-triazolium bromide) [142]. They estimated the melting point to be $360\,K$, again within 10% of the experimental value of $333\,K$. Based on these studies, the authors concluded that molecular simulations can be used to predict the melting point to within 10% of the experimental values.

The problem with the approach just discussed is that it relies upon a kinetic phenomenon, and there is no clear-cut way of knowing how many "voids" to introduce into the crystal to determine the right melting point. The void concentrations used in the simulations are orders of magnitude higher than the defect densities of real crystals. We believe that a better method is to utilise a thermodynamically rigorous approach, whereby the melting temperature is determined by computing crystalline and liquid free energies as a function of temperature and finding the coexistence temperature directly. Such an approach was introduced by Frenkel and Ladd in 1984 [143], but for technical reasons this method has only been applied to relatively simple systems. About five years ago, Grochola introduced an ingenious method whereby the thermodynamic pathway between a crystal and a liquid could be traversed in a reversible manner [144, 145]. The approach was applied to a Lennard-Jones fluid and shown to give equivalent results when compared to the Frenkel–Ladd method. We modified this method and used it to predict the melting point of the Lennard-Jones and NaCl solids [146], benzene and triazole [147], and three alkali nitrate salts, $M[NO_3]$ (M = Li, Na, or K) [148]. We also applied it to compute the melting point of the orthorhombic and monoclinic forms of $[C_4mim]Cl$, as well as the free energy difference between these two polymorphs as a function of temperature [149]. We showed that while reasonable estimates of the melting point are possible, small free energy differences of a few $kJ\cdot mol^{-1}$ can lead to very large differences in the estimated melting point. This suggests that predicting melting points with quantitative accuracy requires both highly accurate methods and very accurate force fields. This is a big challenge for the field.

Recently, Kowsari et al. [150] reported an MD study of an equimolar mixture of an ionic liquid–benzene inclusion crystal. The authors monitored the MSDs and configurational energy to detect the solid-to-liquid transition. The melting of the crystal was characterised by a sudden jump in the MSD and configurational energy. As with the simulations of perfect crystals, superheating was observed and the predicted melting point of 410 K was much higher than the experimental value of 288 K. However, despite the superheating, the authors did observe congruent melting of the crystal; that is, the crystal exhibited melting at a single temperature and the concentrations of the two species in the resulting liquid phase were identical to those in the solid phase.

All the studies involving SLE presented earlier require that a crystal structure be available as initial input to the simulation. In our opinion, this offers a unique opportunity for fruitful collaborations between modellers and crystallographers. In the absence of such structures, one cannot use these methods to predict crystal structures. If a crystal structure exists, then the melting point has probably been measured and there is no need of a prediction! The time is ripe for methods focussed on predicting crystal structures of ionic liquids. This is an extremely active field in the pharmaceutical and protein modelling communities, and translation of these techniques to ionic liquids would be beneficial.

### 6.7.4    Force Fields

It is clear from the foregoing discussion that the results of a simulation are only as good as the force field used. While there are dozens of published force fields, most are for standard imidazolium-based ionic liquids and seldom have the force fields been validated beyond matching a liquid density or two. Lopes and Padua have published the most comprehensive collection of force fields for ionic liquids. Starting with a force field for imidazolium-based ionic liquids [40, 151], these authors have published a series of papers that provide parameters for the trifluoromethylsufate (triflates) and bis[(trifluoromethyl)sulfonyl] amide (bistriflamide) anions [41]; pyridinium, phosphonium, and dicyanamide ions [42]; and 1,2,3-trialkylimidazolium and alkoxycarbonyl imidazolium families of cations, as well as alkylsulfate and alkylsulfonate anions [43]. There are literally dozens of other groups who have developed force fields for different ionic liquids. Some have argued that polarisability is an essential feature of force fields [50, 69], while others have shown that effective partial charges are sufficient for obtaining reliable results [152]. Given the importance of force fields to the effort of modelling ionic liquids, we believe several factors related to force fields should be addressed in future work.

*6.7.4.1    Validation.*    First, force fields need to be *validated* by computing a range of properties and comparing against high-quality experimental data. Simply computing a single liquid density is insufficient to prove that a force field is capable of modelling properties of an ionic liquid. A range of thermodynamic and transport properties should be studied. This includes the density but, perhaps, more important, the volume expansivity (i.e., temperature dependence of the density). Fully flexible isothermal–isobaric simulations of crystals should be carried out, and computed lattice constants compared with X-ray data. Other thermodynamic properties such as surface tensions, heat capacities, and the speed of sound need to be computed. Phase change properties such as enthalpies of fusion, enthalpies of vapourisation, and melting points are extremely sensitive to the quality of a force field and should be used in validation studies whenever the experimental data are available. Transport properties, such as viscosities and self-diffusivities, are also good tests of a force field. Finally, mixture properties such as enthalpies of solution with water and other solvents, and solubilities, should be computed and compared with experiment. While these latter properties involve compounds other than ionic liquids, there are good force fields for many common solutes and experimental mixture data are plentiful.

***6.7.4.2  Extension and Automation.***    While the number of different ionic liquids for which force fields have been developed is increasing, the pace is slow. We need force fields for larger classes of cations and anions if the predictive capabilities of molecular simulations are to come to fruition. In addition, the procedure whereby a force field is developed needs to be made much simpler. The process of developing force fields is tedious and time consuming, and (at least in our experience in the United States and in the United Kingdom) few funding agencies are interested in supporting such work. Inspired by the tools developed in the biological modelling community, our group has been developing a set of automated scripting tools that can greatly simplify and speed up the process of developing force fields. With such tools, an estimate of a set of properties for a completely new ionic liquid can be generated in less than a week — considerably faster than the time it takes to make, characterise, and test a sample experimentally. These procedures need to be expanded and distributed to the broader community, and we hope to be able to do that soon.

***6.7.4.3  Development and Databases.***    Let us say someone has already developed a force field for an ionic liquid of interest. How do you use it? Right now, one has to track down the original paper where it was published and, assuming there are no omissions, typos, or errors in the published force field, manually enter the parameters into the particular piece of software being used to run the simulations. Along each step, there is the possibility of human error, not to mention the fact that often there are one or more errors in published force fields to begin with. We believe that a force field database is needed, where researchers can deposit force fields they have generated and download force fields developed by other researchers. If an error is discovered in a force field, it can be corrected electronically in the database instead of living on forever in the archival literature. The molecular modelling community already has a number of force field databases for other biological and organic liquid systems such as OPLS [30], CHARMM [32], and AMBER [31]. Having a similar repository for the ever-growing ionic liquid force field collection would be extremely beneficial in that it would enable easy access by others, would minimise errors and allow for error corrections, and would provide a single place where validation results could be deposited and compared, enabling "good" and "bad" force fields to be distinguished. The IUPAC and the National Institute of Standards and Technology (NIST) would be ideal organisations for such an effort, following on the heels of their successful ILThermo experimental database [153, 154] and the round-robin experimental benchmark study using [$C_6$mim][NTf$_2$] [10, 155].

### 6.7.5    What Is Not Needed?

We have argued earlier that more work is needed in several research areas. There are dozens of more areas worthy of increased attention that we do not mention here. We do feel that it is important to list the areas of research that are no longer a priority, despite the fact that ever more publications are being churned out examining these questions.

*6.7.5.1    Incremental Changes to Common Ionic Liquids.*    We do not need more MD papers in which a minor tweak is made to an ion (like adding one more carbon atom) and simulations are run to produce a room temperature density and radial distribution function for this "new" ionic liquid. While such papers were noteworthy six to eight years ago, it is no longer enough that a simulation be performed "for the first time" on a particular ionic liquid for it to be worthy of publication. Such simulations do little to advance the field.

*6.7.5.2    Yet Another Refined Force Field.*    We do not need more "refined" force fields for dialkylimidazolium cations. There are many varieties of such force fields already in the literature, and while it would indeed be possible to adjust a parameter here and there to get slightly better agreement with some experimental property, unless a comprehensive set of properties are computed, compared with experiment, and shown to be superior to existing force fields, these activities seem to be of little value.

*6.7.5.3    More of the Same Standard Liquid MD Simulations.*    Unless a particular ionic liquid has potential for use in a specific technical application area, or offers some new interesting characteristics, we would argue that performing "vanilla" MD simulations of the liquid phase and computing a small number of properties such as densities, radial distribution functions, and self-diffusivities is no longer of much interest. Previous simulations have demonstrated some of the underlying characteristics of these liquids, and unless a new ionic liquid deviates from these characteristics, having yet another radial distribution function published that looks like all the others in the literature seems of little merit to us. Instead, researchers should spend time and resources tackling new and interesting problems or searching more deeply for underlying physical phenomena that can help us better understand the physical chemistry of ionic liquids. It is also puzzling to us why groups keep simulating the $[PF_6]^-$ and $[BF_4]^-$ anions, given their instability with water and therefore low probability of practical use.

## 6.8 SUMMARY

We have provided a brief review of how molecular simulations work, and the types of properties related to ionic liquids that can be calculated. We showed that both *physical properties* and *physical insight* can be obtained from these simulations. We also discussed some highlights from the molecular simulation literature over the past 10 years. By no means was our treatment comprehensive; doing so would require perhaps as many pages as this entire volume. Our goal was not to be comprehensive, but to instead show the depth and breadth of the literature and to shed light on the capabilities of this research field. We tried to emphasise the ways in which molecular simulations can help drive innovation in ionic liquid research. We ended by listing some areas to which we think more attention needs to be devoted: phase behaviour studies (vapour–liquid, liquid–liquid, and solid–liquid), as well as advances in force fields. We also listed topics that have run their course, and we hope the modelling community will avoid the temptation to focus on incremental studies, but will instead work on breakthrough techniques and applications. It is absolutely essential that modelling groups form close partnerships with experimental groups. The modelling groups have much to offer in helping drive the discovery of new ionic liquids and in helping develop new application areas. It must also be said that experimental groups will help keep the modelling researchers focussed on important technological areas, thereby avoiding the tendency we often have to explore problems that are fun to model but of little practical interest. There is always a need for exploratory fundamental research, of course, but in ionic liquids research there is also a need to answer questions related to the application of these fascinating materials to solve many of the pressing problems facing us today. In only 10 years, molecular modelling has established itself as a powerful tool within the ionic liquids community. We cannot wait to see what the next 10 years will bring.

## ACKNOWLEDGEMENTS

The authors wish to thank the US Air Force Office of Scientific Research (FA9550-10-1-0244), the US National Science Foundation (CBET-0967458), and Notre Dame's Center for Research Computing for financial support.

## REFERENCES

1 Rogers, R. D., and Seddon, K. R., Ionic liquids: Solvents of the future?, *Science* **302**, 792–793 (2003).

2 Tokuda, H., Hayamizu, K., Ishii, K., Susan, M., and Watanabe, M., Physicochemical properties and structures of room temperature ionic liquids. 2. Variation of alkyl chain length in imidazolium cation, *J. Phys. Chem. B* **109**, 6103–6110 (2005).

3 Urahata, S. M., and Rebeiro, M. C. C., Single particle dynamics in ionic liquids of 1-alkyl-3-methyl-imidazolium cations, *J. Chem. Phys.* **122**, 024511 (2005).

4 Lynden-Bell, R., Atamas, N., Vasilyuk, A., and Hanke, C., Chemical potentials of water and organic solutes in imidazolium ionic liquids: a simulation study, *Mol. Phys.* **100**, 3225–3229 (2002).

5 Shah, J. K., Brennecke, J. F., and Maginn, E. J., Thermodynamic properties of the ionic liquid 1-*n*-butyl-3-methylimidazolium hexafluorophosphate from Monte Carlo simulations, *Green Chem.* **4**, 112–118 (2002).

6 Shah, J. K., and Maginn, E. J., Monte Carlo simulations of gas solubility in the ionic liquid 1-*n*-butyl-3-methylimidazolium hexafluorophosphate, *J. Phys. Chem. B* **109**, 10395–10405 (2005).

7 Shi, W., and Maginn, E. J., Molecular simulation and regular solution theory modelling of pure and mixed gas absorption in the ionic liquid 1-*n*-hexyl-3-methylimidazolium bis(trifluoromethylsulfonyl)amide ([hmim][Tf$_2$N]), *J. Phys. Chem. B* **112**, 16710–16720 (2008).

8 Theodorou, D. N., Progress and outlook in Monte Carlo simulations, *Ind. Eng. Chem. Res.* **49**, 3047–3058 (2010).

9 Maginn, E. J., *Atomistic Simulations of Ionic Liquids*, volume 26 of *Reviews in Computational Chemistry* (John Wiley and Sons, Inc., 2007), and references therein.

10 Chirico, R. D., Diky, V., Magee, J. W., Frenkel, M., and Marsh, K. N., Thermodynamic and thermophysical properties of the reference ionic liquid: 1-hexyl-3-methylimidazolium bis[(trifluoromethyl)sulfonyl]amide (including mixtures). Part 2. Critical evaluation and recommended property values (IUPAC technical report), *Pure Appl. Chem.* **81**, 791–828 (2009).

11 Hert, D. G., Anderson, J. L., Aki, S. N. V. K., and Brennecke, J. F., Enhancement of oxygen and methane solubility in 1-hexyl-3-methylimidazolium bis(trifluoromethylsulfonyl)imide using carbon dioxide, *Chem. Commun.* **20**, 2603–2605 (2005).

12 Lopez-Castillo, Z. K., Kestel, D. E., and Brennecke, J. F., I&EC 150-Solubility of gases in [hmim][Tf$_2$N] and [hmim][eFAP], *Abstracts of Papers of the American Chemical Society* **236** (2008), 236th National Meeting of the American-Chemical-Society, Philadelphia, PA, AUG 17–21, 2008.

13 Bowron, D. T., D'Agostino, C., Gladden, L. F., Hardacre, C., Holbrey, J. D., Lagunas, M. C., McGregor, J., Mantle, M. D., Mullan, C. L., and Youngs, T. G. A., Structure and dynamics of 1-ethyl-3-methylimidazolium acetate via molecular dynamics and neutron diffraction, *J. Phys. Chem. B* **114**, 7760–7768 (2010).

14 Urahata, S. M., and Ribeiro, M. C. C., Structure of ionic liquids of 1-alkyl-3-methylimidazolium cations: A systematic computer simulation study, *J. Chem. Phys.* **120**, 1855–1863 (2004).

15  Wang, Y., Jiang, W., Yan, T., and Voth, G. A., Understanding ionic liquids through atomistic and coarse-grained molecular dynamics simulations, *Acc. Chem. Res.* **40**, 1193–1199 (2007).

16  Canongia Lopes, J. N. A., and Padua, A. A. H., Nanostructural organization in ionic liquids, *J. Phys. Chem. B* **110**, 3330–3335 (2006).

17  Triolo, A., Russina, O., Bleif, H., and Di Cola, E., Nanoscale segregation in room temperature ionic liquids, *J. Phys. Chem. B* **111**, 4641–4644 (2007).

18  Morrow, T. I., and Maginn, E. J., Molecular dynamics study of the ionic liquid 1-*n*-butyl-3-methyl-imidazolium hexafluorophosphate, *J. Phys. Chem. B* **106**, 12807–12813 (2002).

19  Santos, L. M. N. B. F., Canongia Lopes, J. N., Coutinho, J. A. P., Esperança, J. M. S. S., Gomes, L. R., Marrucho, I. M., and Rebelo, L. P. N., Ionic liquids: First direct determination of their cohesive energy, *J. Am. Chem. Soc.* **129**, 284–285 (2007).

20  Köddermann, T., Paschek, D., and Ludwig, R., Molecular dynamics simulations of ionic liquids: A reliable description of structure, thermodynamics and dynamics, *Chem. Phys. Chem.* **8**, 2464–2470 (2007).

21  Liu, Z., Wu, X., and Wang, W., A novel united-atom force field for imidazolium-based ionic liquids, *Phys. Chem. Chem. Phys.* **8**, 1096–1104 (2006).

22  Köddermann, T., Paschek, D., and Ludwig, R., Ionic liquids: Dissecting the enthalpies of vapourization, *Chem. Phys. Chem.*, **9**, 549–555 (2008).

23  Zaitsau, D., Kabo, G., Strechan, A., Paulechka, Y., Tschersich, A., Verevkin, S., and Heintz, A., Experimental vapour pressures of 1-alkyl-3-methylimidazolium bis(trifluoromethylsulfonyl)imides and a correlation scheme for estimation of vaporization enthalpies of ionic liquids, *J. Phys. Chem. A* **110**, 7303–7306 (2006).

24  Earle, M. J., Esperanca, J. M. S. S., Gilea, M. A., Lopes, J. N. C., Rebelo, L. P. N., Magee, J. W., Seddon, K. R., and Widegren, J. A., The distillation and volatility of ionic liquids, *Nature* **439**, 831–834 (2006).

25  Kelkar, M. S., and Maginn, E. J., Calculating the enthalpy of vaporization for ionic liquid clusters, *J. Phys. Chem. B* **111**, 9424–9427 (2007).

26  Ballone, P., Pinilla, C., Kohanoff, J., and Del Pópolo, M. G., Neutral and charged 1-butyl-3-methylimidazolium triflate clusters: Equilibrium concentration in the vapour phase and thermal properties of nanometric droplets, *J. Phys. Chem. B* **111**, 4938–4950 (2007).

27  Leal, J. P., Esperança, J. M. S. S., Minas da Piedade, M. E., Canongia Lopes, J. N., Rebelo, L. P. N., and Seddon, K. R., The nature of ionic liquids in the gas phase, *J. Phys. Chem. A* **111**, 6176–6182 (2007).

28  Strasser, D., Goulay, F., Kelkar, M. S., Maginn, E. J., and Leone, S. R., Photoelectron spectrum of isolated ion-pairs in ionic liquid vapour, *J. Phys. Chem. A* **111**, 3191–3195 (2007).

29  Esperança, J. M. S. S., Canongia Lopes, J. N., Tariq, M., Santos, L. M. N. B. F., Magee, J. W., and Rebelo, L. P. N., Volatility of aprotic ionic liquids-a review, *J. Chem. Eng. Data* **55**, 3–12 (2010).

30  Jorgensen, W. L., Maxwell, D. S., and Tirado-Rives, J., Development and testing of the OPLS all-atom force field on conformational energetics and properties of organic liquids, *J. Am. Chem. Soc.* **118**, 11225–11236 (1996).

31 Cornell, W., Cieplak, P., Bayly, C., Gould, I., Merz, K., Ferguson, D., Spellmeyer, D., Fox, T., Caldwell, J., and Kollman, P., A second generation force field for the simulation of proteins, nucleic acids, and organic molecules, *J. Am. Chem. Soc.* **117**, 5179–5197 (1995).

32 MacKerell, A. D., Bashford, D., Bellott, Dunbrack, R. L., Evanseck, J. D., Field, M. J., Fischer, S., Gao, J., Guo, H., Ha, S., Joseph-McCarthy, D., Kuchnir, L., Kuczera, K., Lau, F. T. K., Mattos, C., Michnick, S., Ngo, T., Nguyen, D. T., Prodhom, B., Reiher, W. E., Roux, B., Schlenkrich, M., Smith, J. C., Stote, R., Straub, J., Watanabe, M., Wiórkiewicz-Kuczera, J., Yin, D., and Karplus, M., All-atom empirical potential for molecular modelling and dynamics studies of proteins, *J. Phys. Chem. B* **102**, 3586–3616 (1998).

33 Hanke, C. G., Price, S. L., and Lynden-Bell, R. M., Intermolecular potentials for simulations of liquid imidazolium salts, *Mol. Phys.* **99**, 801–809 (2001).

34 Williams, D. E., and Cox, S. R., Nonbonded potentials for azahydrocarbons—The importance of the Coulombic interactions, *Acta. Crystallogr. B* **40**, 404 (1984).

35 Hsu, L. Y., and William, D. E., Intermolecular potential-function for crystalline perchlorohydrocarbons, *ACTA Crystallogr. A* **36**, 277–281 (1980).

36 Williams, D. E., and Houpt, D. J., Fluorine nonbonded potential parameters derived from crystalline perflourocarbons, *Acta. Crystallogr. B* **42**, 286 (1986).

37 Cadena, C., Zhao, Q., Snurr, R. Q., and Maginn, E. J., Molecular modelling and experimental studies of the thermodynamic and transport properties of pyridinium-based ionic liquids, *J. Phys. Chem. B* **110**, 2821–2832 (2006).

38 Cadena, C., and Maginn, E. J., Molecular simulation study of some thermophysical and transport properties of triazolium-based ionic liquids, *J. Phys. Chem. B* **110**, 18026–18039 (2006).

39 Gutowski, K. E., and Maginn, E. J., Amine-fuctionalized task specific ionic liquids: A mechanistic explanation for the dramatic increase in viscosity upon complexation with $CO_2$ from molecular simulation, *J. Am. Chem. Soc.* **130**, 14690–14704 (2008).

40 Lopes, J. N. C., Deschamps, J., and Pádua, A. A. H., Modelling ionic liquids using a systematic all-atom force field, *J. Phys. Chem. B* **108**, 2038–2047 (2004).

41 Lopes, J. N. C., and Pádua, A. A. H., Molecular force field for ionic liquids composed of triflate or bistrifylimide anion, *J. Phys. Chem. B* **108**, 16893–16898 (2004).

42 Lopes, J. N. C., and Pádua, A. A. H., Molecular force field for ionic liquids III: Imidazolium, pyridinium, and phosphonium cations; Chloride, bromide, and dicyanamide anions, *J. Phys. Chem. B* **110**, 19586–19592 (2006).

43 Lopes, J. N. C., Pádua, A. A. H., and Shimizu, K., Molecular force field for ionic liquids IV: Trialkylimidazolium and alkoxycarbonyl-imidazolium cations; Alkylsulfonate and alkylsulfate anions, *J. Phys. Chem. B* **112**, 5309–5046 (2008).

44 Shimizu, K., Almantariotis, D., Gomes, M. F. C., Pádua, A. A. H., and Lopes, J. N. C., Molecular force field for ionic liquids V: Hydroxyethylimidazolium, dimethoxy-2-methylimidazolium, and fluoroalkylimidazolium cations and bis(fluorosulfonyl)amide, perfluoroalkanesulfonylamide, and fluoroalkylfluorophosphate anions, *J. Phys. Chem. B* **114**, 3592–3600 (2010).

45  Liu, Z., Huang, S., and Wang, W., A refined force field for molecular simulation of imidazolium-based ionic liquids, *J. Phys. Chem. B* **108**, 12978–12989 (2004).

46  Zhou, G., Liu, X., Zhang, S., Yu, G., and He, H., A force field for molecular simulation of tetrabutylphosphonium amino acid ionic liquids, *J. Phys. Chem. B* **111**, 7078–7084 (2007).

47  Liu, X., Zhou, G., Zhang, S., Wu, G., and Yu, G., Molecular simulation of guanidinium-based ionic liquids, *J. Phys. Chem. B* **111**, 5658–5668 (2007).

48  Liu, Z., Chen, T., Bell, A., and Smit, B., Improved united-atom force field for 1-Alkyl-3-methylimidazolium chloride, *J. Phys. Chem. B* **114**, 4572–4582 (2010).

49  Liu, Z., Chen, T., Bell, A. T., and Smit, B., Improved united-atom force field for 1-Alkyl-3-methylimidazolium chloride (vol 114B, pg 4572, 2010), *J. Phys. Chem. B* **114**, 10692 (2010).

50  Bedrov, D., Borodin, O., Li, Z., and Smith, G. D., Influence of polarization on structural, thermodynamic, and dynamic properties of ionic liquids obtained from molecular dynamics simulations, *J. Phys. Chem. B* **114**, 4984–4997 (2010).

51  Shah, J. K., and Maginn, E. J., A Monte Carlo simulation study of the ionic liquid 1-*n*-butyl-3-methylimidazolium hexafluorophosphate: Liquid structure, volumetric properties and infinite dilution solution thermodynamics of $CO_2$, *Fluid Phase Equilib.* **222**, 195–203 (2004).

52  Canongia Lopes, L. J. N., and Padua, A. A., Nanostructure organization in ionic liquids, *J. Chem. Phys. B* **110**, 3330–3335 (2006).

53  Aparicio, S., Alcalde, R., and Atilhan, M., Experimental and computational study on the properties of pure and water mixed 1-ethyl-3-methylimidazolium l-(+)-lactate ionic liquid, *J. Phys. Chem. B* **114**, 5795–5809 (2010).

54  Bodo, E., Gontrani, L., Triolo, A., and Caminiti, R., Structural determination of ionic liquids with theoretical methods: $C_8$mimBr and $C_8$mimCl. Strengths and weakness of current force fields, *Phys. Chem. Lett.* **1**, 1095–1100 (2010).

55  Dymond, J. H., Malhotra, R., Isdale, J. D., and Glen, N. F., (p,π,T) of n-heptane, toluene, and oct-1-ene in the range 298 to 373 K and 0.1 to 400 MPa and representation by the Tait equation, *J. Chem. Thermodyn.* **20**, 603–614 (1988).

56  Shah, J. K., and Maginn, E. J., Molecular dynamics investigation of biomimetic ionic liquids, *Fluid Phase Equilib.* **294**, 197–205 (2010).

57  Murray, S. M., O'Brien, R. A., Mattson, K. M., Ceccarelli, C., Sykora, R. E., West, K. N., and Davis, J. H., The fluid-mosaic model, homeoviscous adaptation, and ionic liquids: Dramatic lowering of the melting point by side-chain unsaturation, *Angew. Chem.* **49**, 2755–2758 (2010).

58  Tariq, M., Forte, P. A. S., Costa Gomes, M. F., Canongia Lopes, J. N., and Rebelo, L. P. N., Densities and refractive indices of imidazolium and phosphonium-based ionic liquids: Effect of temperature, alkyl chain length, and anion, *J. Chem. Thermodyn.* **41**, 790–798 (2009).

59  Klähn, M., Seduraman, A., and Wu, P., A model for self-diffusion of guanidinium-based ionic liquids: A molecular dynamics simulation study, *J. Phys. Chem. B* **112**, 13849–13861 (2008).

60  Qian, J., Hentschke, R., and Heuer, A., Dynamic heterogeneties of translational and rotational motion of a molecular glass former from computer simulations, *J. Chem. Phys.* **110**, 4514–4522 (1999).

61 Rahman, A., Correlations in motions of atoms in liquid argon, *Phys. Rev.* **136**, A405–A411 (1964).

62 Del Pópolo, M. G., Voth, G. A., On the structure and dynamics of ionic liquids, *J. Phys. Chem. B* **108**, 1744–1752 (2002).

63 Margulis, C. J., Stern, H. A., and Berne, B. J., Computer simulation of a "green chemistry" room-temperature ionic liquid, *J. Phys. Chem. B* **106**, 12107–12121 (2002).

64 Nicotera, I., Oliviero, C., Henderson, W. A., Appetecchi, G. B., and Passerini, S., NMR investigation of ionic liquid-LiX mixtures: Pyrrolidinium cations and TFSI-anions, *J. Phys. Chem. B* **109**, 22814–22819 (2005).

65 Rey-Castro, C., and Vega, L. F., Transport properties of the ionic liquid 1-ethyl-3-methylimidazolium chloride from equilibrium molecular dynamics simulation. The effect of temperature, *J. Phys. Chem. B* **110**, 14426–14435 (2006).

66 Bhargava, B. L., and Balasubramanian, S., Dynamics in a room-temperature ionic liquid: A computer simulation study of 1,3-dimethylimidazolium chloride, *J. Chem. Phys.* **123**, 144505 (2005).

67 Lee, S. U., Jung, J., and Han, Y. K., Molecular dynamics study of the ionic conductivity of 1-*n*-butyl-3-methylimidazolium salts as ionic liquids, *Chem. Phys. Lett.* **406**, 332–340 (2005).

68 Gurkan, B., Goodrich, B. F., Mindrup, E. M., Ficke, L. E., Massel, M., Seo, S., Senftle, T. P., Wu, H., Glaser, M. F., Shah, J. K., Maginn, E. J., Brennecke, J. F., and Schneider, W. F., Molecular design of high capacity, low viscosity, chemically tunable ionic liquids for $CO_2$ capture, *J. Phys. Chem. Lett.* **1**, 3494–3499 (2010).

69 Yan, T. Y., Burnham, C. J., Pópolo, M. G. D., and Voth, G. A., Molecular dynamics simulation of ionic liquids: The effect of electronic polarizability, *J. Phys. Chem. B* **108**, 11877–11881 (2004).

70 Köddermann, T., Ludwing, R., and Paschek, D., On the validity of Stokes-Einstein and Stokes-Einstein-Debye relations in ionic liquids and ionic-liquid mixtures, *Chem. Phys. Chem.* **9**, 1851–1858 (2008).

71 Kelkar, M. S., Shi, W., and Maginn, E. J., Determining the accuracy of classical force fields for ionic liquids: Atomistic simulation of the thermodynamic and transport properties of 1-ethyl-3-methylimidazolium ethylsulfate ([emim][EtSO₄]) and its mixtures with water, *Ind. Eng. Chem. Res.* **47**, 9115–9126 (2008).

72 Harris, K., Relation between the fractional Stokes-Einstein and Nernst-Einstein equations and velocity correlation coefficients in ionic liquids and molten salts, *J. Phys. Chem. B* **114**, 9572–9577 (2010).

73 Kelkar, M. S., and Maginn, E. J., Effect of temperature and water content on the shear viscosity of the ionic liquid 1-ethyl-3-methylimidazolium bis(trifluoro-methanesulfonyl)imide as studied by atomistic simulations, *J. Phys. Chem. B* **111**, 4867–4876 (2007).

74 Zhao, W., Leroy, F., Balasubramanian, S., and Müller-Plathe, F., Shear viscosity of the ionic liquid 1-*n*-butyl-3-methylimidazolium hexafluorophosphate [bmim][PF₆] computed by reverse nonequilibrium molecular dynamics, *J. Phys. Chem. B* **112**, 8129–8133 (2008).

75 Van-Oanh, N.-T., Houriez, C., and Rousseau, B., Viscosity of the 1-ethyl-3-methylimidazolium bis(trifluoromethylsulfonyl)imide ionic liquid from

equilibrium and nonequilibrium molecular dynamics, *Phys. Chem. Chem. Phys.* **12**, 930–936 (2010).

76  Borodin, O., Smith, G. D., and Kim, H., Viscosity of a room temperature ionic liquid: Predictions from nonequilibrium and equilibrium molecular dynamics simulations, *J. Phys. Chem. B* **113**, 4771–4774 (2009).

77  Picálek, J., and Kolafa, J., Molecular dynamics study of conductivity of ionic liquids: The Kohlrausch law, *J. Mol. Liq.* **134**, 29–33 (2007).

78  Kowsari, M. H., Alavi, S., Ashrafizaadeh, M., and Najafi, B., Molecular dynamics simulations of imidazolium-based ionic liquids. II. Transport coefficients, *J. Chem. Phys.* **130**, 014703 (2009).

79  Borodin, O., and Smith, G. D., Structure and dynamics of *n*-methyl-*n*-propyl-pyrrolidinium bis(trifluoromethanesulfonyl)imide ionic liquid from molecular dynamics simulations, *J Phys. Chem. B* **110**, 11481–11490 (2006).

80  MacFarlane, D. R., Forsyth, M., Izgorodina, E. I., Abbott, A. P., Annat, G., and Fraser, K., On the concept of ionicity in ionic liquids, *Phys. Chem. Chem. Phys.* **11**, 4962–4967 (2009).

81  Lynden-Bell, R., Gas–liquid interfaces of room temperature ionic liquids, *Mol. Phys.* **101**, 2625–2633 (2003).

82  Lynden-Bell, R. M., and Del Pópolo, M., Simulation of the surface structure of butylmethylimidazolium ionic liquids, *Phys. Chem. Chem. Phys.* **8**, 949–954 (2006).

83  Bhargava, B., and Balasubramanian, S., Layering at an ionic liquid vapor interface: A molecular dynamics simulation study of [bmim][PF$_6$], *J. Am. Chem. Soc.* **128**, 10073– 10078 (2006).

84  Heggen, B., Zhao, W., Leroy, F., Dammers, A., and Müller-Plathe, F., Interfacial properties of an ionic liquid by molecular dynamics, *J. Phys. Chem. B* **114**, 6954–6961 (2010).

85  Pensado, A., Malfreyt, P., and Pádua, A., Molecular dynamics simulations of the liquid surface of the ionic liquid 1-hexyl-3-methylimidazolium bis(trifluoro-methanesulfonyl) amide: Structure and surface tension, *J. Phys. Chem. B* **113**, 14708–14718 (2009).

86  Perez-Blanco, M., and Maginn, E., Molecular dynamics simulations of $CO_2$ at an ionic liquid interface: Adsorption, ordering, and interfacial crossing, *J. Phys. Chem. B* **114**, 11827–11837 (2010).

87  Yan, T., Li, S., Jiang, W., Gao, X., Xiang, B., and Voth, G., Structure of the liquid–vacuum interface of room-temperature ionic liquids: A molecular dynamics study, *J. Phys. Chem. B* **110**, 1800–1806 (2006).

88  Chang, T. M., and Dang, L. X., Computational studies of structures and dynamics of 1,3-dimethylimidazolium salt liquids and their interfaces using polarizable potential models, *J. Phys. Chem. A* **113**, 2127–2135 (2009).

89  Lynden-Bell, R. M., Kohanoff, J., and Del Pópolo, M. G., Simulation of interfaces between room temperature ionic liquids and other liquids, *Faraday Discuss.* **129**, 57–67 (2005).

90  Sieffert, N., and Wipff, G., Adsorption at the liquid–liquid interface in the biphasic rhodium-catalyzed hydroformylation of 1-hexene in ionic liquids: A molecular dynamics study, *J. Phys. Chem. C* **112**, 6450–6461 (2008).

91  Dou, Q., Sha, M., Fu, H., and Wu, G., Mass distribution and diffusion of [1-butyl-3-methylimidazolium][Y] ionic liquids adsorbed on the graphite surface at 300–800 K, *Chem. Phys. Chem.* **11**, 2438–2443 (2010).

92  Wang, S., Li, S., Cao, Z., and Yan, T., Molecular dynamic simulations of ionic liquids at graphite surface, *J. Phys. Chem. C* **114**, 990–995 (2010).

93  Wang, S., Cao, Z., Li, S., and Yan, T., A molecular dynamics simulation of the structure of ionic liquid (BMIM$^+$/PF$_6^-$)/rutile (110) interface, *Sci. China Ser. B-Chem.* **52**, 1434–1437 (2009).

94  Liu, L., Li, S., Cao, Z., Peng, Y., Li, G., Yan, T., and Gao, X.-P., Well-ordered structure at ionic liquid/rutile (110) interface, *J. Phys. Chem. C* **111**, 12161–12164 (2007).

95  Yan, T., Wang, S., Zhou, Y., Cao, Z., and Li, G., Adsorption of CO$_2$ on the rutile (110) surface in ionic liquid. A molecular dynamics simulation, *J. Phys. Chem. C* **113**, 19389–19392 (2009).

96  Sieffert, N., and Wipff, G., Adsorption at the liquid-liquid interface in the biphasic rhodium-catalyzed hydroformylation of 1-hexene in ionic liquids: A molecular dynamics study, *J Phys Chem C* **112** (2008).

97  Sha, M., Wu, G., Dou, Q., Tang, Z., and Fang, H., Double-layer formation of [Bmim][PF$_6$] ionic liquid triggered by surface negative charge, *Langmuir* **26**, 12667–12672 (2010).

98  Feng, G., Zhang, J. S., and Qiao, R., Microstructure and capacitance of the electrical double layers at the interface of ionic liquids and planar electrodes, *J. Phys. Chem. C* **113**, 4549– 4559 (2009).

99  Chen, S., Wu, G., Sha, M., and Huang, S., Transition of ionic liquid [bmim][PF$_6$] from liquid to high-melting-point crystal when confined in multiwalled carbon nanotubes, *J. Am. Chem. Soc.* **129**, 2416–2417 (2007).

100 Göbel, R., Hesemann, P., Weber, J., Möller, E., Friedrich, A., Beuermann, S., and Taubert, A., Surprisingly high, bulk liquid-like mobility of silica-confined ionic liquids, *Phys. Chem. Chem. Phys.* **11**, 3653–3662 (2009).

101 Néouze, M. A., Bideau, J. L., Gaveau, P., Bellayer, S., and Vioux, A., Ionogels, new materials arising from the confinement of ionic liquids within silica-derived networks, *Chem. Mater.* **18**, 3931–3936 (2006).

102 Bideau, J. L., Gaveau, P., Bellayer, S., Néouze, M.-A., and Vioux, A., Effect of confinement on ionic liquids dynamics in monolithic silica ionogels: $^1$H NMR study, *Phys. Chem. Chem. Phys.* **9**, 5419–5422 (2007).

103 Shim, Y., and Kim, H., Solvation of carbon nanotubes in a room-temperature ionic liquid, *ACS nano* **3**, 1693–1702 (2010).

104 Dong, K., Zhou, G., Liu, X., Yao, X., Zhang, S., and Lyubartsev, A., Structural evidence for the ordered crystallites of ionic liquid in confined carbon nanotubes, *J. Phys. Chem. C* **113**, 10013–10020 (2009).

105 Singh, R., Monk, J., and Hung, F. R., A computational study of the behavior of the ionic liquid [BMIM$^+$][PF$_6^-$] confined inside multiwalled carbon nanotubes, *J. Phys. Chem. C* **114**, 15478–15485 (2010).

106 Pinilla, C., Del Pópolo, M. G., Lynden-Bell, R., and Kohanoff, J., Structure and dynamics of a confined ionic liquid. Topics of relevance to dye-sensitized solar cells, *J. Phys. Chem. B* **109**, 17922–17927 (2005).

107  Sha, M., Wu, G., Liu, Y., Tang, Z., and Fang, H., Drastic phase transition in ionic liquid [Dmim][Cl] confined between graphite walls: New phase formation, *J. Phys. Chem. C* **113**, 4618–4622 (2009).

108  Bhargava, B. L., and Klein, M. L., Aqueous solutions of imidazolium ionic liquids: Molecular dynamics studies, *Soft Matter* **5**, 3475–3480 (2009).

109  Feng, S., and Voth, G. A., Molecular dynamics simulations of imidazolium-based ionic liquid/water mixtures: Alkyl side chain length and anion effects, *Fluid Phase Equilib.* **294**, 148–156 (2010).

110  Raabe, G., and Köhler, J., Thermodynamical and structural properties of binary mixtures of imidazolium chloride ionic liquids and alcohols from molecular simulation, *J. Chem. Phys.* **129**, 144503 (2008).

111  Rodriguez, H., and Brennecke, J. F., Temperature and composition dependence of the density and viscosity of binary mixtures of water plus ionic liquid, *J. Chem. Eng. Data* **51**, 2145–2155 (2006).

112  Del Pópolo, M. G., Mullan, C. L., Holbrey, J. D., Hardacre, C., and Ballone, P., Ion association in [bmim][$PF_6$]/Naphthalene mixtures: An experimental and computational study, *J. Am. Chem. Soc.* **130**, 7032–7041 (2008).

113  Shimizu, K., Gomes, M. F. C., Pádua, A. A. H., Rebelo, L. P. N., and Lopes, J. N. C., On the role of the dipole and quadrupole moments of aromatic compounds in the solvation by ionic liquids, *J. Phys. Chem. B* **113**, 9894–9900 (2009).

114  Cadena, C., Anthony, J. L., Shah, J. K., Morrow, T. I., Brennecke, J. F., and Maginn, E. J., Why is $CO_2$ so soluble in imidazolium-based ionic liquids? *J. Am. Chem. Soc.* **126**, 5300–5308 (2004).

115  Bhargava, B. L., Krishna, A. C., and Balasubramanian, S., Molecular dynamics simulation studies of $CO_2$-[bmim][$PF_6$] solutions: Effect of $CO_2$ concentration, *AIChE J.* **54**, 2971–2978 (2008).

116  Siqueira, L. J. A., Ando, R. A., Bazito, F. F. C., Torresi, R. M., Santos, P. S., and Ribeiro, M. C. C., Shielding of ionic interactions by sulfur dioxide in an ionic liquid, *J. Phys. Chem. B* **112**, 6430–6435 (2008).

117  Widom, B., Some topics in theory of fluids, *J. Chem. Phys.* **39**, 2808–2812 (1963).

118  Lyubartsev, A. P., Martisinovski, A. A., Shevkunov, S. V., and Vorontsov-Velyaminov, P. N., New approach to Monte Carlo calculation of the free energy method of expanded ensembles, *J. Chem. Phys.* **96**, 1776–1783 (1992).

119  Kofke, D. A., Getting the most from molecular simulation, *Mol. Phys.* **102**, 405–420 (2004).

120  Deschamps, J., Gomes, M. F. C., and Pádua, A. A. H., Molecular simulation study of interactions of carbon dioxide and water with ionic liquids, *Chem. Phys. Chem.* **5**, 1049–1052 (2004).

121  Anthony, J. L., Maginn, E. J., and Brennecke, J. F., Solution thermodynamics of imidazolium-based ionic liquids and water, *J. Phys. Chem. B* **105**, 10942–10949 (2001).

122  Urukova, I., Vorholz, J., and Maurer, G., Solubility of $CO_2$, CO, and $H_2$ in the ionic liquid [bmim][$PF_6$] from Monte Carlo simulations, *J. Phys. Chem. B* **109**, 12154–12159 (2005).

123  Panagiotopoulos, A. Z., Direct determination of phase coexistence properties of fluids by Monte Carlo in a new ensemble, *Mol. Phys.* **61**, 813–826 (1987).

124 Shi, W., and Maginn, E. J., Atomistic simulation of the absorption of carbon dioxide and water in the ionic liquid 1-*n*-hexyl-3-methylimidazolium bis(trifluoromethylsulfonyl)imide ([hmim][Tf$_2$N]), *J. Phys. Chem. B*, **112**, 2045–2055 (2008).

125 Anderson, J. L., Dixon, J. K., Maginn, E. J., and Brennecke, J. F., Measurement of SO$_2$ solubility in ionic liquids, *J. Phys. Chem. B* **110**, 15059–15062 (2006).

126 Huang, J., Riisager, A., Wasserscheid, P., and Fehrmann, R., Reversible physical absorption of SO$_2$ by ionic liquids, *Chem. Commun.* 4027–4029 (2006).

127 Shi, W., and Maginn, E. J., Molecular simulation and regular solution theory modelling of pure and mixed gas absorption in the ionic liquid 1-*n*-hexyl-3-methylimidazolium bis(trifluoromethylsulfonyl)amide ([hmim][Tf$_2$N]), *J. Phys. Chem. B* **2008**, 16710–16720 (2008).

128 Chiu, Y.-H., Gaeta, G., Levandier, D. J., Dressler, R. A., and Boatz, J. A., Vacuum electrospray ionization study of the ionic liquid, [Emim][Im], *Int. J. Mass. Spec.* **265**, 146–158 (2007).

129 Arce, P. F., Adrian Robles, P., Graber, T. A., and Aznar, M., Modelling of high-pressure vapour liquid equilibrium in ionic liquids + gas systems using the PRSV equation of state, *Fluid Phase Equilib.* **295**, 9–16 (2010).

130 Tariq, M., Serro, A. P., Mata, J. L., Saramago, B., Esperança, J. M. S. S., Canongia Lopes, J. N., and Rebelo, L. P. N., High-temperature surface tension and density measurements of 1-alkyl-3-methylimidazolium bistriflamide ionic liquids, *Fluid Phase Equilib.* **294**, 131– 138 (2010).

131 Valderrama, J. O., Sanga, W. W., and Lazzús, J. A., Critical properties, normal boiling temperature, and acentric factor of another 200 ionic liquids, *Ind. Eng. Chem. Res.* **47**, 1318–1330 (2008).

132 Shiflett, M. B., Harmer, M. A., Junk, C. R., and Yokozeki, A., Solubility and diffusivity of 1,1,1,2-tetrafluoroethane in room-temperature ionic liquids, *Fluid Phase Equilib.* **242**, 220–232 (2006).

133 Lee, H.-N., and Lodge, T. P., Lower critical solution temperature (LCST) phase behavior of poly(ethylene oxide) in ionic liquids, *J. Phys. Chem. Lett.* **1**, 1962–1966 (2010).

134 Domańska, U., Królikowski, M., and Ślesińska, K., Phase equilibria study of the binary systems (ionic liquid plus thiophene): Desulphurization process, *J. Chem. Thermodyn.* **41**, 1303–1311 (2009).

135 Maia, F. M., Rodriguez, O., and Macedo, E. A., LLE for (water plus ionic liquid) binary systems using [C$_x$mim][BF$_4$] ($x = 6, 8$) ionic liquids, *Fluid Phase Equilib.* **296**, 184–191 (2010).

136 Garcia, J., Garcia, S., Torrecilla, J. S., Oliet, M., and Rodriguez, F., Liquid–liquid equilibria for the ternary systems {heptane + toluene + N-butylpyridinium tetrafluoroborate or N-hexylpyridinium tetrafluoroborate} at T=313.2 K, *J. Chem. Eng. Data* **55**, 2862–2865 (2010).

137 Chaumont, A., Schurhammer, R., and Wipff, G., Aqueous interfaces with hydrophobic room-temperature ionic liquids: A molecular dynamics study, *J. Phys. Chem. B* **109**, 18964–18973 (2005).

138 Sieffert, N., and Wipff, G., The [BMI][Tf$_2$N] ionic liquid/water binary system: A molecular dynamics study of phase separation and of the liquid-liquid interface, *J. Phys. Chem. B* **110**, 13076–13085 (2006).

139  Rooney, D., Jacquemin, J., and Gardas, R. L., Thermophysical properties of ionic liquids, in *Ionic Liquids*, volume 290 of *Topics in Current Chemistry*, 185–212 (2009).

140  Katritzky, A. R., Lomaka, A., Petrukhin, R., Jain, R., Karelson, M., Visser, A. E., and Rogers, R. D., QSPR correlation of the melting point for pyridinium bromides, potential ionic liquids, *J. Chem. Inf. Comput. Sci.* **42**, 71–74 (2002).

141  Alavi, S., and Thompson, D. L., Molecular dynamics studies of melting and some liquid-state properties of 1-ethyl-3-methylimidazolium hexafluorophosphate [emim][PF$_6$], *J. Chem. Phys.* **122**, 154704 (2005).

142  Alavi, S., and Thompson, D. L., Simulations of the solid, liquid and melting of 1-*n*-butyl-4-amino-1,2,4-triazolium bromide, *J. Phys. Chem. B* **109**, 18217–18134 (2005).

143  Frenkel, D., and Ladd, A. J. C., New Monte-Carlo method to compute the free-energy of arbitrary solids -application to the fcc and hcp phases of hard-spheres, *J. Chem. Phys.* **81**, 3188–3193 (1984).

144  Grochola, G., Constrained fluid λ-integration: Constructing a reversible thermodynamic path between the solid and liquid state, *J. Chem. Phys.* **120**, 2122–2126 (2004).

145  Grochola, G., Further application of the constrained fluid lambda-integration method, *J. Chem. Phys.* **122**, 046101 (2005).

146  Eike, D. M., Brennecke, J. F., and Maginn, E. J., Toward a robust and general molecular simulation method for computing solid-liquid coexistence, *J. Chem. Phys.* **122**, 014115 (2005).

147  Eike, D. M., and Maginn, E. J., Atomistic simulation of solid-liquid coexistence for molecular systems: Application to triazole and benzene, *J. Chem. Phys.* **124**, 164503 (2006).

148  Jayaraman, S., Thompson, A. P., von Lilienfeld, O. A., and Maginn, E. J., Molecular simulation of the thermal and transport properties of three alkali nitrate salts, *Ind. Eng. Chem. Res.* **49**, 559–571 (2010).

149  Jayaraman, S., and Maginn, E. J., Computing the melting point and thermodynamic stability of the orthorhombic and monoclinic crystalline polymorphs of the ionic liquid 1-*n*-butyl-3-methylimidazolium chloride, *J. Chem. Phys.* **127**, 214504 (2007).

150  Kowsari, M. H., Alavi, S., Ashrafizaadeh, M., and Najafi, B., Molecular dynamics study of congruent melting of the equimolar ionic-liquid inclusion crystal [emim][NTf$_2$]·C$_6$H$_6$, *J. Chem. Phys.* **132**, 044507 (2010).

151  Lopes, J. N. C., Deschamps, J., and Padua, A. A. H., Modelling ionic liquids using a systematic all-atom force field (vol 104b, pg 2038, 2004), *J. Phys. Chem. B* **108**, 11250 (2004).

152  Bhargava, B. L., and Balasubramanian, S., Refined potential model for atomistic simulations of ionic liquid [bmim][PF$_6$], *J. Chem. Phys.* **127**, 114510 (2007).

153  Dong, Q., Muzny, C.D., Kazakov, A., Diky, V., Magee, J.W., Widegren, J.A., Chirico, R.D., Marsh, K.N., and Frenkel, M., ILThermo: A free-access web database for thermodynamic properties of ionic liquids, *J. Chem. Eng. Data* **52** (4), 1151–1159 (2007).

154 National Institute of Standards and Technology, IUPAC Ionic Liquids Database, ILThermo (NIST Standard Reference Database 147), http://ilthermo.boulder .nist.gov/ILThermo/mainmenu.uix, 2006.

155 Marsh, K.N., Brennecke, J.F., Chirico, R.D., Frenkel, M., Heintz, A., Magee, J.W., Peters, C.J., Rebelo, L.P.N., and Seddon, K.R., Thermodynamic and thermophysical properties of the reference ionic liquid: 1-hexyl-3-methylimidazolium bis[(trifluoromethyl)sulfonyl]amide (including mixtures). Part 1. Experimental methods and results (IUPAC technical report), *Pure Appl. Chem.* **81** (5), 781–790 (2009).

# 7 Biocatalytic Reactions in Ionic Liquids

FLORIAN STEIN and UDO KRAGL

Department of Chemistry, University of Rostock, Rostock, Germany

## ABSTRACT

Today biocatalytic reactions are widely used in industrial processes. The spectrum of applications has grown since the discovery that enzymes are active even in organic solvents. The main advantages of organic solvents are increased solubility of hydrophobic substrates, easy product purification, possible enzyme stabilising effects, and suppression of side reactions such as hydrolysis. But biocatalysts show a lack of activity and stability in protic and polar solvents, which is a common problem using enzymes. The hope of scientists for ionic liquids is to close this gap. Many studies have been carried out since early reports published in 2000. Many applications show good activities, yields, and/or selectivities for single and multiphase systems. In addition to that, many researchers deal with the question about influence of ionic liquids and impurities such as water on the biocatalytic reactions. Another field of interest is the use of whole-cell systems, as they allow the use of a greater variety of enzymes because of their integrated cofactor regeneration. In order to use a whole-cell system, several important points have to be taken into consideration, such as finding an appropriate solvent, catalyst, and downstream process. In addition to these points, the toxicity of ionic liquids against cells is also very important, and several reports give an overview of which ionic liquids are biocompatible.

## 7.1 INTRODUCTION

Biocatalysts, such as enzymes and whole-cell systems, have been used for a long time in human history. In ancient Mesopotamia, China, and Japan,

*Ionic Liquids Further UnCOILed: Critical Expert Overviews*, First Edition.
Edited by Natalia V. Plechkova and Kenneth R. Seddon.
© 2014 John Wiley & Sons, Inc. Published 2014 by John Wiley & Sons, Inc.

biocatalysts were used to refine food or to produce alcoholic drinks. Researchers in the nineteenth century investigated enzymes and carried out the first syntheses, Emil Fischer developed the lock-and-key principle, and Croft Hill synthesised isomaltose with the aid of an enzyme [1, 2]. In the twentieth century, enzymes were mostly used and investigated in aqueous solution, but the first studies about enzymes in organic solvents were published in the late 1980s [3–5]. The main advantages of organic solvents are increased solubility of hydrophobic substrates, easy product purification, possible enzyme-stabilising effects, and suppression of side reactions such as hydrolysis. Nowadays biocatalysts are widely used in varying industrial processes. $\beta$-Tyrosinase is applied as biocatalyst in the production of L-dopa, and dextrine is synthesised by an amylase [6, 7]. Biocatalysts work excellently in non-polar and aqueous solvents but show a lack of activity and stability in protic and polar solvents, which is a common problem using enzymes. The hope of scientists is for ionic liquids to close this gap. The first promising studies showed comparable activity and stability with conventional media [8, 9]. To the present day, many studies and reviews are dealing with biocatalytic synthesis in ionic liquids, although there is still a need to clarify the relationships between structure, stability, and activity.

## 7.2 ENZYMES IN IONIC LIQUIDS

In 2000, Erbeldinger et al. reported the synthesis of aspartame from carboxybenzoxy-L-aspartate and L-phenylalanine methyl ester chloride. The special feature of this reaction was the unusual choice of solvent using a thermolysin-catalysed reaction in $[C_2mim][PF_6]$. The initial enzyme activity and yield were of the same order of magnitude as in conventional solvents. Surprisingly, the stability of the enzyme was much higher than in the commonly used ethyl ethanoate [8]. The study was ground-breaking for research in the field of biocatalysis in ionic liquids and led to a rapid increase of publications.

The first enzymes that were applied in research were lipases. This Enzyme Class (EC 3.1.1.3) already showed good activity and stability in organic solvents and established new ways of synthesis, for instance, esterification of fatty acids with alcohols. A fundamental problem of commonly used organic solvents is insufficient solubility of the substrate. Polar substances dissolve better in polar than in non-polar media. In contrast, enzymes show only small or no activity in polar solvents because the enzyme competes for the needed water in order to sustain its conformation [10]. This effect depends on the nature of both the enzyme and the solvent. High substrate solubility for polar substances was achieved by using ionic liquids as solvent, which additionally were believed to offer good enzyme activity. Park and Kazlauskas discovered a possibility to improve and control selectivity [11]. They solvated glucose in the ionic liquid $[C_2mim][BF_4]$, and acetylated it with the aid of *Candida antarctica* Lipase B (CaLB) and vinyl acetate. The problem with using common organic solvents is the higher solubility of the product than of the precursor, resulting in a further acetylation and a decreasing yield. With the use of 1-(2-methoxyethyl)-3-methyl-3$H$-imidazolium chloride ([$C_1OC_2$mim][$BF_4$]), a mono-acylated product was isolated with a yield of 99% and a selectivity of 93%. The main advantage, in contrast to propanone as solvent, is the 100 times higher solubility of glucose. In general, lipases showed the highest tolerance to ionic liquids among all researched enzymes. Ionic liquids with the general structure [$C_n$mim]Y and [$C_n$py]Y ($n$ = 2–10, $Y^-$ = [$BF_4$]$^-$ or [$PF_6$]$^-$) were similarly efficient as the organic solvents methyl *tert*-butyl ether, toluene, and dioxane. For these investigations, lipases CaLB and PCL (*Pseudomonas cepacia* lipase) were used [12–14]. Lipases are able to catalyse a broad range of common and uncommon types of reactions in ionic liquids. Esterification, trans-esterification, acylation of amines, and the aminolysis have been reported [13, 15–19]. Some of these reactions are characterised by good enantioselectivities and yields.

Besides lipases, other enzyme classes have also been investigated. Among others, glycosidases, esterases, phosphatases, laccases, glucooxidases, and dehydrogenases have been utilised as biocatalysts in ionic liquids [18, 20–24]. The very first described biocatalysis in an ionic liquid was carried out with the protease thermolysin, as already mentioned before. Studies with $\alpha$-chymotrypsin showed a very good conversion for a trans-esterification of *N*-acetyl-L-phenylalanine ethyl ester with 1-propanol in different ionic liquids. The enzyme activity in the ionic liquid [$C_2$mim][N(SO$_2$CF$_3$)$_2$] had a value of 13.6 U mg$^{-1}$ and hence was in the same range as found in the organic solvent 1-propanol (23.2 U mg$^{-1}$). Half-life and conversion increased for all ionic liquids used, as opposed to a system containing 1-propanol as solvent in which denaturation took place within 30 minutes after addition of enzyme. Moreover, the amount of product formed was only 25% [25]. Table 7.1 shows a small selection of important biocatalytic syntheses in ionic liquids out of a large number of published papers.

New interesting possibilities open up with two-phase systems for biocatalysis in ionic liquids, since dehydrogenases, for example, are dependent on cofactors and show no activity in pure ionic liquid systems. By using a biphasic system, it is possible to overcome this restriction. While enzyme and cofactors remain in the aqueous phase, product and substrate are present in the ionic liquid phase, a good distribution coefficient provided [26, 27]. Further advantages are simplified catalyst separation, recycling of enzyme solutions, and improved conversions. In two-phase systems composed of H$_2$O/[$C_4$mim][$PF_6$], thioanisol was oxidised by peroxidase and glucose oxidase. The stereoselectivity was comparable with purely aqueous systems, accompanied by a largely improved solubility of the substrates. The isolation of products and substrate from the ionic liquid phase is essential, and a principal problem. The often proposed distillation works only for volatile compounds. Otherwise, remaining involatile compounds limit the recyclability of the ionic liquids.

**TABLE 7.1    A Small Selection of Important Biocatalytic Syntheses Reported in Ionic Liquids**

| Enzyme Class | Enzyme | Ionic Liquid[a] | Comment | Reference |
|---|---|---|---|---|
| *Oxidoreductase* | | | | |
| 1.1.1.1 | Alcohol dehydrogenase | 1 | Biphasic, improved yield | [24] |
| 1.10.3.2 | Laccase | 4-9 | Good activity and stability | [88, 89] |
| 1.11.1.7 | Horseradish peroxidase | 3,6 | Used in sol–gel and w/IL emulsion | [90, 91] |
| *Hydrolase* | | | | |
| 3.1.1.3 | *C. antarctica* Lipase B | 1,2,6,10,11 | Good stability, activity, and yield | [13, 16, 46] |
| 3.1.3.1 | Alkaline phosphatase | 12,salts | Influence of salts on activity | [21, 92] |
| 3.2.1.21 | Cellulase | 13,14 | Used on crop samples | [85, 93] |
| 3.2.1.23 | $\beta$-Galactosidase | 1,2,6,15,16 | Whole-cell-system good possibility | [20, 52] |
| 3.4.21.1 | $\alpha$-Chymotrypsin | 2,6,10,11 | Substrate and alkyl-length influence | [25] |
| 3.4.21.62 | Subtilisin | 17 | Increased yield and ee% | [94] |
| 3.4.24.27 | Thermolysin | 2 | First published synthesis in ionic liquid | [8] |
| 3.5.1.4 | Peptid amidase | 15,18 | Amidation of H-Ala-Phe-OH | [95] |
| *Lyase* | | | | |
| 4.1.2.37 | Hydroxynitril lyase | 6,10,19 | High yield with high ee% | [96, 97] |
| *Isomerase* | | | | |
| 5.1.2.2 | Mandelate racemase | 15,16,20 | Influence of $a_w$ on synthesis | [98] |

[a] Ionic liquids are designated: (1) [C$_4$mim][N(SO$_2$CF$_3$)$_2$], (2) [C$_4$mim][PF$_6$], (3) [C$_8$mim][N(SO$_2$CF$_3$)$_2$], (4) [C$_4$mim]Br, (5) [C$_4$mim]Br, (6) [C$_4$mim][BF$_4$], (7) [C$_2$mim][C$_2$H$_5$SO$_4$], (8) [C$_2$mim][CH$_3$SO$_4$], (9) [C$_2$mim][C$_1$OC$_2$OC$_2$OSO$_3$], (10) [C$_2$mim][BF$_4$], (11) [C$_2$mim][N(SO$_2$CF$_3$)$_2$], (12) [C$_2$H$_5$NH$_3$][NO$_3$], (13) [C$_2$mim][CH$_3$COO], (14) [C$_2$mim][(C$_2$H$_5$)$_2$PO$_4$], (15) [C$_1$mim][CH$_3$SO$_4$], (16) [C$_4$mim][C$_8$H$_{17}$SO$_4$], (17) [C$_2$py][CF$_3$COO], (18) [C$_4$mim][CH$_3$SO$_4$], (19) [C$_5$mim][BF$_4$], and (20) [C$_8$mim][PF$_6$].

Additional purification steps may overcompensate the positive effects, leading to higher costs at the end [23].

Many enzymes tolerate ionic liquids in their reactions only to a certain amount. Beyond this, most of them will become deactivated or denaturated. It has to be tested, for the given combination of enzyme and ionic liquid, whether or not this is a reversible or irreversible process. But even when denaturation may occur, a better process performance may be achieved, for example, higher yields by suppression of side reactions. For a final evaluation, a careful investigation of thermodynamic and kinetic properties of the reaction system is mandatory.

## 7.3    SINGLE-PHASE AND MULTIPHASE SYSTEMS FOR BIOCATALYSIS IN IONIC LIQUIDS

There are many options using an ionic liquid in a biocatalytic system. First of all, they can be employed as a single-phase system by adding pure ionic liquid. Then, it is possible to create a biphasic system by combining a water-immiscible ionic liquid, such as $[C_4mim][PF_6]$, with water or buffer. Ionic liquids can also be used as additives and solubilising agents, and as water-in-ionic liquid (w/IL) micro-emulsions.

Not many enzymes tolerate high concentrations of ionic liquids if they are utilised in a single-phase system. Mainly lipases are suitable for such reactions, as already discussed earlier. In addition, enzymes are scarcely dissolved in ionic liquids and form suspensions. But advantages of a single-phase system include the good solubility of polar substances and the suppression of side reactions. Disadvantages include the problems of product isolation and recycling of ionic liquids. However, products with a low boiling point can be evaporated from the reaction mixture and product isolation can be eased, implicating another two advantages: the reusability of the reaction mixture is provided for, and removal of the reaction products leads to a shift of the chemical equilibrium towards the products [28, 29].

By using a two-phase system, it is possible to simplify downstream processes to a further extent. A biphasic system is generally assembled from an aqueous phase and an ionic liquid phase containing a water-immiscible ionic liquid such as $[C_4mim][PF_6]$ or $[C_4mim][N(SO_2CF_3)_2]$. Organic substrates and products should dissolve better in ionic liquids, while the enzymes remain in the aqueous buffer solution. By this means, high product yields and good enzyme activities are achieved and inhibition by-products or substrates are avoided. The application to different classes of enzymes besides hydrolases and lipases is an additional advantage, since cofactor regeneration is easier and the partial inactivating effect of the ionic liquids can be reduced severely in a two-phase systems. The mass transfer limitation can be considered a negative aspect in biphasic systems. Due to the high viscosity of ionic liquids, insufficient mixing can lead to a limitation of mass transfer, inducing decreased enzyme activity

[9]. But in an experiment with a well-mixed system, the ionic liquid [C$_4$mim] [PF$_6$], having a high viscosity of 397 mPa s, showed higher initial reaction rates than [C$_4$mim][N(SO$_2$CF$_3$)$_2$], which has a viscosity of only 27 mPa s [30].

A promising new possibility of increasing the solubility of enzymes in ionic liquids is w/IL. The enzyme is stabilised in this system, containing buffer and water-immiscible ionic liquid, with the aid of surfactants. Nano- or micro-sized water domains are formed [31]. This layer of water protects the enzymes from the damaging impact of solvent while maintaining good stability and activity. Currently, a major problem is the low solubility of surfactants in ionic liquids. This limitation can be overcome by producing new surfactants with better solubility properties or by adding an organic solvent as solubiliser. Moniruz-zaman et al. reported the first example for an enzymatic reaction in a micro-emulsion of [C$_8$mim][N(SO$_2$CF$_3$)$_2$] with the help of sodium bis(2-ethylhexyl) sulfosuccinate (AOT). The hydrolysis of 4-nitrophenyl butanoate by the lipase *Burkholderia cepacia* was investigated, and a higher activity than in a micro-emulsion of isooctane (water-in-oil [w/o] emulsion) was achieved [32]. In a more recent study, ternary mixtures of [C$_4$mim][PF$_6$]/H$_2$O/Tween20 were the centre of research, and their stability and activity were determined. A conversion of nearly 100% was obtained in a mixture of 80% IL, 15% Tween20, and 5% water. Furthermore, different compositions of the ternary system were compared with each other and with a w/o emulsion consisting of AOT and hexane. All tested w/IL emulsions provided higher yields and mostly better activities while showing vastly improved stabilities compared with a purely aqueous system [33]. Still, the use of surfactants could lead to additional efforts necessary for downstream processing.

Hussain et al. demonstrated another application for ionic liquids in a bio-catalytic system. To gain higher substrate solubility, the work group added 20% of water-miscible or water-immiscible ionic liquids to a fermentation of 6-bromo-$\beta$-tetralone to (S)-6-bromo-$\beta$-tetralol. Especially interesting was the fact that water soluble [C$_2$mim][4-CH$_3$C$_6$H$_4$SO$_3$] allows, besides a high yield, a good initial activity and also an excellent substrate solubility. The prevented aggregation of yeast cells and thus better suspension is the proposed reason for obtaining improved results (compared with a 10% ethanol solution) [34].

## 7.4 INFLUENCE OF IONIC LIQUIDS ON ENZYME AND SUBSTRATE

Ionic liquids have the extraordinary ability to solubilise natural materials such as cellulose, carbohydrates, and amino acids [8, 35–37]. In particular, the dis-solution of cellulose and other biopolymers makes it possible to find new or enhanced ways of utilisation and production of basic chemicals from nature.

Enzymes do not often dissolve in pure ionic liquids, but form a suspension in the case of a well-mixed system. Although enzyme solvation in hydrophilic ionic liquids such as [C$_4$mim][NO$_3$], [C$_4$mim][MeSO$_4$], or [C$_4$mim][lactate] can be achieved, no observable catalytic activity with lipases has been found. These ionic liquids break up protein–protein bonds, resulting in a denaturation of the enzymes. So far the understanding is that the deactivation is mainly due to the anion. A few publications have dealt with the potential of ionic liquids for renaturation of proteins. The reversible denaturation of lysozyme in [EtNH$_3$] [NO$_3$], for example, has been investigated by fluorescence spectroscopy [38].

In general, a solvent should interact as little as possible with substrates, products, and the biocatalyst itself. But ionic liquids cannot be considered completely inert. A weak acidity at the H2-proton within the ring was discovered in 1-alkyl-3-methylimidazolium ionic liquids [39]. Most of the commonly used anions showed neutral or weak basic characteristics. These have the ability to interact with acidic or basic elements of reaction mixtures and can affect parameters such as activity and stability. To explain and understand the interaction between protein and ionic liquid, many researchers refer to the Hofmeister series, since ionic liquids dissociate in sufficient amounts of water into cations and anions. For 40 years, it has been known that enzyme activity follows the series developed by Hofmeister. Kosmotropic anions and chaotropic cations have been discovered as protein stabilising agents (see Figure 7.1). Kosmotropic anions compete for water that is associated with the enzyme. Thus, the surface area of the enzyme decreases and unfolding of the enzyme structure is inhibited. Chaotropic anions, on the other hand, bind to the protein–water interface due to their high polarisability, and consequentially destabilise the enzyme [40, 41].

The situation becomes more complex when ionic liquids are used as pure solvent with a small amount of water. Chaotropic anions such as [BF$_4$]$^-$ or [PF$_6$]$^-$, in contrast to kosmotropic anions, compete only weakly with the water at the protein–water interface [40, 42, 43]. In this case, the enzyme needs the water to maintain its conformation, meaning that kosmotropic anions are destabilising. Generally, the Hofmeister series shows only poor correlation when using a pure ionic liquid: for example, the chaotropic anion [NO$_3$]$^-$ has a negative impact on lipase activity. Obviously, the hydrophobicity is much

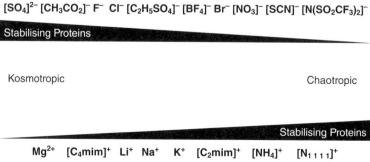

**Figure 7.1** Hofmeister series to describe the influence of ionic liquids on proteins [40, 42, 43].

more dominant, resulting in retained activity of enzyme suspensions in ionic liquids. Nevertheless, there is a publication reporting the impact of protein denaturants and stabilisers on water structure, claiming that the Hofmeister effects are more or less a coincidence [44]. However, it can be concluded that the Hofmeister series is much more suitable for understanding the influence of ionic liquids on enzyme activity than the log $P$ value. The often used polarity scales for ionic liquids using Reichhardt's dye or other tools seem not to be useful to describe ionic liquid–protein interactions.

## 7.5   WATER CONTENT AND WATER ACTIVITY

Water content plays a major role in enzyme-catalysed reactions in organic media. Every enzyme needs a minimum of water to maintain its conformation and unfold their ternary structures. A sphere of protecting water molecules is discussed in different publications. Of importance, a complete removal of water from the organic solvent induces a severely reduced enzyme activity [10, 17, 45]. For a long time, no relation between water content in different organic solvents and enzyme activity could be detected, until Peter Halling and his colleagues described a correlation with water activity. It was found that an enzyme showed the same activity when, besides the temperature, the same water activity was used in different organic solvents instead of the water content [10]. The thermodynamic water activity $a_w$ is the free amount of water in a substance, which is in direct contact with the surrounding atmosphere. This can be described and measured by the quotient of water vapour pressure of the substance and of pure water:

$$a_w = \frac{p_w}{p_{ws}},$$

where $a_w$ is the water activity, $p_w$ is the water vapour pressure of the substance (in pascals), and $p_{ws}$ is the saturated vapour pressure of the pure water (in pascals).

In a similar way, the water content of ionic liquids should be also controlled carefully, when used as reaction media for biocatalysts. However, in order to compare ionic liquids with each other and with organic media, researchers should work with water activity instead of water content. This will lead to more consistent data and reliable parameters for enzyme activity, stability, and conversion [46]. Garcia et al. published a comparative study about the influence of non-conventional media on biocatalysis and used defined water activities for comparing the results (see Figure 7.2) [47]. In addition, it has to be considered that ionic liquids contain some residual water, even after drying, due to their relative hydrophobicities. Furthermore, many ionic liquids are hygroscopic and attract water as long as they are not kept under water-free conditions [48, 49].

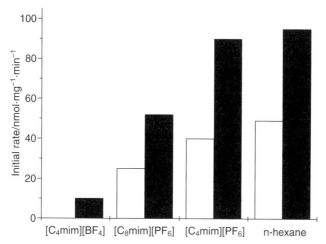

**Figure 7.2** Initial rates of trans-esterification for immobilised cutinase in different solvents, $a_w = 0.2$ (white), $a_w = 0.7$ (black) [47].

While Karl Fischer titration is an easy and fast opportunity to receive information on the current water content in a liquid system, there is no adequate method for the determination of water activity. Devices for the measurement of water activity often need large sample volumes and also a considerable amount of time, since the ionic liquid must equilibrate with its vapour phase above. It is possible to avoid this problem by plotting the measured water activity versus water content for a solvent at a given temperature. By means of these data, it is possible to determine the water activity in this solvent by measuring only its water content. It has to be kept in mind that all compounds, not only the solvent, in a mixture contribute to the water content. Another problem, which has yet to be solved, is adjusting a defined water activity in a system. Equilibration with a saturated salt solution is a commonly used technique (see Figure 7.3). A vessel with the ionic liquid or organic solvent is placed in another larger sealed vessel, containing an oversaturated salt solution. The vapour phase over the saturated salt solution equilibrates with the vapour phase above the solvent. The equilibrating vapour phase is in contact with the solvent and is also subject to equilibration, resulting in a solvent with a specific water activity, which is dependent on the different salts used in the saturated salt solution. Still, this method needs a great amount of time for equilibration.

## 7.6 IMPURITIES

Especially at the beginning of research into the area of ionic liquids, many results were difficult to reproduce. A possible reason was, and still is today,

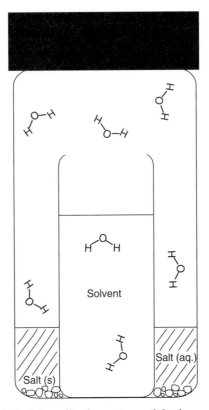

**Figure 7.3** Method for adjusting water activity in a solvent system.

impurities in ionic liquids. An excellent example is the varying of melting temperature from 5.8 to 15 °C of the ionic liquid [$C_4$mim][$BF_4$] [50]. Widegren et al. also report that viscosities vary around 30% at different temperatures [51]. These impurities can consist of solvents and reagents used in the synthesis of ionic liquids, remaining salts, as well as water. They decrease the viscosity and influence the melting point. Ionic liquids, polluted with starting substances, can also have a significant impact on the biocatalytic reaction since they could affect both the activity and the stability of the enzyme. Consequently, every synthesis should be followed by a purification step. Evaporation in vacuum or extraction may be a sufficient method to remove impurities. Another possibility lies in a change of the synthetic procedure. Conversion in metathesis reactions, for example, is strongly dependent on the salt used for precipitation [50]. As long as all groups work with the same high quality of ionic liquids, reproducibility and comparability of results can be achieved.

## 7.7    BIOCATALYSIS IN WHOLE-CELL SYSTEMS

While lipases and hydrolases easily catalyse reactions in pure ionic liquids, enzymes such as dehydrogenases can only be active in two-phase systems or in one-phase systems with water-miscible ionic liquids. These enzymes require cofactors for their regeneration. But cofactors are expensive and should not inhibit the reaction in any way. This problem can be avoided by using whole cells as biocatalysts since they have integrated cofactor regeneration. Whole cells are used in many examples to produce fine chemicals, often with high regio- and enantioselectivities. Besides, wild-type organism recombinant cells with specific properties are used, which could be resistant to solvents or able to achieve improved yields and activities. Despite the many advantages, there are still problems to be solved in whole-cell biocatalysis. In general, similar rules should be applied for whole-cell systems and isolated enzymes. However, organisms have additional possibilities for interaction with the surrounding environment.

To set up a successful whole-cell biotransformation, the following points have to be addressed:

(i)   For the desired reaction, a suitable whole-cell biocatalyst has to be found or designed with the help of genetic engineering.

(ii)  Other enzymes present in the cell may cause undesirable side reactions.

(iii) Downstream processing of the reaction is complex and depends on whether the product is sequestered inside the cells or excreted from the cells into the solvent. Moreover, a contamination with cell residue, biosurfactans, polynucleotides, and polysaccharides is possible.

(iv)  Accumulation of substrates and products in the cells leads to inhibition of the reaction. Overall cell catalytic activity declines subsequently.

(v)   As long as product concentration is low, either due to limited solubility or due to low productivity, large volumes of aqueous solution have to be processed with all associated negative aspects, such as waste and energy consumption.

To overcome these problems, a biphasic system can be applied. Products and precursors will be extracted into the organic phase, while fermentation takes place in the aqueous phase. *In situ* extraction is strongly dependent on the solvent; it should provide a high log $P$ value, dissolve products and precursors better than the buffer phase, and should not be cytotoxic or harmful to humans and the environment. Few solvents come into consideration because of this specification. Long-chained alkyl alcohols, aldehydes, and ketones are possible media for fermentation. By making the correct choice, substrate and products remain in the organic layer, while the biocatalyst is located in the aqueous phase. A limitation by catalyst inhibition is avoided, and additional side reactions such as hydrolysis in esterifications are suppressed.

The potential of ionic liquids for use with whole cells has also been tested quite early. However, a prediction as to which ionic liquid is more suitable for a catalytic system is difficult since results in several studies are contradictory. Ionic liquids should fulfil the same requirements as organic solvents. First, high distribution coefficients for substrates and products are desired to overcome inhibition. With a better distribution into the ionic liquid phase, better yields and selectivity can be expected. As opposed to organic solvents such as toluene aggregation into the interphase of a two-phase system was not observed. This characteristic is beneficial for downstream processing. In this context, well-mixed phases are necessary to eliminate mass transfer limitations due to the high viscosity of ionic liquids. Otherwise, poorer conversions and slower initial reaction rates than in organic systems will be observed [9]. If the ionic liquid is dispersed in the system as fine spheres, the mass transfer will not be limited. Another possibility to decrease this limitation is the use of more hydrophilic substrates resulting in a higher concentration of substrate in the aqueous phase [30].

Experiences with whole-cell systems in ionic liquids are mostly based on empirical data. Scientists report biocatalysis with a good conversion but a poor enantiomeric excess, or the other way around. But there are also counter-examples. In a system containing buffer phase, $[C_4mim][N(SO_2CF_3)_2]$, and cells of *Lactobacillus kefir*, twice the yield with constant high enantiomeric excess was obtained. This system was scaled up to a 200-cm$^3$ system, with comparable results [28, 52].

Lou et al. used a whole-cell system from *Saccharomyces cerevisiae* for the reduction of acetyltrimethylsilane, $CH_3C(O)SiMe_3$, in order to produce enantiopure (S)-1-trimethylsilylethanol, $CH_3CH(OH)SiMe_3$ (see Figure 7.4). The hydrophobic $[C_4mim][PF_6]$ and the hydrophilic $[C_4mim][BF_4]$ were used. In contrast to a hexane/buffer system and pure aqueous buffer system, better initial activities, yields, and enantiomeric excess were achieved (see Table 7.2). Both ionic liquid systems could be recycled at least six times without losing noteworthy activity [53].

Alongside extractive fermentation, extraction is also a possible downstream process. Important for an effective extraction is the distribution coefficient of product in the ionic liquid, $K_d$:

$$K_d = \frac{m_a([A_i] - [A_f])}{m_s[A_f]},$$

where $m_a$ is the mass of the aqueous phase (in grams), $m_s$ is the mass of the solvent phase, $[A_i]$ is the initial concentration of solute in aqueous phase (in mol·l$^{-1}$), and $[A_f]$ is the final concentration of solute in the aqueous phase (in mol·l$^{-1}$).

Ionic liquids have also been tested for selective extraction of natural compounds among others [54]. The ionic liquid $[P_{6\,6\,6\,14}][N(SO_2CF_3)_2]$ has relatively

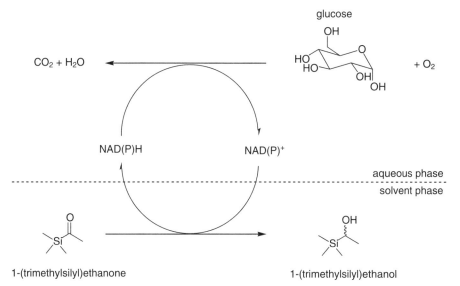

1-(trimethylsilyl)ethanone                                1-(trimethylsilyl)ethanol

**Figure 7.4**  Conversion of prochiral acetyltrimethylsilane to (S)-1-trimethylsilylethanol with *Saccharomyces cerevisiae* [53].

**TABLE 7.2    Reduction in Acetyltrimethylsilane (ATMS) to (S)-1-Trimethylsilylethanol with *Saccharomyces cerevisiae*, Buffer: Tris-HCl [53]**

| Solvent | pH | $c$(ATMS)/ mM | Initial rate/ $mM\,h^{-1}$ | Yield/% | ee/% |
|---|---|---|---|---|---|
| Buffer (aq.) | 7.3 | 14 | 4.88 | 84.1 | 82.7 |
| Hexane:buffer $(2:1)^a$ | 8.0 | 42 | 1.45 | 97.4 | 95.4 |
| [C₄mim][PF₆]:buffer $(1:6)^a$ | 7.3 | 84 | 63.4 | 99.9 | 99.9 |
| [C₄mim][BF₄]:buffer (10%) | 7.3 | 77 | 74.5 | 99.2 | 99.9 |

$^a$ Formation of a biphasic system.

high distribution coefficients at pH 11 for ethanoic, lactic, and succinic acid, with $K_d = 6, 14$, and 23, respectively [55]. Amino acids were also extracted well from different ionic liquids, adding dicyclohexano-18-crown-6 to the ionic liquid phase formed by [C₄mim][PF₆]. The distribution coefficient for tryptophan was 76: 96% of amino acid was recovered. Similar results were obtained with a model solution for a fermentation broth [56].

The extraction of penicillin G from an aqueous two-phase system containing ionic liquid, water, and Na[H₂PO₄] showed an extraction yield of 93% with [C₄mim]Cl, and 93.7% with [C₄mim][BF₄]. The yield from fermentation at pH

5–6 was 91.5% [57]. Furthermore, aniline, benzoic acid, 4-toluic acid, and erythromycin were extracted with the help of [C$_4$mim][PF$_6$] [9, 58]. However, some of the examples also suffer from the problem of isolation of the product from the ionic liquid phase, and for its recycling.

## 7.8   ENVIRONMENTAL IMPACT OF IONIC LIQUIDS

If ionic liquids are to be used in biocatalytic synthesis, information about their toxicity must be gathered. Toxicity data are necessary to evaluate survivability of the microorganisms in the fermentation broth, but additionally to allow a risk assessment for the environment. Objects of investigation are, besides the intended or unintended contamination of the environment, the behaviour and deposition of ionic liquids within the environmental cycle of matter. Synthesis of ionic liquids is the first problematic step in evaluation of toxicity. Basically, the very low vapour pressure of ionic liquids has the benefit of producing less air pollution than any organic solvent. Many different solvents are used during the synthesis and recovery of ionic liquids, resulting in atmospheric loading with pollutants and reducing the advantage of low vapour pressure [59]. Investigations concerning toxicity and biodegradability discovered a massive threat for aquatic organism from several ionic liquids [60–62]. The toxicity of imidazolium- and pyridinium-based ionic liquids was found to be dependent on the length of the alkyl chain. In conclusion, the longer the side chain, the higher is the toxicity of an ionic liquid (see Table 7.3). 1-Octyl-3-methylimidazolium bromide, for example, shows toxicity towards *Vibrio fischerii* exceeding the EC$_{50}$ value of toluene and benzene, whereas 1-butyl-3-methylimidazolium shows almost no toxicity at all [63].

It should be kept in mind that, similar to the work with enzymes, impurities may be responsible for the toxicity effect. This might be especially important when reading early studies because producers of ionic liquids had to learn how

**TABLE 7.3   Toxicity of ionic liquids with a culture from *V. fischerii* [63]**

| Ionic Liquid | log $P$ | EC$_{50}$/ppm |
|---|---|---|
| 1-Octyl-3-methylimidazolium bromide | 0.80 | 1.77 |
| 1-Hexyl-3-methylimidazolium bromide | 0.15 | 6.44 |
| Toluene | 2.73 | 31.7 |
| 1-Butyl-3-methylpyridinium dicyanamide | −2.40 | 98.0 |
| Benzene | 2.13 | 108 |
| 1-Butylpyridinium chloride | – | 440 |
| 1-Butyl-3-methylimidazolium chloride | −2.40 | 897 |
| Trichloromethane | 1.97 | 1,199 |
| 1-Methylimidazole | – | 1,218 |
| 1-Butyl-3-methylimidazolium bromide | −2.48 | 2,248 |
| Methanol | −0.74 | 101,068 |

to improve purity and had to install appropriate tools for quality control. Ionic liquids show a particular behaviour that influences their environmental fate, as well as the analysis of ionic liquids. Depending on the concentration and nature of solvent, ionic liquids form aggregates or cluster. Of course, the type of cation and anion influences this as well. The property could be verified by mass spectrometry, conductometry, and simulation [64–66]. These clusters might be able to explain the solubility effect of hydrophilic compounds when present in aqueous phase, especially since addition of salt normally leads to decreasing solubility (salting out).

These disadvantages led to an ecological revision of the handling of ionic liquids, with the result that more and more scientists try to focus their attention on exploring environmental friendly ionic liquids. Successes were achieved in ecological balance by changing to a microwave-based solvent-free reaction control for the synthesis of 1-alkyl-3-methyl imidazolium salts [67]. Regarding the regeneration of ionic liquids from reaction systems, Keskin et al. described methods and results for the extraction of ionic liquids with supercritical $CO_2$ [68]. In order to reduce the toxicity of ionic liquids themselves, the application of ions found in nature or gained from natural substances is beneficial. Several publications report the successful synthesis of ionic liquids in which either cations or anions are represented by amino acids [69, 70]. Additionally there are attempts to increase the biodegradability of ionic liquids [71]. In the area of environmental sustainability, more research has yet to be done, especially with respect to the unknown toxicity of numerous compounds. The Centre for Environmental Research and Technology of the University of Bremen (UFT) created a comprehensive database that provides information, among other things, about the hazardous behaviour of various ionic liquids [72].

Speaking of environmental aspects, it is also essential to turn attention towards the biocatalytic reactions in ionic liquids. Many studies just cover the influence on the environment, but not the fermentation of ionic liquids. Biocompatibility of solvents in a two-phase system is an important factor. It is important to differentiate between molecular toxicity and phase toxicity. If a solvent has a direct impact on the cell or the enzyme, it is defined as molecular toxicity. The solvent can modify cell walls, inhibit enzymes, denaturise proteins, or expand membranes by inclusions. Phase toxicity occurs when fundamental nutrients or substrates, necessary for the function of an organism, are extracted into the solvent phase, leading to an inhibition of cell growth. By forming a solvent layer around the biocatalyst, mass transfer can be reduced. Another possibility of phase toxicity is extraction of essential cell components from the cell, leading to cell death [73, 74].

Prediction of cell toxicity in ionic liquids is rather difficult as well. Some organisms within a species tolerate ionic liquids better than others. *Lactobacillus delbruekii* showed, in the presence of [$C_6$mim][$PF_6$], a relatively high activity of 94% (in relation to the aqueous system), while other *Lactobacillus* species indeed showed activity, but only in range of 4–80% [75]. An interesting concept in whether an organism tolerates a solvent is the critical log $P$ value (log $P_{crit}$). This parameter has to be determined for individual organisms. A log $P$ value higher than the log $P_{crit}$ of the cell indicates a biocompatible solvent for this organism. For instance, a cell with a log $P_{crit}$ of 2 needs a more hydrophilic medium than a cell with a log $P_{crit}$ of 3. This concept could be useful in finding evidence for compatible ionic liquids [76].

But aside from all behaviour of cells or ionic liquids, processes are also responsible for toxicity. Ionic liquids such as [C$_2$mim][BF$_4$] and [C$_4$mim][PF$_6$] show a different behaviour depending on process conditions. If a solid agar plate is used, [C$_4$mim][PF$_6$] will be less toxic than [C$_2$mim][BF$_4$], whereas, in a suspended medium, [C$_4$mim][PF$_6$] is more toxic. An explanation for this phenomenon is based on the diffusion of ionic liquid to the cell cultures. Diffusion of hydrophobic [C$_4$mim][PF$_6$] to the solid agar plate is hindered, but in a well-mixed suspension a large contact area can be established between organism and ionic liquid so that the toxic effect of [C$_4$mim][PF$_6$] becomes more important [77].

As a general rule, the toxicity of ionic liquids in fermentations has a similar effect to that in the environmentally relevant toxicity. Long alkyl chains are more toxic than short chains [78]. Cornmell et al. wanted to know where exactly ionic liquids affect the cell, and used Fourier transform infrared spectroscopy (FT-IR). Cells were exposed to different ionic liquids, lysed, and afterwards separated and analysed. It was experimentally verified that toxic ionic liquids accumulate in the lipid membrane of the cell and cause membrane disruption. The process seems to proceed slower with biocompatible ionic liquids [79].

Closely related to the evaluation of toxicity and environmental behaviour is the abiotic and biotic degradation of ionic liquids. There has been very little work done so far, and published data indicate that, depending on their nature, ionic liquids might not be easily degradable [62, 71]. Studies are again hampered by aggregate formation and the difficulty for exact analysis. For widening industrial applications, these data must be generated by the producers of ionic liquids under the regulations of REACH (**R**egistration, **E**valuation, **A**uthorisation and restriction of **CH**emical substances) [80]. For a final judgment of the beneficial application for ionic liquids in biotechnology, a complete life cycle analysis or eco-efficiency analysis has to be done. BASF has published this analysis for their BASIL process [81]. So far, only limited progress has been made for the use of ionic liquids in biotechnology. This is due to the fact that there are insufficient data available on the early development step in the laboratory. Nevertheless, first attempts have been published [82].

## 7.9 CONCLUDING REMARKS AND FUTURE ASPECTS

In the past 10 years, a large number of ionic liquids and biocatalysts have been tested for applicability in reactions. The partially inconsistent results give an ample scope for future research in that field. Understanding the interaction of

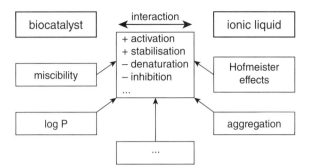

**Figure 7.5**    Interaction between ionic liquids and biocatalyst and possible influence.

ionic liquids with biocatalysts will be the preferred focus in the future (see Figure 7.5). Nevertheless, some advantages of ionic liquids compared with organic solvents can be crystallised from all of these studies:

(i) Ionic liquids are able to dissolve polar and hydrophilic substrates, which are not soluble in organic liquids, such as amino acids and carbohydrates.

(ii) High thermal stability (up to 400 °C) allows higher temperatures for syntheses.

(iii) Improved stability for many enzymes.

(iv) Separation of substrate and products by extraction with organic solvents enables a recycling of ionic liquids that still contains the catalyst.

New approaches for reducing limitations, such as activity loss in ionic liquids, include the immobilisation of enzymes in micro-emulsions in w/IL. Another possibility is preparation of functionalised ionic liquids that are designed to the requirements of the biocatalytic system. An alternative path is screening for new enzyme types and the subsequent change in protein structure in order to find ideal enzymes for a specific ionic liquid. Ilmberger et al. cloned a cellulose, which already showed activity in the wild type, multiple times, and increased enzyme activity [83]. Furthermore, ionic liquids are used to improve existing syntheses selectively, since their positive properties (such as substrate solubility) are very useful. The enhancement of a synthesis for enantiopure alcohols from different racemic phenyl ethanols, for instance, is described in the literature [84]. Ionic liquids could play a central role if their strength is utilised. Datta et al.'s work shows a hyperthermophilic cellulase tolerating up to 15% of $[C_2mim][O_2CCH_3]$ and still retaining high enzymatic activity in Avicel and a natural matrix. This cellulase could now be used for the hydrolysis of cellulose which was extracted with the help of ionic liquids. This way, expensive purification steps in the downstream processing of ionic liquid-dissolved cellulose could be eliminated [85]. A successful evaluation of the impact of the environmental behaviour of ionic liquids will be crucial. But a strong exchange with other disciplines might also lead to a better understanding of their implications for biotechnology. Examples are simple models being able to predict the properties of ionic liquids, such as melting point or distribution behaviour [86, 87].

## REFERENCES

1   Bornscheuer, U. T., and Buchholz, K., Highlights in biocatalysis: Historical landmarks and current trends, *Eng. Life Sci.* **5**, 309–323 (2005).

2   Fischer, E., Nobel Lecture: Syntheses in the purine and sugar group. http://www.nobelprize.org/nobel_prizes/chemistry/laureates/1902/fischer-lecture.html (2012).

3   Alvarez-Macarie, E., and Baratti, J., Short chain flavour ester synthesis by a new esterase from Bacillus licheniformis, *J. Mol. Catal. B: Enzym.* **10**, 377–383 (2000).

4   Goderis, H. L., Ampe, G., Feyten, M. P., Fouwé, B. L., Guffens, W. M., Vancauwenbergh, S. M., and Tobback, P. P., Lipase-catalyzed ester exchange-reactions in organic media with controlled humidity, *Biotechnol. Bioeng.* **30**, 258–266 (1987).

5   Zaks, A., and Klibanov, A. M., Enzyme-catalyzed processes in organic-solvents, *Proc. Natl. Acad. Sci. U. S. A.* **82**, 10, 3192–3196 (1985).

6   Ho, P. Y., Chiou, M. S., and Chao, A. C., Production of L-DOPA by tyrosinase immobilized on modified polystyrene, *Appl. Biochem. Biotechnol.* **111**, 139–152 (2003).

7   Hänsel, R., and Sticher, O., *Pharmakognosie-Phytopharmazie* (Springer Verlag, Berlin, Heidelberg, New York, 2007).

8   Erbeldinger, M., Mesiano, A. J., and Russell, A. J., Enzymatic catalysis of formation of Z-aspartame in ionic liquid: An alternative to enzymatic catalysis in organic solvents, *Biotechnol. Prog.* **16**, 1129–1131 (2000).

9   Cull, S. G., Holbrey, J. D., Vargas-Mora, V., Seddon, K. R., and Lye, G. J., Room-temperature ionic liquids as replacements for organic solvents in multiphase bioprocess operations, *Biotechnol. Bioeng.* **69**, 227–233 (2000).

10   Valivety, R. H., Halling, P. J., and Macrae, A. R., Reaction-rate with suspended lipase catalyst shows similar dependence on water activity in different organic-solvents, *Biochim. Biophys. Acta. Protein. Struct. Mol. Enzymol.* **1118**, 218–222 (1992).

11   Park, S., and Kazlauskas, R. J., Improved preparation and use of room-temperature ionic liquids in lipase-catalyzed enantio- and regioselective acylations, *J. Org. Chem.* **66**, 8395–8401 (2001).

12   Kim, K. W., Song, B., Choi, M. Y., and Kim, M. J., Biocatalysis in ionic liquids: Markedly enhanced enantioselectivity of lipase, *Org. Lett.* **3**, 1507–1509 (2001).

13   Lau, R. M., van Rantwijk, F., Seddon, K. R., and Sheldon, R. A., Lipase-catalyzed reactions in ionic liquids, *Org. Lett.* **2**, 4189–4191 (2000).

14 Schöfer, S. H., Kaftzik, N., Wasserscheid, P., and Kragl, U., Enzyme catalysis in ionic liquids: Lipase catalysed kinetic resolution of 1-phenylethanol with improved enantioselectivity, *Chem. Commun.* **5**, 425–426 (2001).

15 Bélafi-Bakó, K., Dormo, N., Ulbert, O., and Gubicza, L., Application of pervaporation for removal of water produced during enzymatic esterification in ionic liquids, *Desalination* **149**, 267–268 (2002).

16 Lozano, P., De Diego, T., Carrié, D., Vaultier, M., and Iborra, J. L., Over-stabilization of *Candida antarctica* lipase B by ionic liquids in ester synthesis, *Biotechnol. Lett.* **23**, 18, 1529–1533 (2001).

17 Nara, S. J., Harjani, J. R., and Salunkhe, M. M., Lipase-catalysed transesterification in ionic liquids and organic solvents: A comparative study, *Tetrahedron Lett.* **43**, 16, 2979–2982 (2002).

18 Persson, M., and Bornscheuer, U. T., Increased stability of an esterase from Bacillus stearothermophilus in ionic liquids as compared to organic solvents, *J. Mol. Catal. B: Enzym.* **22**, 21–27 (2003).

19 Irimescu, R., and Kato, K., Investigation of ionic liquids as reaction media for enzymatic enantioselective acylation of amines, *J. Mol. Catal. B: Enzym.* **30**, 5–6, 189–194 (2004).

20 Kaftzik, N., Wasserscheid, P., and Kragl, U., Use of ionic liquids to increase the yield and enzyme stability in the beta-galactosidase catalysed synthesis of N-acetyllactosamine, *Org. Process Res. Dev.* **6**, 553–557 (2002).

21 Yang, Z., Liu, X. J., Chen, C., and Halling, P. J., Hofmeister effects on activity and stability of alkaline phosphatase, *Biochim. Biophys. Acta: Proteins Proteomics* **1804**, 821–828 (2010).

22 Hinckley, G., Mozhaev, V. V., Budde, C., and Khmelnitsky, Y. L., Oxidative enzymes possess catalytic activity in systems with ionic liquids, *Biotechnol. Lett.* **24**, 2083–2087 (2002).

23 Okrasa, K., Guibé-Jampel, E., and Therisod, M., Ionic liquids as a new reaction medium for oxidase-peroxidase-catalyzed sulfoxidation, *Tetrahedron: Asymmetry* **14**, 2487–2490 (2003).

24 Eckstein, M., Villela, M., Liese, A., and Kragl, U., Use of an ionic liquid in a two-phase system to improve an alcohol dehydrogenase catalysed reduction, *Chem. Commun.*, 1084–1085 (2004).

25 Lozano, P., de Diego, T., Guegan, J. P., Vaultier, M., and Iborra, J. L., Stabilization of alpha-chymotrypsin by ionic liquids in transesterification reactions, *Biotechnol. Bioeng.*, **75**, 563–569 (2001).

26 Eckstein, M., Daußmann, T., and Kragl, U., Recent developments in NAD(P)H regeneration for enzymatic reductions in one- and two-phase systems, *Biocatal. Biotransform.* **22**, 89–96 (2004).

27 Walker, A. J., and Bruce, N. C., Cofactor-dependent enzyme catalysis in functionalized ionic solvents, *Chem. Commun.* 2570–2571 (2004).

28 Howarth, J., James, P., and Dai, J. F., Immobilized baker's yeast reduction of ketones in an ionic liquid, [bmim]PF$_6$ and water mix, *Tetrahedron Lett.* **42**, 7517–7519 (2001).

29 Itoh, T., Akasaki, E., and Nishimura, Y., Efficient lipase-catalyzed enantioselective acylation under reduced pressure conditions in an ionic liquid solvent system, *Chem. Lett.* **2**, 154–155 (2002).

30 Pfruender, H., Jones, R., and Weuster-Botz, D., Water immiscible ionic liquids as solvents for whole cell biocatalysis, *J. Biotechnol.* **124**, 182–190 (2006).

31 Moniruzzaman, M., Nakashima, K., Kamiya, N., and Goto, M., Recent advances of enzymatic reactions in ionic liquids, *Biochem. Eng. J.* **48**, 295–314 (2010).

32 Moniruzzaman, M., Kamiya, N., Nakashima, K., and Goto, M., Water-in-ionic liquid microemulsions as a new medium for enzymatic reactions, *Green Chem.* **10**, 497–500 (2008).

33 Pavlidis, I. V., Gournis, D., Papadopoulos, G. K., and Stamatis, H., Lipases in water-in-ionic liquid microemulsions: Structural and activity studies, *J. Mol. Catal. B: Enzym.* **60**, 50–56 (2009).

34 Hussain, W., Pollard, D. J., and Lye, G. J., The bioreduction of a beta-tetralone to its corresponding alcohol by the yeast Trichosporon capitatum MY1890 and bacterium Rhodococcus erythropolis MA7213 in a range of ionic liquids, *Biocatal. Biotransform.* **25**, 443–452 (2007).

35 Swatloski, R. P., Spear, S. K., Holbrey, J. D., and Rogers, R. D., Dissolution of cellose with ionic liquids, *J. Am. Chem. Soc.* **124**, 4974–4975 (2002).

36 Kimizuka, N., and Nakashima, T., Spontaneous self-assembly of glycolipid bilayer membranes in sugar-philic ionic liquids and formation of ionogels, *Langmuir* **17**, 6759–6761 (2001).

37 Eckstein, M., Sesing, M., Kragl, U., and Adlercreutz, P., At low water activity alpha-chymotrypsin is more active in an ionic liquid than in non-ionic organic solvents, *Biotechnol. Lett.* **24**, 867–872 (2002).

38 Summers, C. A., and Flowers, R. A., Protein renaturation by the liquid organic salt ethylammonium nitrate, *Protein Sci.* **9**, 2001–2008 (2000).

39 Dupont, J., and Spencer, J., On the noninnocent nature of 1,3-dialkylimidazolium ionic liquids, *Angew. Chem.* **43**, 5296–5297 (2004).

40 Yang, Z., Hofmeister effects: An explanation for the impact of ionic liquids on biocatalysis, *J. Biotechnol.* **144**, 12–22 (2009).

41 Zhao, H., Effect of ions and other compatible solutes on enzyme activity, and its implication for biocatalysis using ionic liquids, *J. Mol. Catal. B: Enzym.* **37**, 16–25 (2005).

42 Constantinescu, D., Weingärtner, H., and Herrmann, C., Protein denaturation by ionic liquids and the Hofmeister series: A case study of aqueous solutions of ribonuclease A, *Angew. Chem.* **46**, 8887–8889 (2007).

43 Zhao, H., Are ionic liquids kosmotropic or chaotropic? An evaluation of available thermodynamic parameters for quantifying the ion kosmotropicity of ionic liquids, *J. Chem. Technol. Biotechnol.* **81**, 877–891 (2006).

44 Batchelor, J. D., Olteanu, A., Tripathy, A., and Pielak, G. J., Impact of protein denaturants and stabilizers on water structure, *J. Am. Chem. Soc.* **126**, 1958–1961 (2004).

45 Bornscheuer, U. T., *Hydrolases in Organic Synthesis: Regio-and Stereoselectivebiotransformations* (Wiley, Weinheim, 2006).

46 Berberich, J. A., Kaar, J. L., and Russell, A. J., Use of salt hydrate pairs to control water activity for enzyme catalysis in ionic liquids, *Biotechnol. Prog.* **19**, 1029–1032 (2003).

47 Garcia, S., Lourenço, N. M. T., Lousa, D., Sequeira, A. F., Mimoso, P., Cabral, J. M. S., Afonso, C. A. M., and Barreiros, S., A comparative study of biocatalysis in non-conventional solvents: Ionic liquids, supercritical fluids and organic media, *Green Chem.* **6**, 466–470 (2004).

48 Sheldon, R. A., Lau, R. M., Sorgedrager, M. J., van Rantwijk, F., and Seddon, K. R., Biocatalysis in ionic liquids, *Green Chem.* **4**, 147–151 (2002).

49 Anthony, J. L., Maginn, E. J., and Brennecke, J. F., Solution thermodynamics of imidazolium-based ionic liquids and water, *J. Phys. Chem. B* **105**, 10942–10949 (2001).

50 Seddon, K. R., Stark, A., and Torres, M. J., Influence of chloride, water, and organic solvents on the physical properties of ionic liquids, *Pure Appl. Chem.* **72**, 2275–2287 (2000).

51 Widegren, J. A., Saurer, E. M., Marsh, K. N., and Magee, J. W., Electrolytic conductivity of four imidazolium-based room-temperature ionic liquids and the effect of a water impurity, *J. Chem. Thermodyn.* **37**, 569–575 (2005).

52 Pfruender, H., Amidjojo, M., Kragl, U., and Weuster-Botz, D., Efficient whole-cell biotransformation in a biphasic ionic liquid/water system, *Angew. Chem.* **43**, 4529–4531 (2004).

53 Lou, W. Y., Zong, M. H., and Smith, T. J., Use of ionic liquids to improve whole-cell biocatalytic asymmetric reduction of acetyltrimethylsilane for efficient synthesis of enantiopure (S)-1-trimethylsilylethanol, *Green Chem.* **8**, 147–155 (2006).

54 Oppermann, S., Stein, F., and Kragl, U., Ionic liquids for two phase systems and their application for purification, extraction and biocatalysis, *Appl. Microbiol. Biotechnol.* **89**, 493–499 (2011).

55 Klasson, K., Ridenour, W., Davison, B., and McFarlane, J., *Direct Capture of Organic Acids from Fermentation Media Using Ionic Liquids,* Report No. ORNL/TM-2004/192, Nuclear Science and Technology Division, Oak Ridge National Laboratory, Oak Ridge, Tennessee (2004).

56 Smirnova, S. V., Torocheshnikova, I. I., Formanovsky, A. A., and Pletnev, I. V., Solvent extraction of amino acids into a room temperature ionic liquid with dicyclohexano-18-crown-6, *Anal. Bioanal. Chem.* **378**, 1369–1375 (2004).

57 Liu, Q. F., Yu, J., Li, W. L., Hu, X. S., Xia, H. S., Liu, H. Z., and Yang, P., Partitioning behavior of penicillin G in aqueous two phase system formed by ionic liquids and phosphate, *Sep. Sci. Technol.* **41**, 2849–2858 (2006).

58 Huddleston, J. G., Willauer, H. D., Swatloski, R. P., Visser, A. E., and Rogers, R. D., Room temperature ionic liquids as novel media for "clean" liquid-liquid extraction, *Chem. Commun.* 1765–1766 (1998).

59 Zhang, Y., Bakshi, B. R., and Demessie, E. S., Life cycle assessment of an ionic liquid versus molecular solvents and their applications, *Environ. Sci. Technol.* **42**, 1724–1730 (2008).

60 Ranke, J., Stolte, S., Störmann, R., Arning, J., and Jastorff, B., Design of sustainable chemical products: The example of ionic liquids, *Chem. Rev.* **107**, 2183–2206 (2007).

61 Ropel, L., Belveze, L. S., Aki, S., Stadtherr, M. A., and Brennecke, J. F., Octanol-water partition coefficients of imidazolium-based ionic liquids, *Green Chem.* **7**, 83–90 (2005).

62  Wells, A. S., and Coombe, V. T., On the freshwater ecotoxicity and biodegradation properties of some common ionic liquids, *Org. Process Res. Dev.* **10**, 794–798 (2006).

63  Docherty, K. M., and Kulpa, C. F., Toxicity and antimicrobial activity of imidazolium and pyridinium ionic liquids, *Green Chem.* **7**, 185–189 (2005).

64  Dorbritz, S., Ruth, W., and Kragl, U., Investigation on aggregate formation of ionic liquids, *Adv. Synth. Catal.* **347**, 1273–1279 (2005).

65  D'Anna, F., Frenna, V., La Marca, S., Noto, R., Pace, V., and Spinelli, D., On the characterization of some [bmim][X]/co-solvent binary mixtures: A multidisciplinary approach by using kinetic, spectrophotometric and conductometric investigations, *Tetrahedron*, **64**, 672–680 (2008).

66  Carper, W. R., Langenwalter, K., Nooruddin, N. S., Kullman, M. J., Gerhard, D., and Wasserscheid, P., Aggregation models of potential cyclical trimethylsulfonium dicyanamide ionic liquid clusters, *J. Phys. Chem. B*, **113**, 2031–2041 (2009).

67  Varma, R. S., and Namboodiri, V. V., An expeditious solvent-free route to ionic liquids using microwaves, *Chem. Commun.* 643–644 (2001).

68  Keskin, S., Kayrak-Talay, D., Akman, U., and Hortacsu, O., A review of ionic liquids towards supercritical fluid applications, *J. Supercrit. Fluids* **43**, 150–180 (2007).

69  Ohno, H., and Fukumoto, K., Amino acid ionic liquids, *Acc. Chem. Res.* **40**, 1122–1129 (2007).

70  Tao, G. H., He, L., Sun, N., and Kou, Y., New generation ionic liquids: Cations derived from amino acids, *Chem. Commun.* 3562–3564 (2005).

71  Gathergood, N., Scammells, P. J., and Garcia, M. T., Biodegradable ionic liquids. Part III. The first readily biodegradable ionic liquids, *Green Chem.* **8**, 156–160 (2006).

72  Stolte, S., and Störmann, R., The UFT/Merck Ionic Liquids Biological Effects Database, http://www.il-eco.uft.uni-bremen.de/ (2011).

73  Leon, R., Fernandes, P., Pinheiro, H. M., and Cabral, J. M. S., Whole-cell biocatalysis in organic media, *Enzyme Microb. Technol.* **23**, 483–500 (1998).

74  Bar, R., Effect of interphase mixing on a water organic-solvent 2-liquid phase microbial system: Ethanol fermentation, *J. Chem. Technol. Biotechnol.* **43**, 49–62 (1988).

75  Matsumoto, M., Mochiduki, K., and Kondo, K., Toxicity of ionic liquids and organic solvents to lactic acid-producing bacteria, *J. Biosci. Bioeng.* **98**, 344–347 (2004).

76  Baumann, M. D., Daugulis, A. J., and Jessop, P. G., Phosphonium ionic liquids for degradation of phenol in a two-phase partitioning bioreactor, *Appl. Microbiol. Biotechnol.* **67**, 131–137 (2005).

77  Ganske, F., and Bornscheuer, U. T., Growth of Escherichia coli, Pichia pastoris and Bacillus cereus in the presence of the ionic liquids [BMIM][BF$_4$] and [BMIM][PF$_6$] and organic solvents, *Biotechnol. Lett.* **28**, 465–469 (2006).

78  Lee, S. M., Chang, W. J., Choi, A. R., and Koo, Y. M., Influence of ionic liquids on the growth of Esherichia coli, *Korean J. Chem. Eng.* **22**, 687–690 (2005).

79  Cornmell, R. J., Winder, C. L., Tiddy, G. J. T., Goodacre, R., and Stephens, G., Accumulation of ionic liquids in Escherichia coli cells, *Green Chem.* **10**, 836–841 (2008).

80  European Commission, REACH, http://ec.europa.eu/enterprise/sectors/chemicals/reach/index_en.htm (2012).

81  Saling, P., Label Eco-Efficiency Analysis: Acid quench with the ionic liquid BASIL®
    http://www.basf.com/group/corporate/en_GB/function/conversions:/publish/
    content/sustainability/eco-efficiency-analysis/images/BASF_Eco-Efficiency
    _Label_Basil_2005.pdf (2005).

82  Kholiq, M. A., and Heinzle, E., Modelling process and lastingness evaluation of a
    complete cell biotransformation in two-phase ionic liquid/water system, *Chem. Ing.
    Tech.* **78**, 307–316 (2006).

83  Ilmberger, N., Pottkamper, J., and Streit, W. R., Enzymes from the metagenome for
    the biocatalysis of ionic fluids, *Chem. Ing. Tech.* **82**, 77–80 (2009).

84  Singh, M., Singh, R. S., and Banerjee, U. C., Enantioselective transesterification of
    racemic phenyl ethanol and its derivatives in organic solvent and ionic liquid using
    Pseudomonas aeruginosa lipase, *Process Biochem.* **45**, 25–29 (2010).

85  Datta, S., Holmes, B., Park, J. I., Chen, Z. W., Dibble, D. C., Hadi, M., Blanch, H.
    W., Simmons, B. A., and Sapra, R., Ionic liquid tolerant hyperthermophilic cellu-
    lases for biomass pretreatment and hydrolysis, *Green Chem.* **12**, 338–345 (2009).

86  Preiss, U., Bulut, S., and Krossing, I., In silico prediction of the melting points of
    ionic liquids from thermodynamic considerations: A case study on 67 salts with a
    melting point range of 337°C, *J. Phys. Chem. B* **114**, 11133–11140 (2010).

87  Krossing, I., and Slattery, J. M., Semi-empirical methods to predict the physical
    properties of ionic liquids: An overview of recent developments, *Z. Phys. Chemie-
    Int. J. Res. Phys. Chem. Chem. Phys.* **220**, 1343–1359 (2006).

88  Shipovskov, S., Gunaratne, H. Q. N., Seddon, K. R., and Stephens, G., Catalytic
    activity of laccases in aqueous solutions of ionic liquids, *Green Chem.* **10**, 806–810
    (2008).

89  T'Avares, A. P. M., Rodriguez, O., and Macedo, E. A., Ionic liquids as alternative
    co-solvents for laccase: Study of enzyme activity and stability, *Biotechnol. Bioeng.*
    **101**, 201–207 (2008).

90  Moniruzzaman, M., Kamiya, N., and Goto, A., Biocatalysis in water-in-ionic liquid
    microemulsions: A case study with horseradish peroxidase, *Langmuir* **25**, 977–982
    (2009).

91  Liu, Y., Wang, M. J., Li, J., Li, Z. Y., He, P., Liu, H. T., and Li, J. H., Highly active
    horseradish peroxidase immobilized in 1-butyl-3-methylimidazolium tetrafluorob-
    orate room-temperature ionic liquid based sol-gel host materials, *Chem. Commun.*
    1778–1780 (2005).

92  Magnuson, D. K., Bodley, J. W., and Evans, D. F., The activity and stability of
    alkaline phosphatase in solutions of water and the fused salt ethylammonium
    nitrate, *J. Solution Chem.* **13**, 583–587 (1984).

93  Li, Q., He, Y. C., Xian, M., Jun, G., Xu, X., Yang, J. M., and Li, L. Z., Improving
    enzymatic hydrolysis of wheat straw using ionic liquid 1-ethyl-3-methyl imidazo-
    lium diethyl phosphate pretreatment, *Bioresour. Technol.* **100**, 3570–3575 (2009).

94  Zhao, H., and Malhotra, S. V., Enzymatic resolution of amino acid esters using ionic
    liquid N-ethyl pyridinium trifluoroacetate, *Biotechnol. Lett.* **24**, 1257–1260 (2002).

95  Kaftzik, N., Neumann, S., Kula, M. R., and Kragl, U., Enzymatic condensation reac-
    tions in ionic liquids, in *Ionic Liquids as Green Solvents: Progress and Prospects*,
    eds. R. D. Rogers and K. R. Seddon, Vol. 856 (American Chemical Society, Wash-
    ington, 2003), pp. 206–211.

96   Lou, W. Y., Xu, R., and Zong, M. H., Hydroxynitrile lyase catalysis in ionic liquid-containing systems, *Biotechnol. Lett.* **27**, 1387–1390 (2005).

97   Gaisberger, R. P., Fechter, M. H., and Griengl, H., The first hydroxynitrile lyase catalysed cyanohydrin formation in ionic liquids, *Tetrahedron-Asymmetry* **15**, 2959–2963 (2004).

98   Kaftzik, N., Kroutil, W., Faber, K., and Kragl, U., Mandelate racemase activity in ionic liquids: Scopes and limitations, *J. Mol. Catal. A: Chem.* **214**, 107–112 (2004).

# 8 Ionicity in Ionic Liquids: Origin of Characteristic Properties of Ionic Liquids

MASAYOSHI WATANABE and HIROYUKI TOKUDA

Department of Chemistry and Biotechnology, Yokohama National University, Yokohama, Japan

## ABSTRACT

Physicochemical properties of 1-ethyl-3-methylimidazolium ionic liquids having $[CF_3CO_2]^-$, $[CH_3SO_3]^-$, $[CH_3CO_2]^-$, $[CF_3SO_3]^-$, $[BF_4]^-$, and $[N(CF_3SO_2)_2]^-$ anions were measured precisely. The ionicity, in other words the dissociativity of the ionic liquids, was estimated from the molar conductivity ratios $(\Lambda_{imp}/\Lambda_{NMR})$, where $\Lambda_{imp}$ is molar conductivity from ionic conductivity measurements and $\Lambda_{NMR}$ is that from ionic diffusivity by pulsed-gradient spin-echo nuclear magnetic resonance (NMR) measurements. The ionicity, $\Lambda_{imp}/\Lambda_{NMR}$, was controlled by the magnitude and balance of Coulombic forces, which were mainly altered by the anionic Lewis basicities, and van der Waals forces. The $\Lambda_{imp}/\Lambda_{NMR}$ values of the ionic liquids ranged from 0.5 to 0.8 and followed the anion-dependent order $[BF_4]^- > [N(CF_3SO_2)_2]^- > [CF_3SO_3]^- > [CF_3CO_2]^- > [CH_3SO_3]^- > [CH_3CO_2]^-$, which is the inverse order of the anionic basicities or the order of acidities of their conjugate acids. Calibration curves for the Lewis basicity–$\Lambda_{imp}/\Lambda_{NMR}$ relationship are presented for the 1-ethyl-3-methylimidazolium, $[C_2mim]^+$, and 1-butyl-3-methylimidazolium, $[C_4mim]^+$, ionic liquids, from which either $\Lambda_{imp}/\Lambda_{NMR}$ or Lewis basicity can be estimated if either of them is known. Viscosities of various $[C_2mim]^+$ and $[C_4mim]^+$ ionic liquids were correlated with the effective ionic concentration $C_{eff}$, which was estimated from the ionicity and molar concentration of these ionic liquids. With increasing $C_{eff}$, the viscosities tended to increase, although at the same $C_{eff}$, $[C_4mim]^+$ ionic liquids showed higher viscosities than $[C_2mim]^+$ ionic liquids because of the enhanced van der Waals forces.

*Ionic Liquids Further UnCOILed: Critical Expert Overviews*, First Edition.
Edited by Natalia V. Plechkova and Kenneth R. Seddon.
© 2014 John Wiley & Sons, Inc. Published 2014 by John Wiley & Sons, Inc.

## 8.1   INTRODUCTION

The growing interest in room temperature ionic liquids is a result of their unusual liquid properties, such as

(1) low vapour pressure and low flammability,
(2) wide liquid temperature range in conjunction with thermal stability,
(3) fast ion transport,
(4) unusual solubility, and
(5) structure-forming properties on the nanoscale.

The designability of ionic liquids by various combinations of available cations and anions, and further by preparing new cations and anions, allows significant flexibility in material design and in the optimisation of the properties of ionic liquids for specific purposes. However, the complexity of the inter-ionic interactions in ionic liquids makes the prediction of their properties difficult and, consequently, choice and design of the cations and anions are not always successful.

It is generally accepted that ionic liquids consist entirely of ions. On the basis of their high ionic conductivity, it can also be speculated that ionic liquids self-dissociate into ions (charged species) even in the absence of molecular solvents. However, the quantitative magnitude of the dissociativity was unknown prior to our studies [1–6]. Our interest lies in self-dissociativity, hereafter referred to as "ionicity," of ionic liquids [1–6] because the self-dissociative nature of ionic liquids plays a definite role in bringing about their characteristic properties. For instance, if the ionicity is low, the ionic liquid appears to have neither low vapour pressure nor high thermal stability, nor high ionic conductivity [6]. We have succeeded in the quantitative estimation of the ionicity of ionic liquids by measuring the ratio ($\Lambda_{imp}/\Lambda_{NMR}$) of the molar conductivity from ionic conductivity measurements ($\Lambda_{imp}$) to the molar conductivity ($\Lambda_{NMR}$) estimated from cationic and anionic diffusivities [1]. The $\Lambda_{NMR}$ value assumes that every diffusing species detected by pulsed-gradient spin-echo (PGSE)–nuclear magnetic resonance (NMR) measurements contributes to the molar conductivity. In contrast, $\Lambda_{imp}$ depends on the migration of charged species in an electric field. The ratio $\Lambda_{imp}/\Lambda_{NMR}$, thus, indicates the proportion of ions (charged species) that contribute to ionic conduction from all the diffusing species in the time- and space-averaged measurements. It has been revealed that, if the ionic charge is transported by the so-called vehicle mechanism [7], the ionicity of ionic liquids becomes lower than unity, which indicates that not all of the diffusive species in the ionic liquids give rise to ionic conduction; that is, ion aggregates or clusters are formed [1–6]. The ionicity is affected by the Lewis basicity of anions [2], and the Lewis acidity of cations [4], as well as van der Waals interactions between constituent ions [3]. The lower the Lewis basicity and acidity and the van der Waals interaction, the higher is the ionicity of the ionic liquids [6]. Typical ionic liquids have been shown to have ionicities of *ca.* 0.5–0.8 [5].

It is known that salts of $[C_2mim]^+$ form a range of ionic liquids with different anions and that the resulting ionic liquids have relatively low viscosity. We have systematically reported the ionicity of 15 different ionic liquids [1–6]; however, reports on $[C_2mim]^+$ ionic liquids are rather limited, that is, $[C_2mim][N(CF_3SO_2)_2]$ [3] and $[C_2mim][BF_4]$ [1]. Here we add $[C_2mim][CF_3SO_3]$, $[C_2mim][CF_3CO_2]$, $[C_2mim][CH_3SO_3]$, and $[C_2mim][CH_3CO_2]$ to the entries of our ionicity study. Fundamental physicochemical properties, particularly transport properties of these ionic liquids, are presented. We also discuss how the ionicity is affected by the ionic structure of ionic liquids, and how the ionicity affects the physicochemical properties of ionic liquids by taking into account the ionicity data that we have collected.

## 8.2  METHODOLOGY

### 8.2.1  Synthesis

The ionic liquids $[C_2mim][CF_3CO_2]$, $[C_2mim][CH_3SO_3]$, and $[C_2mim][CH_3CO_2]$ were synthesised by the $CO_2$-releasing reactions of 1-ethyl-3-methylimidazolium-2-carboxylate with $CF_3CO_2H$, $CH_3SO_3H$, and $CH_3CO_2H$, respectively. 1-Ethyl-3-methylimidazolium-2-carboxylate was prepared according to the previously reported procedure [8]. 1-Ethylimidazole, dimethyl carbonate, and anhydrous methanol were put into a pressure vessel and mixed at 120 °C for 36 h. Methanol was removed from the reaction mixture under reduced pressure to yield a white solid, which was purified by recrystallisation from an ethanenitrile–tetrahydrofuran mixture. Into methanolic solutions of $CF_3CO_2H$, $CH_3SO_3H$, and $CH_3CO_2H$, a methanolic solution of 1-ethyl-3-methylimidazolium-2- carboxylate was added dropwise, stirred for several hours, and dried under a reduced pressure at 60 °C to remove the methanol, yielding $[C_2mim][CF_3CO_2]$, $[C_2mim][CH_3SO_3]$, and $[C_2mim][CH_3CO_2]$, respectively. $[C_2mim][CF_3SO_3]$ was prepared by a metathetic reaction of $[C_2mim]Cl$ and $Na[CF_3SO_3]$. The $[C_2mim][CF_3SO_3]$ was finally dehydrated under high vacuum with heating over 48 h. The $[C_2mim]^+$ ionic liquids obtained were stored in an argon atmosphere glovebox ($[O_2] < 1$ ppm; $[H_2O] < 1$ ppm). The structures of the ionic liquids were identified by $^1H$ and $^{13}C$ NMR spectroscopy and fast atom bombardment mass spectra (FAB-MS). Chloride content in $[C_2mim][CF_3SO_3]$ was maintained at least below the solubility limit of AgCl in water ($1.4\,mg\,l^{-1}$), which was checked by adding an $AgNO_3$ solution. The water content of all of the ionic liquids, as determined by Karl–Fischer titration, was below 40 ppm. All of the data for $[C_2mim][N(CF_3SO_2)_2]$ [3] and $[C_2mim][BF_4]$ [1] were obtained from our previous work.

### 8.2.2  Thermal Analysis

Differential scanning calorimetry (DSC) was carried out under a dinitrogen atmosphere. The samples were tightly sealed in aluminium pans in a dry glovebox. The samples were heated up to 80 °C, and subsequently cooled to −150 °C and reheated, and cooled at a rate of 10 °C·min$^{-1}$, unless otherwise noted. The glass transition temperature ($T_g$) and melting point ($T_m$) were determined from the DSC thermograms during the programmed reheating steps. Thermogravimetric measurements were conducted from room temperature to 550 °C at a heating rate of 10 °C min$^{-1}$ under a dinitrogen atmosphere.

### 8.2.3  Density

The density measurements were performed using a thermoregulated density metre. The measurements were conducted in the range of 15–40 °C.

### 8.2.4  Conductivity

The bulk ionic conductivity was measured by complex impedance measurements. The ionic liquids were introduced into a conductivity cell with platinised platinum electrodes, with a cell constant of *ca.* 1 cm$^{-1}$. The measurement was carried out at controlled temperatures with cooling from +100 to −10 °C.

### 8.2.5  Viscosity

The viscosity measurements were carried out with a cone-plate viscometer under a dinitrogen atmosphere. The temperature was controlled in the range from +80 to −10 °C.

### 8.2.6  Self-Diffusion Coefficients

The PGSE-NMR measurements were conducted according to our previously reported procedure [2]. The sine gradient pulse, providing a gradient strength of up to 12 T m$^{-1}$, was used throughout the measurements in this study. The self-diffusion coefficients were measured using a simple Hahn spin-echo sequence, (i.e., 90°–$\tau$–180°–$\tau$–acquisition), with a gradient pulse incorporated in each $\tau$ period. Measurements of the cationic and anionic self-diffusion coefficients in each ionic liquid were performed for either $^1$H (399.7 MHz) or $^{19}$F (376.1 MHz) nuclei. The self-diffusion coefficients of the cations and anions were calculated using the Stejskal–Tanner equation [9]. The measurements were performed in a range of temperatures with gradual cooling from 80 to −10 °C; samples were allowed to remain in thermal equilibrium at each temperature for 30 minutes prior to the measurements.

### 8.2.7   Solvent Polarity Parameter

The solvatochromic polarity scale of the ionic liquids was investigated using a dye, the copper(II) complex [Cu(acac)(tmen)][BPh$_4$] (Hacac = pentane-2,4-dione, trivially acetylacetone, tmen = $N,N,N',N'$-tetramethylethylenediamine, Me$_2$NCH$_2$CH$_2$NMe$_2$) [10]. It is well known that [Cu(acac)(tmen)][BPh$_4$] gives a good correlation between the donor number of a solvent and the maximum absorption wavelength ($\lambda_{Cu}$) [10]. It was confirmed that the anionic donor ability of ionic liquids strongly affected the $\lambda_{Cu}$ values [5].

## 8.3   PHYSICOCHEMICAL, PROPERTIES OF [C$_2$mim]$^+$-BASED IONIC LIQUIDS

Thermal properties of the [C$_2$mim]$^+$-based ionic liquids are shown in Table 8.1, together with their formula weights. Glass transition temperatures ($T_g$) were observed for all ionic liquids except for [C$_2$mim][N(CF$_3$SO$_2$)$_2$] and [C$_2$mim][CF$_3$SO$_3$], which indicated relatively fast crystallisation kinetics of these two ionic liquids. The $T_g$ values of [C$_2$mim][N(CF$_3$SO$_2$)$_2$] and [C$_2$mim][CF$_3$SO$_3$] could be obtained by means of a rapid temperature change (50 °C min$^{-1}$). The $T_g$ values of the ionic liquids with fluorinated anions ([N(CF$_3$SO$_2$)$_2$]$^-$, [BF$_4$]$^-$, [CF$_3$SO$_3$]$^-$, and [CF$_3$CO$_2$]$^-$) were lower than those with alkyl anions ([CH$_3$CO$_2$]$^-$ and [CH$_3$SO$_3$]$^-$). The $T_m$ values of the ionic liquids with fluorinated anions, except for [C$_2$mim][BF$_4$], were lower than –10 °C. The [BF$_4$]$^-$ anion has lower conformational flexibility than other fluorinated anions, which might explain the high $T_m$ value. The variation of the thermal stability ($T_d$) of the [C$_2$mim]$^+$ ionic liquids was largely dependent on the anionic structures, and followed the order

$$[N(CF_3SO_2)_2]^- > [BF_4]^- > [CF_3SO_3]^- > [CH_3SO_3]^- > [CH_3CO_2]^- > [CF_3CO_2]^-.$$

**TABLE 8.1   Formula Weight (FW) and Thermal Properties of 1-Ethyl-3-Methylimidazolium Ionic Liquids**

|  | FW/g mol$^{-1}$ | $T_g$/°C$^a$ | $T_m$/°C$^a$ | $T_d$/°C$^b$ |
|---|---|---|---|---|
| [C$_2$mim][(CF$_3$SO$_2$)$_2$N] | 391.3 | –87$^c$ | –18 | 439 |
| [C$_2$mim][CH$_3$COO] | 170.2 | –68 |  | 215 |
| [C$_2$mim][BF$_4$] | 198.0 | –89 | 15 | 427 |
| [C$_2$mim][CH$_3$SO$_3$] | 206.3 | –77 | 37 | 343 |
| [C$_2$mim][CF$_3$COO] | 224.2 | –87 | –14 | 177 |
| [C$_2$mim][CF$_3$SO$_3$] | 260.2 | –93$^c$ | –10 | 406 |

$^a$ Onset temperature of heat capacity change ($T_g$) and an endothermic peak ($T_m$) determined by DSC.
$^b$ Onset temperature of mass loss ($T_d$).
$^c$ Detected by rapid cooling.

**TABLE 8.2 Density Equation Parameters from Equation (8.1) and Molar Concentration at 30 °C ($M_{30}$)**

|  | $a/10^{-4}\,g\,cm^{-3}\,K^{-1}$ | $b/g\cdot cm^{-3}$ | $M_{30}/10^{-3}\,mol\,cm^{-3}$ |
|---|---|---|---|
| [$C_2$mim][N(CF$_3$SO$_2$)$_2$] | 10.0 | 1.82 | 3.87 |
| [$C_2$mim][CH$_3$COO] | 6.51 | 1.29 | 6.44 |
| [$C_2$mim][BF$_4$] | 8.40 | 1.53 | 6.44 |
| [$C_2$mim][CH$_3$SO$_3$] | 6.80 | 1.45 | 6.01 |
| [$C_2$mim][CF$_3$COO] | 7.71 | 1.52 | 5.74 |
| [$C_2$mim][CF$_3$SO$_3$] | 8.46 | 1.64 | 5.30 |

This order is consistent with the inverse order of the Lewis basicity of the anions [5]. The exceptionally low thermal stability of [$C_2$mim][CF$_3$CO$_2$] implies thermal instability of the [CF$_3$CO$_2$]$^-$ anion.

The temperature dependence of the density of the ionic liquids showed a linear decrease with increasing temperature, which can be represented by Equation (8.1),

$$\rho = b - aT. \tag{8.1}$$

The best-fit parameters for the linear fitting are listed in Table 8.2. The observed trend indicated that the larger the formula weight of the anions, the higher is the density of the ionic liquids. Ionic liquids with fluorinated anions have higher densities than those with the corresponding non-fluorinated anions. The calculated molar concentration of the ionic liquids at 30 °C ($M_{30}$) increased with a decrease in the formula weight of the anions and became higher than 6.0 M for [$C_2$mim][CH$_3$SO$_3$], [$C_2$mim][BF$_4$], and [$C_2$mim][CH$_3$CO$_2$].

The temperature dependences of the viscosity ($\eta$) and conductivity ($\sigma$) of the ionic liquids are shown in Figure 8.1 and Figure 8.2, respectively. The profiles fitted to the Vogel–Fulcher–Tammann (VFT) equations are also depicted in Figure 8.1 and Figure 8.2, and the corresponding VFT equations are Equation (8.2) and Equation (8.3),

$$\eta = \eta_0 \exp[B/(T - T_0)], \tag{8.2}$$

$$\sigma = \sigma_0 \exp[-B/(T - T_0)], \tag{8.3}$$

where the constants $\eta_0$ (mPa s), $\sigma_0$ (S cm$^{-1}$), $B$ (K), and $T_0$ (K) are also adjustable parameters. The best-fit parameters for the viscosities and conductivities are listed in Table 8.3 and Table 8.4, respectively.

The viscosity values (Figure 8.1) showed a marked dependence on whether or not the ionic liquids comprised fluorinated anions, with the observed trend that the ionic liquids of fluorinated anions had lower viscosity. The conductivity values (Figure 8.2), on the other hand, were affected not only by the viscosity values but also by the molar concentrations (Table 8.2). The ionic

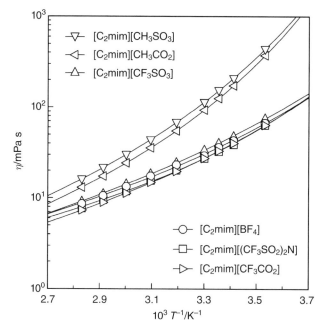

**Figure 8.1**    Temperature dependence of viscosity for [C₂mim]⁺ ionic liquids.

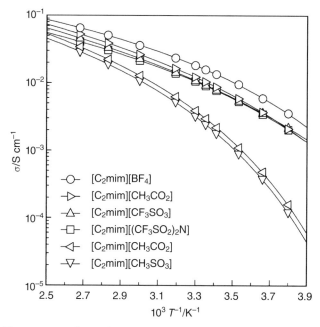

**Figure 8.2**    Temperature dependence of ionic conductivity for [C₂mim]⁺ ionic liquids.

**TABLE 8.3 VFT Parameters of Viscosity Data**

| | $\eta_0/10^{-1}\,\mathrm{mPa\,s}$ | $B/10^2\,\mathrm{K}$ | $T_0/\mathrm{K}$ |
|---|---|---|---|
| $[C_2mim][(CF_3SO_2)_2N]$ | $4.0 \pm 1.3$ | $5.09 \pm 0.81$ | $182 \pm 10$ |
| $[C_2mim][CH_3COO]$ | $1.7 \pm 0.1$ | $6.98 \pm 0.09$ | $192 \pm 1$ |
| $[C_2mim][BF_4]$ | $2.5 \pm 0.9$ | $6.82 \pm 1.00$ | $161 \pm 10$ |
| $[C_2mim][CH_3SO_3]$ | $1.7 \pm 0.1$ | $7.55 \pm 0.20$ | $187 \pm 2$ |
| $[C_2mim][CF_3COO]$ | $1.9 \pm 0.3$ | $6.89 \pm 0.52$ | $165 \pm 5$ |
| $[C_2mim][CF_3SO_3]$ | $1.8 \pm 0.5$ | $7.98 \pm 0.81$ | $151 \pm 8$ |

**TABLE 8.4 VFT Equation Parameters of Ionic Conductivity Data**

| | $\sigma_0/10^{-1}\,\mathrm{S\,cm^{-1}}$ | $B/10^2\,\mathrm{K}$ | $T_0/\mathrm{K}$ |
|---|---|---|---|
| $[C_2mim][(CF_3SO_2)_2N]$ | $5.8 \pm 0.2$ | $5.54 \pm 0.13$ | $165 \pm 2$ |
| $[C_2mim][CH_3COO]$ | $10.1 \pm 0.4$ | $6.07 \pm 0.09$ | $194 \pm 1$ |
| $[C_2mim][BF_4]$ | $6.6 \pm 0.4$ | $4.55 \pm 0.18$ | $176 \pm 3$ |
| $[C_2mim][CH_3SO_3]$ | $8.7 \pm 0.4$ | $6.06 \pm 0.12$ | $195 \pm 1$ |
| $[C_2mim][CF_3COO]$ | $9.8 \pm 0.3$ | $6.12 \pm 0.08$ | $163 \pm 1$ |
| $[C_2mim][CF_3SO_3]$ | $11.0 \pm 0.3$ | $7.16 \pm 0.10$ | $148 \pm 1$ |

liquids with lower viscosity and higher molar concentrations exhibited higher conductivity.

The PGSE-NMR method for the determination of the self-diffusion coefficient allows the evaluation of the diffusivity of NMR-sensitive nuclei. Since the ionic liquids used in this study include NMR-sensitive $^1$H and $^{19}$F (or $^1$H) nuclei in the cation and anion, respectively, each self-diffusion coefficient could be independently determined and was defined as self-diffusion coefficients of the cation ($D_{cation}$) and anion ($D_{anion}$). The temperature dependence of the diffusion coefficients in each case exhibited convex, curved profiles; therefore, experimental data were fitted with the VFT equation for diffusivity, Equation (8.4),

$$D = D_0 \exp[-B/(T-T_0)], \tag{8.4}$$

where the constants $D_0$ ($\mathrm{cm^2\,s^{-1}}$), $B$ (K), and $T_0$ (K) are adjustable parameters. The best-fit parameters of the ionic diffusivity are summarised in Table 8.5. The simple sum of the cationic and anionic self-diffusion coefficients ($D_{cation} + D_{anion}$) for these ionic liquids at each temperature was calculated and the best-fit parameters for the VFT equation were estimated. The results are also listed in Table 8.5. The ($D_{cation} + D_{anion}$) values for the $[C_2mim]^+$ ionic liquids followed the order

$$[BF_4]^- \approx [CF_3CO_2]^- \approx [N(CF_3SO_2)_2]^- > [CF_3SO_3]^- > [CH_3CO_2]^- > [CH_3SO_3]^-.$$

**TABLE 8.5    VFT Equation Parameters of Self-Diffusion Coefficient Data**

|  | $D_0/10^{-4} \text{cm}^2\text{s}^{-1}$ | $B/10^2 \text{K}$ | $T_0/\text{K}$ |
|---|---|---|---|
| [C₂mim][N(CF₃SO₂)₂] | | | |
| Cation | $1.1 \pm 0.1$ | $8.16 \pm 0.41$ | $147 \pm 4$ |
| Anion | $0.9 \pm 0.1$ | $8.57 \pm 0.30$ | $147 \pm 3$ |
| Cation + anion | $2.1 \pm 0.2$ | $8.36 \pm 0.35$ | $146 \pm 3$ |
| [C₂mim][CH₃COO] | | | |
| Cation | $1.6 \pm 0.3$ | $8.88 \pm 0.47$ | $174 \pm 4$ |
| Anion | $2.0 \pm 0.4$ | $9.65 \pm 0.56$ | $170 \pm 4$ |
| Cation + anion | $3.6 \pm 0.6$ | $9.24 \pm 0.50$ | $172 \pm 4$ |
| [C₂mim][BF₄] | | | |
| Cation | $1.2 \pm 0.1$ | $8.56 \pm 0.40$ | $144 \pm 4$ |
| Anion | $1.4 \pm 0.2$ | $9.34 \pm 0.50$ | $141 \pm 4$ |
| Cation + anion | $2.6 \pm 0.3$ | $8.93 \pm 0.42$ | $143 \pm 4$ |
| [C₂mim][CH₃SO₃] | | | |
| Cation | $1.4 \pm 0.4$ | $8.96 \pm 0.85$ | $174 \pm 7$ |
| Anion | $1.1 \pm 0.2$ | $8.59 \pm 0.62$ | $180 \pm 5$ |
| Cation + anion | $2.5 \pm 0.6$ | $8.80 \pm 0.74$ | $177 \pm 6$ |
| [C₂mim][CF₃COO] | | | |
| Cation | $0.9 \pm 0.2$ | $7.25 \pm 0.64$ | $159 \pm 7$ |
| Anion | $0.9 \pm 0.2$ | $7.59 \pm 0.63$ | $159 \pm 6$ |
| Cation + anion | $1.7 \pm 0.4$ | $7.41 \pm 0.62$ | $159 \pm 6$ |
| [C₂mim][CF₃SO₃] | | | |
| Cation | $2.0 \pm 0.4$ | $10.62 \pm 0.71$ | $126 \pm 6$ |
| Anion | $1.9 \pm 0.4$ | $11.36 \pm 0.72$ | $126 \pm 6$ |
| Cation + anion | $3.9 \pm 0.7$ | $10.97 \pm 0.63$ | $125 \pm 5$ |

This order is roughly consistent with the inverse order of the magnitude of the viscosity, which indicates that the translational dynamics of each ion is governed by the macroscopic viscosity.

Since the molar concentration of these ionic liquids is strongly dependent on the formula weight (anionic structures) (Table 8.2), the molar conductivity was calculated from the ionic conductivity and molar concentration. The VFT equation for the molar conductivity is given in Equation (8.5),

$$\Lambda = \Lambda_0 \exp[-B/(T-T_0)], \tag{8.5}$$

where $\Lambda_0$ ($\text{S cm}^2 \text{mol}^{-1}$), $B$ (K), and $T_0$ (K) are constants.

The best-fit parameters for the temperature dependence ($\Lambda_{\text{imp}}$) are shown in Table 8.6. The molar conductivity of the ionic liquids can also be calculated from the self-diffusion coefficients ($\Lambda_{\text{NMR}}$), as determined by the PGSE-NMR measurements, using the Nernst–Einstein equation, Equation (8.6),

$$\Lambda_{\text{NMR}} = N_A e^2 (D_{\text{cation}} + D_{\text{anion}})/kT = F^2(D_{\text{cation}} + D_{\text{anion}})/\text{RT}, \tag{8.6}$$

**TABLE 8.6    VFT Parameters of Molar Conductivity Data Based on Impedance Measurements ($\Lambda_{imp}$)**

|  | $\Lambda_0/10^2\,S\,cm^2\,mol^{-1}$ | $B/10^2\,K$ | $T_0/K$ |
|---|---|---|---|
| $[C_2mim][(CF_3SO_2)_2N]$ | $1.9 \pm 0.1$ | $6.04 \pm 0.15$ | $161 \pm 2$ |
| $[C_2mim][CH_3COO]$ | $1.8 \pm 0.1$ | $6.36 \pm 0.09$ | $192 \pm 1$ |
| $[C_2mim][BF_4]$ | $1.3 \pm 0.1$ | $4.86 \pm 0.18$ | $172 \pm 3$ |
| $[C_2mim][CH_3SO_3]$ | $1.7 \pm 0.1$ | $6.33 \pm 0.12$ | $193 \pm 1$ |
| $[C_2mim][CF_3COO]$ | $2.1 \pm 0.1$ | $6.57 \pm 0.09$ | $160 \pm 1$ |
| $[C_2mim][CF_3SO_3]$ | $2.6 \pm 0.1$ | $7.70 \pm 0.10$ | $144 \pm 1$ |

**TABLE 8.7    VFT Parameter of Molar Conductivity Based on Diffusivity Measurements ($\Lambda_{NMR}$)**

|  | $\Lambda_0/10^2\,S\,cm^2\,mol^{-1}$ | $B/10^2\,K$ | $T_0/K$ |
|---|---|---|---|
| $[C_2mim][N(CF_3SO_2)_2]$ | $3.1 \pm 0.3$ | $6.41 \pm 0.25$ | $159 \pm 3$ |
| $[C_2mim][CH_3COO]$ | $5.6 \pm 0.8$ | $7.60 \pm 0.39$ | $180 \pm 4$ |
| $[C_2mim][BF_4]$ | $3.9 \pm 0.3$ | $6.90 \pm 0.27$ | $155 \pm 3$ |
| $[C_2mim][CH_3SO_3]$ | $4.0 \pm 0.8$ | $7.22 \pm 0.55$ | $185 \pm 5$ |
| $[C_2mim][CF_3COO]$ | $2.8 \pm 0.5$ | $5.78 \pm 0.49$ | $169 \pm 6$ |
| $[C_2mim][CF_3SO_3]$ | $5.3 \pm 0.7$ | $8.51 \pm 0.46$ | $138 \pm 5$ |

where $N_A$ is the Avogadro number, $e$ is the electric charge on each ionic carrier, $k$ is the Boltzmann constant, $F$ is the Faraday constant, and $R$ is the universal gas constant.

The best-fit parameters of the VFT equation for the temperature dependence of the molar conductivity calculated from the ionic diffusion coefficient and Equation (8.6) are listed in Table 8.7.

The data and parameters in Table 8.1, Table 8.2, Table 8.3, Table 8.4, Table 8.5, Table 8.6, and Table 8.7 represent a precise and reliable compilation of the physicochemical properties, particularly transport properties, of these 1-ethyl-3-methylimidazolium ionic liquids.

## 8.4    TRANSFERENCE NUMBER AND IONICITY

Figure 8.3 shows the apparent cationic transference number, $D_{cation}/(D_{cation} + D_{anion})$, for each of the 1-ethyl-3-methylimidazolium ionic liquids, calculated from the PGSE-NMR diffusivity data as a function of temperature. The cationic transference number represents the relative diffusivity of the cation. The transference numbers followed the approximate order

$$[N(CF_3SO_2)_2]^- > [CF_3SO_3]^- > [CH_3SO_3]^- > [CF_3CO_2]^- > [BF_4]^- > [CH_3CO_2]^-,$$

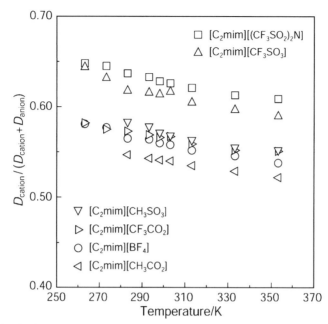

**Figure 8.3** Cationic transference numbers for 1-ethyl-3-methylimidazolium ionic liquids determined from PGSE-NMR diffusivity measurements as a function of temperature.

which clearly indicates that the relative diffusivity of the cations decreases with a decrease in the size of the anions.

However, the cationic self-diffusion coefficient was larger than the anionic diffusion coefficient, even for small anions such as $[BF_4]^-$ and $[CH_3CO_2]^-$. The cationic and anionic radii of $[C_2mim][BF_4]$ have been reported to be 0.303 nm and 0.227 nm, respectively [11, 12]. Nonetheless, the cationic transference number of $[C_2mim][BF_4]$ was estimated to be 0.55–0.58. It should be noted that all of the ionic liquids that we have reported exhibit cationic transference numbers higher than 0.5, irrespective of the ionic sizes [1–4]. There appears to be a certain electrostatic effect that causes faster diffusivity of the cations than of the anions in these ionic liquids. A recent molecular dynamics simulation of the self-diffusion coefficients of 1,3-dialkylimidazolium ionic liquids also demonstrates higher diffusivity of the cations than of the anions [13]. In this study, the cationic transference number decreased almost linearly with increasing temperature in all the ionic liquids, indicating the higher thermal acceleration of the anionic diffusion than the $[C_2mim]^+$ diffusion. This is consistent with the previously observed tendency in the $[C_4mim]^+$ ionic liquids [1–4].

For all the ionic liquids over the entire temperature range considered, the electrochemical molar conductivity ($\Lambda_{imp}$) was lower than the calculated molar conductivity ($\Lambda_{NMR}$; obtained from the PGSE-NMR diffusivity data and

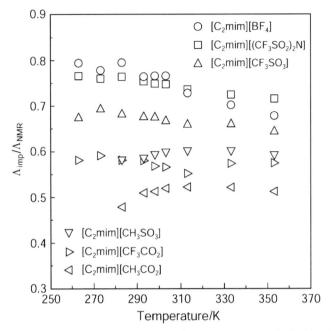

**Figure 8.4** $\Lambda_{imp}/\Lambda_{NMR}$ (ionicity, dissociativity) of 1-ethyl-3-methylimidazolium ionic liquids as a function of temperature.

Equation (8.6)). Thus, the ratio $\Lambda_{imp}/\Lambda_{NMR}$ (ionicity) was lower than unity (Figure 8.4), which is consistent with observations reported in our series of previous studies [1–4], in which the observed ratios were shown to correlate with the ionic nature of the ionic liquids [5, 6]. It is clear that not all of the diffusive species in the ionic liquids lead to ionic conduction; that is, ion aggregates or clusters are formed. The NMR measurements could not distinguish between associated non-charged species and charged ionic species on the NMR timescale. The timescales for any association or dissociation of the ionic species are much shorter than those of dc conductivity measurements. In the conductivity measurements, an individual charged ionic species migrates in an electric field for a characteristic time during which it exists as a charged species, but it may associate with other ions to form a neutral aggregate for another characteristic length of time. When the ions exist in such a neutral form, they do not contribute to conduction.

The variation of the anions in the 1-ethyl-3-methylimidazolium ionic liquids led to the different $\Lambda_{imp}/\Lambda_{NMR}$ values, and the ionicity followed the order

$$[BF_4]^- > [N(CF_3SO_2)_2]^- > [CF_3SO_3]^- > [CF_3CO_2]^- > [CH_3SO_3]^-$$
$$> [CH_3CO_2]^- \text{ at } 30°C,$$

and ranged from 0.5 to 0.8. Since the cationic structure is fixed as $[C_2mim]^+$, the order of $\Lambda_{imp}/\Lambda_{NMR}$ depends entirely on the nature of the anion, which we will discuss in the following section. It was rather surprising for us that even $[C_2mim][CH_3CO_2]$, consisting of the strongly Lewis basic $[CH_3CO_2]^-$ anion, exhibited an ionicity value higher than 0.5.

## 8.5   CORRELATION OF IONICITY WITH IONIC STRUCTURES AND PHYSICOCHEMICAL PROPERTIES

The ionicity of the ionic liquids should be controlled by the types, magnitudes, and balances of the inter-ionic forces, which vary depending on the cationic and anionic structures. Figure 8.5 shows the relationship between $\lambda_{Cu}$ and $\Lambda_{imp}/\Lambda_{NMR}$ for the 1-ethyl-3-methylimidazolium and 1-butyl-3-methylimidazolium ionic liquids with different anionic structures.

It can be seen from Figure 8.5 that the relationships between $\lambda_{Cu}$ and $\Lambda_{imp}/\Lambda_{NMR}$ for $[C_2mim]^+$ and $[C_4mim]^+$ ionic liquids give two different lines and that a decrease in $\lambda_{Cu}$, that is, a decrease in the donor ability of the ionic liquids, induces an increase in $\Lambda_{imp}/\Lambda_{NMR}$. Since the cation was kept constant for each series of ionic liquids, the difference in $\lambda_{Cu}$ is a function of the donor ability of the anions. Insensitivity of $\lambda_{Cu}$ towards the cationic structures was also confirmed by observation of the same $\lambda_{Cu}$ values for the $[C_2mim]^+$ and $[C_4mim]^+$ ionic liquids when the anion was kept constant. It should be noted that when

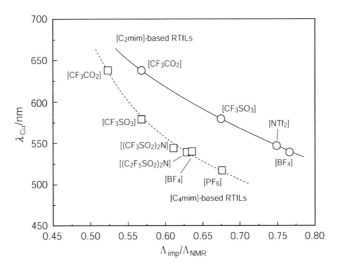

**Figure 8.5**   The relationship between $\Lambda_{Cu}$ and $\Lambda_{imp}/\Lambda_{NMR}$ (ionicity, dissociativity) for $[C_2mim]^+$ and $[C_4mim]^+$ ionic liquids with different anionic structures at 30 °C. The solid and broken lines in the plots are guides for the eyes.

the donor ability of the anion is the same, $[C_4mim]^+$ ionic liquids show lower ionicity values than $[C_2mim]^+$ ionic liquids. We have demonstrated that the ionicity values are controlled not only by the Lewis basicity and acidity of anions and cations, respectively [2, 4], but also by van der Waals interaction [3] between ions, including inductive and dispersion forces. Higher Lewis basicity and acidity and stronger van der Waals interactions result in lower ionicity [6]. The two independent lines in Figure 8.5 are attributed to stronger van der Waals interaction for the 1-butyl-3-methylimidazolium ionic liquids owing to the longer alkyl chain length. Although it was impossible to measure $\lambda_{Cu}$ for $[C_2mim][CH_3CO_2]$ and $[C_2mim][CH_3SO_3]$ because of a shortage of the obtained samples, we could estimate the Lewis basicity (donor ability) of these anions from the $\lambda_{Cu}-\Lambda_{imp}/\Lambda_{NMR}$ relationship for the $[C_2mim]^+$ ionic liquids and the $\Lambda_{imp}/\Lambda_{NMR}$ values of $[C_2mim][CH_3CO_2]$ and $[C_2mim][CH_3SO_3]$. The obtained Lewis basicity order is as follows:

$$[CH_3CO_2]^- > [CF_3CO_2]^- > [CH_3SO_3]^- > [CF_3SO_3]^- > [N(CF_3SO_2)_2]^- > [BF_4]^-.$$

The $pK_a$ values of the conjugated acids of $[CH_3CO_2]^-$, $[CF_3CO_2]^-$, and $[CH_3SO_3]^-$ are 4.8, 0.5, and $-2.0$, respectively, and they confirm the order mentioned earlier. If we use these $\lambda_{Cu}-\Lambda_{imp}/\Lambda_{NMR}$ relationships as calibration curves, we can estimate either $\lambda_{Cu}$ or $\Lambda_{imp}/\Lambda_{NMR}$ when either of them is known. The $\lambda_{Cu}$ values of $[C_2mim][N(CN)_2]$ and $[C_2mim][SCN]$ were measured at 652 nm and 702 nm, respectively, which gave the respective ionicity values of 0.55 and 0.50 from the calibration curve. $[C_2mim][PF_6]$ is a solid at room temperature; however, its ionicity is roughly estimated as 0.8, assuming that it gives the same $\lambda_{Cu}$ as $[C_4mim][PF_6]$.

It is interesting to note that the order of ionicity of the 1-ethyl-3-methylimidazolium ionic liquids corresponds well with the observed trend for $T_d$ (Table 8.1). This correlation indicates that the more dissociative ionic liquids exhibit the higher thermal stability. The decomposition or pyrolysis of the 1,3-dialkylimidazolium salts was reported to proceed via an $S_N2$ mechanism [14, 15], which indicates that an increase in the Lewis basicity (nucleophilicity) of the anions accelerates the $S_N2$ decomposition reactions of the ionic liquids. The observed variation of $T_d$ in response to the change in the nucleophilicity of the anions in this study is therefore understandable and is in agreement with the literature [14, 15].

From the results presented in Figure 8.5, it is clear that the ionicity is affected not only by the Lewis basicity of the anions but also by van der Waals interactions between the ions. It has also been clarified, from *ab initio* calculations [16], that the difference in the former parameter mainly influences the Coulombic interaction between cations and anions. Therefore, simply put, ionicity is determined by a subtle balance between Coulombic and van der Waals interactions between ions. As we proposed earlier [5], values of the effective ionic concentration $C_{eff} = M_{30} \times \Lambda_{imp}/\Lambda_{NMR}$ become a useful indicator of the strength of Coulombic interactions since $(C_{eff})^{-1/3}$ is a measure of

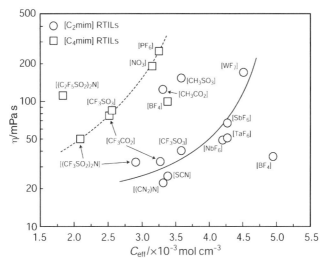

**Figure 8.6** $C_{\text{eff}}$ dependency of viscosity for 1-ethyl-3-methylimidazolium and 1-butyl-3-methylimidazolium ionic liquids with different anionic structures at 30 °C. The solid and broken lines are guides for the eyes.

the average inter-ionic distance. Figure 8.6 shows the $C_{\text{eff}}$ dependence of the $\eta$ values at 30 °C for the 1-ethyl-3-methylimidazolium and 1-butyl-3-methylimidazolium ionic liquids with different anionic structures. $C_{\text{eff}}$ of [C$_4$mim][NO$_3$] was calculated using an estimate of $\Lambda_{\text{imp}}/\Lambda_{\text{NMR}}$ (0.55) and the reported density data at 30 °C (1.1497 g cm$^{-3}$) [17]. The reported data for $\eta$ [17] was then plotted against the calculated value of $C_{\text{eff}}$. For [C$_2$mim][N(CN)$_2$] and [C$_2$mim][SCN], $C_{\text{eff}}$ was similarly estimated from $\Lambda_{\text{imp}}/\Lambda_{\text{NMR}}$ obtained from Figure 8.5 and their densities [18, 19]. The literature value of $\eta$ for [C$_2$mim][SCN] at 25 °C [19] was used in Figure 8.6 because of lack of data at 30 °C. Since conjugated acids of [NbF$_6$]$^-$, [SbF$_6$]$^-$, [TaF$_6$]$^-$, and [WF$_7$]$^-$ are super-strong acids, and their strengths appear to be similar to or stronger than that of HPF$_6$, the $\Lambda_{\text{imp}}/\Lambda_{\text{NMR}}$ values of these ionic liquids were assumed to be 0.8, except for [C$_2$mim][WF$_7$], for which a value of 0.85 was used. The reported $\eta$ values [20] of these ionic liquids were plotted against the estimated $C_{\text{eff}}$ values.

From Figure 8.6, it is apparent that the $\eta$–$C_{\text{eff}}$ relationships for the 1-ethyl-3-methylimidazolium and 1-butyl-3-methylimidazolium ionic liquids are different. The [C$_4$mim]$^+$ ionic liquids exhibited a higher $\eta$ value at constant $C_{\text{eff}}$. Although some amount of scatter of data was observed, the $\eta$ values generally tended to increase with increasing $C_{\text{eff}}$ for each series of the ionic liquids. Since $\eta$ reflects inter-ionic frictions, an increase in the Coulombic or van der Waals interaction induces an increase in $\eta$. If similar van der Waals interactions can be assumed in the 1-ethyl-3-methylimidazolium and 1-butyl-3-methylimidazolium ionic liquids, an increase in $C_{\text{eff}}$ induces an increase in the

Coulombic interaction, thereby resulting in an increase in the $\eta$ values. When the anions were the same ($[N(CF_3SO_2)_2]^-$, $[CF_3CO_2]^-$, and $[CF_3SO_3]^-$), 1-butyl-3-methylimidazolium ionic liquids gave higher $\eta$ values than 1-ethyl-3-methylimidazolium and ionic liquids, which again confirms stronger van der Waals interaction in the 1-butyl-3-methylimidazolium ionic liquids. The $\eta$–$C_{eff}$ relationship for the 1-butyl-3-methylimidazolium ionic liquids showed an increase in $\eta$ values as $C_{eff}$ increased, the exceptions being $[C_4mim][N(C_2F_5SO_2)_2]$ and $[C_4mim][BF_4]$. $[C_4mim][N(C_2F_5SO_2)_2]$ exhibited a high $\eta$ value, despite low $C_{eff}$, which appeared to be because of the enhanced van der Waals interaction originating from $[N(C_2F_5SO_2)_2]^-$, as seen in $[C_nmim][N(CF_3SO_2)_2]$, whose $\eta$ values increase with the alkyl chain length [3]. The exceptionally low $\eta$–value for $[BF_4]^-$ ionic liquids was also observed in the $\eta$–$C_{eff}$ relationship for the 1-ethyl-3-methylimidazolium ionic liquids. It is noteworthy that a change in the anionic structure from $[CF_3CO_2]^-$ to $[CH_3CO_2]^-$ and from $[CF_3SO_3]^-$ to $[CH_3SO_3]^-$ brought about a large increase in the $\eta$ values in spite of the similar $C_{eff}$ values. The perfluorinated anionic structures appeared to have a viscosity-lowering effect.

## 8.6   CONCLUSIONS

The major conclusions of this study can be summarised as follows:

1. By the neutralisation of 1-ethyl-3-methylimidazolium-2-carboxylate with protic acids, which have higher acidity than carbonic acid, ionic liquids having the conjugated bases of the protic acids as counteranions could be easily prepared via $CO_2$-releasing reactions. Special purification procedures were not necessary for the obtained ionic liquids. This method was successfully applied to the preparation of $[C_2mim][CF_3CO_2]$, $[C_2mim][CH_3SO_3]$, and $[C_2mim][CH_3CO_2]$.

2. Physicochemical properties, particularly transport properties, of 1-ethyl-3-methylimidazolium ionic liquids having $[CF_3CO_2]^-$, $[CH_3SO_3]^-$, $[CH_3CO_2]^-$, $[CF_3SO_3]^-$, $[BF_4]^-$, and $[N(CF_3SO_2)_2]^-$ anions were precisely measured. The data for the former four ionic liquids are presented for the first time in this study.

3. The ionicity of the ionic liquids was controlled by the magnitude and balance of Coulombic forces, which, in this study, were mainly altered by the anionic Lewis basicities, and the van der Waals forces. The ionicity, $\Lambda_{imp}/\Lambda_{NMR}$, of the ionic liquids ranged from 0.5 to 0.8 and followed the anion-dependent order $[BF_4]^- > [N(CF_3SO_2)_2]^- > [CF_3SO_3]^- > [CF_3CO_2]^- > [CH_3SO_3]^- > [CH_3CO_2]^-$, which is the inverse order of the anionic donor ability or the order of acidity of the conjugated acids.

4. Calibration curves for the $\lambda_{Cu}$–$\Lambda_{imp}/\Lambda_{NMR}$ relationship were obtained for the 1-ethyl-3-methylimidazolium and 1-butyl-3-methylimidazolium ionic

liquids, from which either $\Lambda_{imp}/\Lambda_{NMR}$ or $\lambda_{Cu}$ can be estimated if either of them is known.

5. Viscosities of various 1-ethyl-3-methylimidazolium and 1-butyl-3-methylimidazolium ionic liquids were correlated with the effective ionic concentration $C_{eff}$, which was estimated from the ionicity and molar concentration of these ionic liquids. With increasing $C_{eff}$, the viscosities tended to increase, although at the same $C_{eff}$ values, 1-butyl-3-methylimidazolium ionic liquids showed higher viscosities than 1-ethyl-3-methylimidazolium ionic liquids because of the enhanced van der Waals force effect.

## ACKNOWLEDGEMENT

This research was supported by a Grant-in-Aid for Scientific Research from the MEXT of Japan in the priority area "Science of Ionic Liquids" (No. 452-17073009) and for basic research (No. B-20350104).

## REFERENCES

1  Noda, A., Hayamizu, K., and Watanabe, M., Pulsed-gradient spin-echo [1]H NMR and [19]F NMR ionic diffusion coefficient, viscosity and ionic conductivity of non-chloroaluminate room temperature ionic liquids, *J. Phys. Chem. B* **105**, 4603–4610 (2001).

2  Tokuda, H., Hayamizu, K., Ishii, K., Susan, M. A. B. H., and Watanabe, M., Physicochemical properties and structures of room temperature ionic liquids. 1. Variation of anionic species, *J. Phys. Chem. B* **108**, 16593–16600 (2004).

3  Tokuda, H., Hayamizu, K., Ishii, K., Susan, M. A. B. H., and Watanabe, M., Physicochemical properties and structures of room temperature ionic liquids. 2. Variation of alkyl chain length of the imidazolium cation, *J. Phys. Chem. B* **109**, 6103–6110 (2005).

4  Tokuda, H., Ishii, K., Susan, M. A. B. H., Tsuzuki, S., Hayamizu, K., and Watanabe, M., Physicochemical properties and structures of room temperature ionic liquids. 3. Variation of cationic species, *J. Phys. Chem. B* **110**, 2833–2839 (2006).

5  Tokuda, H., Tsuzuki, S., Susan, M. A. B. H., Hayamizu, K., and Watanabe, M., How ionic are room temperature ionic liquids? An indicator of the physicochemical properties, *J. Phys. Chem. B* **110**, 19593–19600 (2006).

6  Ueno, K., Tokuda, H., and Watanabe, M., Ionicity in ionic liquids: correlation with ionic structure and physicochemical properties, *Phys. Chem. Chem. Phys.* **12**, 1649–1658 (2010).

7  Noda, A., Susan, M. A. B. H., Kudo, K., Mitsushima, S., Hayamizu, K., and Watanabe, M., Brønsted acid-base ionic liquids as proton conducting non-aqueous electrolytes, *J. Phys. Chem. B* **107**, 4024–4033 (2003).

8  Holbrey, J. D., Reichert, W. M., Tkatchenko, I., Bouajila, E., Walter, O., Tommasi I., and Rogers, R. D., 1,3-Dimethylimidazolium-2-carboxylate: The unexpected

synthesis of an ionic liquid precursor and carbene-$CO_2$ adduct, *Chem. Commun.* 28–29 (2003).

9   Stejskal, E. O., and Tanner, J. E., Spin diffusion measurements: spin echoes in the presence of a time-dependent field gradient, *J. Chem. Phys.* **42**, 288–292 (1965)

10  Soukup, R. W., and Sone, K., (Acetylacetonato)(*N,N,N′,N′*-tetramethylethylene-diamine) copper(II) tetraphenylborate as a solvent basicity indicator, *Bull. Chem. Soc. Jpn.* **60**, 2286–2288 (1987).

11  Ue, M., Mobility and ionic association of lithium and quaternary ammonium salts in propylene carbonate and $\gamma$-butyrolactone, *J. Electrochem. Soc.* **141**, 3336–3342 (1994).

12  Ue, M., Murakami, A., and Nakamura, S., A convenient method to estimate ion size for electrolyte materials design, *J. Electrochem. Soc.* **149**, A1385–A1388 (2002).

13  Tsuzuki, S., Shinoda, W, Saito, H., Mikami, M., Tokuda, H., and Watanabe, M., Molecular dynamic simulations of ionic liquids: cation and anion dependence of self-Diffusion coefficients of ions, *J. Phys. Chem. B.* **113**, 10641–10649 (2009).

14  Awad, W. H., Gilman, J. W., Nyden, M., Harris, R. H., Sutto, T. E., Callahan, J., Trulove, P. C., DeLong, H. C., and Fox, D. M., Thermal degradation studies of alkyl-imidazolium salts and their application in nanocomposites, *Thermochimica Acta* **409**, 3–11 (2004).

15  Fox, D. M., Awad, W. H., Gilman, J. W., Maupin, P. H., DeLong, H. C., and Trulove, P. C., Flammability, thermal stability, and phase change characteristics of several trialkylimidazolium Salts, *Green Chem.* **5**, 724–727 (2003).

16  Tsuzuki, S., Tokuda, H., Hayamizu, K., and Watanabe, M., Magnitude and directionality of interaction in ion pairs of ionic liquids: Relationship with ionic conductivity, *J. Phys. Chem. B* **109**, 16474–16481 (2005).

17  Seddon, K. R., Stark, A., and Torres, M. J., Viscosity and density of 1-alkyl-3-methylimidazolium ionic liquids, in *Clean Solvents: Alternative Media for Chemical Reactions and Processing*, eds. M Abraham and L. Moens, ACS Symp. Ser., Vol. 819 (American Chemical Society, Washington D.C., 2002) pp. 34–49.

18  MacFarlane, D. R., Golding, J., Forsyth, S., Forsyth, M., and Deacon, G. B., Low viscosity ionic liquids based on organic salts of the dicyanaamide anion, *Chem. Commun.* 1430–1431 (2001).

19  McHale, G., Hardacre, C., Ge, R., Doy, N. Allen, R. W. K., MacInnes, J. M., Bown, M. R., and Newton, M. I., Density-viscosity product of small-volume ionic liquid samples using quartz crystal impedance analysis, *Anal. Chem.* **80**, 5806–5811 (2008).

20  Matsumoto, K., Hagiwara, R., Yoshida, R., Ito, Y., Mazej, Z., Benkic, P., Zemva, B., Tamada, O., Yoshino, H., and Matsubara, S., Synthesis, structure and properties of 1-ethyl-3-methyl- imidazolium salts of fluorocomplex anions, *Dalton Trans.* 144–149 (2004).

# 9 Dielectric Properties of Ionic Liquids: Achievements So Far and Challenges Remaining

HERMANN WEINGÄRTNER

Physical Chemistry II, Faculty of Chemistry and Biochemistry,
Ruhr-University Bochum, Bochum, Germany

## ABSTRACT

The static dielectric constant of an ionic liquid is not measurable by conventional methods because the samples are largely short-circuited by their intrinsic electrical conductance. It is, however, possible to determine this quantity by recording the frequency-dependent dielectric dispersion curve in the microwave regime, followed by extrapolation to quasi-static conditions. This review compiles the information on static dielectric constants available from such experiments, and discusses trends in the cation and anion dependence. The results classify most aprotic ionic liquids as moderately polar solvents with dielectric constants of the order of $\varepsilon = 10$–$12$; protic ionic liquids, however, exhibit much higher values.

## 9.1 INTRODUCTION

Many innovative applications of ionic liquids benefit from the possibility of optimising their solvent properties by cation and anion variation [1–3]. Such modifications can dramatically affect the outcome and rate of chemical reactions. An understanding of these phenomena requires the characterisation and understanding of solvation, solvation dynamics, and their effects on reactants and transition states. The frequency-dependent dielectric response and its

*Ionic Liquids Further UnCOILed: Critical Expert Overviews*, First Edition.
Edited by Natalia V. Plechkova and Kenneth R. Seddon.
© 2014 John Wiley & Sons, Inc. Published 2014 by John Wiley & Sons, Inc.

zero-frequency limit—the relative static dielectric permittivity or "static dielectric constant"—reveal important facets of these properties [4, 5].

The dielectric response can be probed by dielectric relaxation spectroscopy (DRS), which measures the response of the dielectric polarisation of a sample to an electric field [6, 7]. In uncharged volatile organic compounds (VOCs), dielectric relaxation is driven by rotational motions of dipolar species, henceforth briefly denoted as "dipolar" processes. In electrically conducting systems, the dielectric response is also affected by translational motions of the charged species, henceforth denoted as "ionic" processes. In the case of complex ions, which possess an electric dipole moment, the dielectric response is partly dipolar and partly ionic.

In the case of VOCs and ionic liquids of low viscosity, a large portion of the dipolar and ionic dynamics driving dielectric relaxation occurs on the nanosecond to sub-picosecond timescale, which corresponds to the microwave and far-infrared (FIR) regions of the dielectric spectrum. While the microwave region is captured by well-established experimental techniques [6], FIR methods are currently only available in a few specialised laboratories.

The range of broadband microwave spectra roughly extends from 1 MHz to several tens of gigahertz. The location of the dielectric modes in this regime usually depends on the viscosity $\eta$ of the sample. Because high viscosities form barriers for most applications, interest usually focusses on low-viscosity ionic liquids [1–3]. For low-viscosity VOCs at ambient conditions, $\eta$ is typically of the order of 1 mPa s (=1 cP), and the slowest component of dipolar dynamics occurs on the timescale of a few picoseconds. Thus, the relevant dielectric modes are usually centred at the upper edge of the microwave regime, say, above 10 GHz. By contrast, under ambient conditions, the lowest viscosities of ionic liquids are of the order of several tens of mPa s [1, 3], which is one to two orders of magnitude higher than the viscosities of simple VOCs. Accordingly, modes in ionic liquids are shifted to lower frequencies.

Although the basic DRS techniques are the same for VOCs and ionic liquids, two features render applications to ionic liquids particularly difficult [4, 5]: first, the direct current (DC) electrical conductivity contributes to the dielectric spectrum and superimposes upon the processes of interest. Second, conducting samples give rise to polarisations at the interface between the liquid and the sample cell, usually called "electrode polarisations." Both effects rapidly increase with decreasing frequency, obscuring low-frequency spectra [4, 5, 8]. Despite these aberrant perturbations, DRS has for a long time provided unique information on the structure and dynamics of conducting electrolyte solutions [5]. We have shown in a pilot study of ethylammonium nitrate [9] that DRS can be beneficially applied to ionic liquids as well.

As a particularly interesting aspect, the low-frequency tail of the dielectric spectrum of an ionic liquid provides experimental access to its static dielectric constant, $\varepsilon$ [10, 11]. For conducting media, standard capacitance methods for determining $\varepsilon$ fail because the ionic liquid short-circuits the sample cell. If a sufficiently large segment of the frequency-dependent dielectric response is covered, $\varepsilon$ can be determined by zero-frequency extrapolation [10].

This overview spotlights the achievements made so far in characterising and understanding dielectric properties of ionic liquids, and on the challenges remaining, with a major focus on studies from the author's laboratory.

## 9.2    A GLANCE AT DIELECTRIC THEORY OF ELECTRICALLY CONDUCTING SYSTEMS

For VOCs, the interpretation of dielectric spectra benefits from experience over many years [7], and the outcomes of experiments usually match quite well with intuition. The dielectric behaviour of ionic liquids is more subtle. To avoid misconceptions, such as the belief that the static dielectric constant of an ionic liquid is infinite [12], a glance at the dielectric theory of electrically conducting systems is mandatory before discussing the dielectric properties of ionic liquids in detail.

DRS measures the interaction of a sample with an oscillating, low-amplitude electric field. The field-induced dielectric polarisation reflects the fluctuation of the total, that is, macroscopic electric dipole moment, $\mathbf{M}(t)$, of the sample (bold quantities denote vectors). The fluctuations of $\mathbf{M}(t)$ can be of different origins:

1. In VOCs, $\mathbf{M}(t)$ fluctuates due to orientational motions of the molecular dipoles, while translational motions do not contribute directly. Note that there are indirect translational contributions because collision-induced and interaction-induced high-frequency contributions will affect the macroscopic electric dipole moment. These features render DRS as a key method for studying the orientational dynamics of VOCs [7, 13]. The results are readily described in terms of a frequency-dependent dielectric permittivity.

2. In simple molten salts, such as NaCl, the individual ions lack a permanent electric dipole moment, so that orientational polarisation is absent. In the case of charged species, translational dynamics, however, affect $\mathbf{M}(t)$, leading to a frequency-dependent electrical conductivity [14]. Moreover, one expects interaction-induced high-frequency contributions to $\mathbf{M}(t)$.

3. Ionic liquids with dipolar ions share properties of VOCs and simple molten salts; that is, relaxation is partly dipolar and partly ionic [15].

Because the dielectric modes lack specificity to these mechanisms, their correct assignment is highly challenging. Molecular dynamics (MD) simulations can assist interpretation, but adequate simulations are still scarce [15–18]. Dielectric theory [6, 7] starts with Maxwell's equations, which have to be supplemented by materials equations, namely the frequency-dependent relative dielectric permittivity, Equation (9.1),

$$\varepsilon^*(v) = \varepsilon_r'(v) - i\varepsilon_r''(v) \; (i^2 = -1),$$    (9.1)

and the frequency-dependent electrical conductivity, Equation (9.2),

$$k^*(v) = \kappa'(v) - i\kappa''(v).$$    (9.2)

The real parts of the two quantities reflect the in-phase response of the sample to the oscillating field; the imaginary parts reflect the out-of-phase response. At finite frequencies, the dipolar and ionic contributions, $\kappa^*(v)$ and $\varepsilon^*(v)$, are coupled and merge to a generalised dielectric function, Equation (9.3),

$$\Sigma^*(v) = \Sigma'(v) - i\Sigma''(v).$$    (9.3)

Experiments probe the propagation of electromagnetic waves in the sample, which can be expressed in terms of $\Sigma^*(v)$. The real part, $\Sigma'(v)$, of the complex dielectric function is denoted as *dielectric dispersion*. In essence, $\Sigma'(v)$ indicates how far the sample polarisation is able to follow the oscillating field. The imaginary part, $\Sigma'(v)$, denoted as *dielectric loss*, reflects the absorption of electromagnetic radiation by the sample. Dispersion and loss are linked through the Kramers–Kronig relation [6, 7]. In principle, therefore, it is possible to obtain the complete information either from the dispersion curve $\Sigma'(v)$ or from the loss signal $\Sigma''(v)$. However, for reasons of experimental accuracy, it is often preferable to synchronously record and analyse both $\Sigma'(v)$ and $\Sigma''(v)$.

At zero-frequency, $\varepsilon^*(0)$ and $\kappa^*(0)$ become decoupled, and are real quantities. The latter property enables the definition of the static dielectric constant in terms of the zero-frequency limit, Equation (9.4),

$$\varepsilon^*(0) = \Sigma'(0) \equiv \varepsilon,$$    (9.4)

of the dispersion curve [6, 7]. The corresponding zero-frequency limit of the conductivity, Equation Equation (9.5),

$$\kappa^*(0) = \kappa'(0) \equiv \kappa_{DC},$$    (9.5)

yields the DC conductivity, which is measurable by standard conductance methods. Because $\kappa_{DC}$ is independently measurable, it is convenient to decompose $\kappa^*(v)$ according to Equation (9.6):

$$\kappa^*(v) = \kappa_{DC} + \kappa_{ex}^*(v).$$    (9.6)

The excess term, $\kappa_{ex}^*(v)$, captures all the frequency-dependent contributions to the conductance. This separation yields the working Equation (9.7) [6]:

$$\Sigma^*(\nu) = \varepsilon'(\nu) - i\varepsilon''(\nu) - i\kappa_{DC}/2\pi\nu\varepsilon_0. \tag{9.7}$$

In the limit $\nu \to 0$, the $\nu^{-1}$-dependent term in Equation (9.7), called *Ohmic loss*, causes the measured imaginary part $\Sigma''(\nu)$ to diverge. $\varepsilon_0$ is the permittivity of the vacuum. The Ohmic loss can be experimentally corrected for because $\kappa_{DC}$ is independently measurable. Alternatively, it can be extracted from the low-frequency branch of $\Sigma''(\nu)$, where only the $\nu^{-1}$-dependent contribution survives.

The non-divergent parts, $\varepsilon'(\nu)$ and $\varepsilon''(\nu)$, in Equation (9.7) are of the form defined in Equations (9.8):

$$\varepsilon'(\nu) = \varepsilon_r'(\nu) - i\kappa_{ex}''(\nu)/2\pi\nu\varepsilon_0 \tag{9.8a}$$

and

$$\varepsilon''(\nu) = \varepsilon_r''(\nu) - i\kappa_{ex}'(\nu)/2\pi\nu\varepsilon_0. \tag{9.8b}$$

Despite the conductance correction in Equation (9.7), the residual terms in Equation (9.8a) and Equation (9.8b) still involve conductance contributions that are related to frequency-dependent deviations from the DC conductance, usually denoted as *conductance dispersion*. To a first approximation, it is often assumed that in the megahertz/gigahertz regime, the conductivity is a frequency-independent constant equal to $\kappa_{DC}$. In the latter case, the real part, $\Sigma'(\nu) \cong \varepsilon'(\nu)$, and the conductance-corrected imaginary part, $\varepsilon''(\nu) \cong \Sigma''(\nu) - \kappa_{DC}/2\pi\nu\varepsilon_0$, involve only dipolar contributions.

It is mandatory to note alternative, but fully interchangeable, representations of the experimental spectra in terms of the complex conductivity, Equation (9.9),

$$\kappa^*(\nu) = i2\pi\nu\varepsilon_0\varepsilon^*(\nu), \tag{9.9}$$

or in terms of the dielectric modulus, defined as the inverse permittivity [19], Equation (9.10),

$$M^*(\nu) = 1/\varepsilon^*(\nu). \tag{9.10}$$

These representations are widely used for describing relaxation in solid and glassy ionic conductors [6, 20]. In the latter case, the permittivity $\varepsilon^*(\nu)$ and, in particular, its real part, $\varepsilon'(\nu)$, are seldom considered, although they can shed light on interesting facets of the data not captured by the alternative representations [21]. The preference of the permittivity formalism in most studies of low-viscosity ionic liquids is founded in the key role ascribed to dipolar, as opposed to ionic, relaxation mechanisms in the microwave regime [4, 5]. In a few cases, the conductance [22, 23] and modulus representations [24] have been applied to ionic liquids as well.

In concluding this theoretical digression, some key features may be pinpointed:

1. By contrast to VOCs on the one hand, and simple molten salts on the other, the dielectric response of an ionic liquid has both dipolar and ionic components.

2. In the limit of $\nu \to 0$, the measured imaginary part, $\Sigma''(\nu)$, of the complex dielectric function shows a divergence founded in the contribution from the DC conductivity. This divergence can be removed by correction for the Ohmic loss, but the corrected spectrum is still affected by conductance dispersion.

3. The real part, $\Sigma'(\nu)$, is not affected by the DC conductivity and is non-divergent. Because the static dielectric constant is defined by the zero-frequency limit of $\Sigma'(\nu)$, it is well defined and finite, while in VOCs only dipolar processes contribute to $\varepsilon$, for a conducting liquid $\varepsilon$ involves ionic contributions.

## 9.3 PHENOMENOLOGICAL DESCRIPTION OF DIELECTRIC SPECTRA OF IONIC LIQUIDS

### 9.3.1 Microwave Spectra

The vast majority of DRS studies concern the microwave regime, where experimental methods are well established and now largely benefit from general progress in measurement technology. Above all, the availability of vectorial network analysers up to several tens of gigahertz facilitates methods such as the coaxial reflection technique, applied by us [4]. Although the experimental methods are the same for VOCs and ionic liquids, the Ohmic loss and electrode polarisation render dielectric studies of conducting ionic liquids challenging.

As a prototypical example for low-viscosity ionic liquids [4, 25–29], Figure 9.1 shows the measured real and imaginary parts of the complex permittivity of [C$_2$mim][OTf] at 25 °C [30]. Visual inspection of $\Sigma''(\nu)$ reveals the $\nu^{-1}$ divergence of the Ohmic loss. Distinct relaxation processes only become visible in the conductance-corrected spectrum, $\varepsilon''(\nu) = \Sigma''(\nu) - \kappa_{DC}/2\pi\nu\varepsilon_0$. At low frequencies, $\varepsilon''(\nu)$ is a very small fraction of $\Sigma''(\nu)$, thus imposing a low-frequency limit for an accurate extraction of $\varepsilon''(\nu)$.

In the simplest case, dielectric relaxation is exponential in the time domain, as for example predicted by the well-known Debye model of diffusive reorientational dynamics. Exponential relaxation is commonly denoted as the "Debye process" [6, 7]. Fourier–Laplace transformation of an exponential relaxation process in the time domain leads to a Lorentzian spectral shape of the corresponding permittivity contribution in the frequency domain, $\Delta\varepsilon^*(\nu)$, which is characterised by the relaxation amplitude $S$ and the Debye relaxation time $\tau_D$, Equation (9.11),

$$\Delta\varepsilon^*(\nu) = S/(1 + i2\pi\nu\tau_D).\tag{9.11}$$

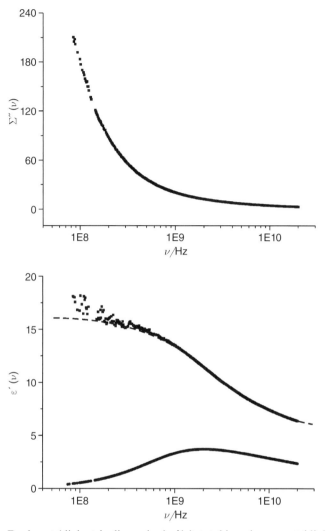

**Figure 9.1**   Real part (dielectric dispersion) $\varepsilon'(\nu)$, total imaginary part (dielectric loss) $\Sigma''(\nu)$, and conductance-corrected imaginary part $\varepsilon''(\nu) = \Sigma''(\nu) - \kappa_{DC}/2\pi\nu\varepsilon_0$ of the frequency-dependent complex dielectric function of $[C_2mim][OTf]$ at $25\,°C$. The increase in $\varepsilon'(\nu)$ at low frequencies relative to the dashed line reflects the onset of electrode polarisation effects.

However, compared with the predictions from Equation (9.11), the decay of $\varepsilon'(\nu)$ and the loss signal $\varepsilon''(\nu)$ are largely broadened. For low-viscosity VOCs, such a broadening is not typical.

Another special feature of dielectric spectra of ionic liquids concerns the existence of pronounced high-frequency processes. For $\nu \to \infty$, the intermolecular dynamics ceases, but dielectric dispersion is still affected by non-relaxing intramolecular contributions, mainly due to ion polarisability. Izgorodina et al. [31] have shown that calculated ion polarisabilities correlate well with the optical refractive index of the sample, $n$, so that in case of ionic liquids, the polarisability contribution does not differ substantially from that of VOCs.

In most VOCs, high-frequency extrapolation of the measured dispersion curves $\varepsilon'(\nu)$ yields a high-frequency limit of $\varepsilon_\infty \cong n^2 \cong 2$, as predicted by the Maxwell relation [7]. By contrast, microwave spectra of ionic liquids usually extrapolate to values well above $n^2$ [4, 9, 25–27]. The difference $\varepsilon_\infty - n^2$ indicates that microwave experiments do not capture the complete intermolecular dynamics, presumably due to processes in the terahertz regime. This conjecture [9] is now well confirmed by experiments in the terahertz regime [32–37]. Taken together, microwave spectra of ionic liquids are rationalised by a broad low-frequency mode, superimposed by weak wings of terahertz processes, and in some cases by weak modes at intermediate frequencies.

In analogy to well-established procedures for describing spectral broadenings in viscous VOCs [6, 7, 13], in initial work on ionic liquids [4, 5] it was tempting to parameterise the low-frequency mode by an asymmetrical distribution of relaxation times, adopting the Cole–Davidson (CD) model [6, 7, 13]. There is, however, mounting evidence [27, 33] that, after proper correction for high-frequency processes, a symmetrical distribution of relaxation times according to the Cole–Cole (CC) model [6, 7, 13] is more apt. The CC function, Equation (9.12),

$$\Delta\varepsilon^*(\nu) = S/\{1 + (i2\pi\nu\tau)^{1-\alpha}\}, \tag{9.12}$$

implies a symmetrical broadening of the Lorentzian signal characterised by an exponent $0 \le \alpha < 1$. In the Debye limit, $\alpha = 0$. Such a spectral shape is mainly suggested by a comparative analysis of dielectric and optical Kerr effect (OKE) spectra [27, 33].

There are other approaches for spectral parameterisation of broadened processes. Non-exponential relaxation in the time domain is often described by the Kohlrausch–Williams–Watts (KWW) stretched exponential function [6, 7, 13]. The KWW function has, for example, been used for representing time-dependent relaxation functions of ionic liquids deduced from MD simulations [16, 38]. Unfortunately, it does not possess an analytical Fourier–Laplace transform, which complicates the analysis of spectra in the frequency domain. For ionic liquids, correlations of the KWW parameters with those of the CC/CD distributions were considered by Schröder and Steinhauser [38].

Some comments on the chosen parameterisation are mandatory:

1. Because both the CD and CC models are empirical, their application is neither founded in theory nor does the spectral shape guide molecular interpretation.
2. The statistical quality of the CC and CD fits does not allow a clear-cut conclusion in favour of one of the models [27]. The rationale is founded in the fact that the wings of high-frequency processes obscure the spectral shapes.
3. The change in parameterisation indicates the importance of low-frequency contributions that are not accounted for by the CD model, which approaches Debye behaviour at the low-frequency side. It is likely that these long-time processes are founded in the micro-heterogeneous structure of the ionic liquids [33].
4. The two models exhibit different asymptotic behaviour as $\nu \to 0$. As an aberrant consequence, some reported dielectric constants need re-evaluation.

### 9.3.2   Terahertz Spectra

Only a few groups have so far studied the sub-picosecond part of the dielectric response. Sub-picosecond dynamics can be probed by conventional FIR spectroscopy [36, 37], but the experimental capabilities now greatly benefit from advances in femtosecond laser pulse technology [39], which permit probing of the femtosecond dielectric response by terahertz time domain spectroscopy (TDS) [32–35].

The terahertz spectra exhibit a wide range of behaviour. As an example, Figure 9.2 shows the conductance-corrected dielectric loss spectrum of ethylammonium nitrate, where a strong microwave process [9] is supplemented by a mode near 1.7 THz [35]. At higher frequencies, FIR spectra identify further processes [35, 37]. The comparatively simple spectrum contrasts, for example, to the spectrum of $[C_4mim][BF_4]$, where many overlapping relaxation processes render the dielectric loss remarkably persistent up to the terahertz regime [5, 32].

## 9.4   MOLECULAR PROCESSES AFFECTING THE DIELECTRIC RESPONSE

### 9.4.1   Dipolar Processes

Depending on the timescale, the dielectric response of an ionic liquid can be driven by dipolar as well as ionic mechanisms. The lack of specificity to these mechanisms renders the assignment of the dielectric modes difficult. Interpretations typically resort to general knowledge about the structure and dynamics

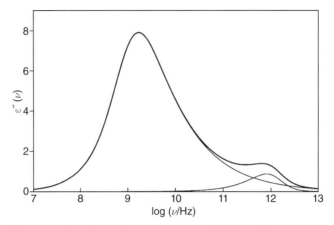

**Figure 9.2** The "complete" conductance-corrected dielectric loss spectrum $\varepsilon''(\nu)$ of ethylammonium nitrate at $25\,°C$ constructed from microwave [9] and terahertz-TDS data [35]. The spectrum is deconvoluted into a broad-end low-frequency process and a high-frequency process. The low-frequency process is fitted by a symmetrical CC relaxation time distribution. The high-frequency process is fitted by a damped harmonic oscillator.

of ionic liquids [1–3], specifically designed MD simulations [15–18], comparison with other spectroscopic data [28, 33, 40], and long-standing experience on dielectric processes in VOCs [7] and electrolyte solutions [5].

On very short timescales, dipolar processes are founded in the librational dynamics of the molecular ions. The spectral assignment of terahertz modes to librational processes is, however, difficult because a multitude of (translational) intermolecular vibrations of the ions in the cage of their neighbours can contribute to the same spectral region. Compared with these intermolecular vibrations, contributions of librational motions to the dielectric spectrum seem less relevant.

There is consensus that the dominant mode in the megahertz/gigahertz regime mainly reflects the co-operative reorientation of dipolar cations [4, 25–29], as confirmed by comparison with magnetic relaxation data [28, 40], OKE spectra [33], and MD simulations [16–18]. At a first glance, this dominant mode resembles the reorientational relaxation of dipolar species in VOCs, but some observations make a strong case against a detailed analogy:

- Comparison of dielectric and OKE spectra [33] indicates that cation reorientation occurs via large-angle jumps.
- Hydrodynamic approaches, which successfully describe dipole reorientation in VOCs, fail to rationalise the observed rotational modes in ionic liquids [4, 27, 41], although they reasonably account for translational diffusion of the ions [42, 43].

- For VOCs, non-exponential relaxation is usually observed in viscous and glassy states, where it signals relaxation in a spatially heterogeneous environment. In ionic liquids, non-exponential relaxation also occurs in low-viscosity systems.

In ionic liquids, non-exponential relaxation seems to reflect spatial heterogeneity resulting from the existence of meso-scale hydrophilic and hydrophobic domains in the ionic liquid structure [1–3]. The existence of such domains is key to the highly amphiphilic properties of ionic liquids, which can, for example, incorporate apolar molecules in hydrophobic domains and polar molecules in hydrophilic domains. By variation of the hydrophobic side chains of 1,3-dialkylimidazolium ionic liquids, Mizoshiri et al. [29] have estimated the local dielectric constant of hydrophilic domains in aprotic ionic liquids to be of the order of $\varepsilon = 20$, and that of hydrophobic domains to be of the order of $\varepsilon = 2.5$.

It is an open question how far these heterogeneous structures are related to the observed jump reorientation and to the failure of hydrodynamic approaches for ion reorientation. In low-viscosity VOCs, the low-frequency modes normally reflect small-angle diffusive reorientation and are well described by hydrodynamic approaches.

The failure of hydrodynamic models was noted in several papers [4, 26, 27] and recently studied in detail for ionic liquids containing the tetra(hexafluoroisopropoxy)aluminate(III) ($[\mathrm{Al(hfip)_4}]^-$) anion [41]. Because the viscosities of $[\mathrm{Al(hfip)_4}]^-$ ionic liquids are low, and cation modification does not substantially change the viscosity [44], this novel class of ionic liquids provides excellent candidates for testing hydrodynamic theories.

Hydrodynamics links the single-particle reorientation time $\tau_{\mathrm{rot}}$ of a particle at temperature $T$ to the bulk viscosity $\eta$ of the surrounding medium, resulting in the well-known Stokes–Einstein–Debye (SED) equation, and its descendants [45]:

$$\tau_{\mathrm{rot}} = 3V_{\mathrm{eff}}\eta f/k_{\mathrm{B}}T, \tag{9.13}$$

where $V_{\mathrm{eff}}$ is the effective ("hydrodynamic") volume of the rotating particle, $k_{\mathrm{B}}$ is the Boltzmann constant, and $f$ is a coupling factor that accounts for the hydrodynamic boundary conditions at the surface of the rotating particle. The effective volumes, $V_{\mathrm{eff}}$, obtained from the dielectric relaxation times (after correction for co-operative effects) are at least two orders of magnitude smaller than the geometric volumes of the cations [41]. NMR relaxation data [40] show that this failure is founded in the rotational dynamics of the cations, and not special to dielectric relaxation. Instead of discussing this effect in terms of the anomalously small volume, it could also be seen as a local viscosity that is much smaller than that of the bulk. The results imply an almost frictionless cation reorientation, which is consistent with the meso-structure, and may be related to ion reorientation by large-angle jumps.

Spectral contributions by dipolar anions are far more difficult to assess. Halide ions and anions of high molecular symmetry, such as $[BF_4]^-$ or $[PF_6]^-$, are dielectrically inactive. Other dipolar anions result in unexpectedly weak dielectric modes, if these are detectable at all [4, 5]. An example is $[C_2mim]$ [OTf], where a weak mode at the upper edge of the spectrum in Figure 9.1 accounts, at least in part, for the asymmetric shape of the loss curve $\varepsilon''(\nu)$ [30].

Notable exceptions are alkyl sulfates such as $[C_2mim][C_2H_5SO_4]$, which exhibit unusually high static dielectric constants [11]. Spectral analysis shows that these are caused by a strong mode near 10 GHz, which is not present in other ionic liquids [46]. While it seems inescapable to attribute this mode to anion reorientation, the fast relaxation and the low amplitudes of anion contributions in other ionic liquids are not clear.

Other candidates for dipolar relaxation are long-lived ion pairs (and higher dipolar ion clusters). Ion pairing in ionic liquids is a much debated issue [1–3]. By analysis of the conductance–ion diffusion relationship, Watanabe and coworkers [42, 43] have concluded that a notable fraction of ions contribute to mass transport, but not to charge transport, as is expected for cations and anions forming neutral ion pairs

In dielectric spectra, ion pairs should reveal themselves by intense modes because oppositely charged ions at contact give rise to very high electric dipole moments. Ion-pair modes are well documented for salts in solvents of low and moderate polarity, and make DRS a powerful tool for ion-pair spectroscopy [5, 47, 48]. By contrast, there is no evidence for such modes in neat ionic liquids [4, 5]. Obviously, cation–anion configurations, which cause the observed reduction in the conductance of neat ionic liquids, do not survive on the timescale of molecular reorientation, signalling lifetimes of less than $\sim$10 ps. Ion-pair modes have, however, indeed been identified on short timescales in the terahertz regime [36]. There is also no convincing evidence for contributions of larger ion clusters. Note that such aggregates are dielectrically invisible, if the ions are involved in symmetric clusters with low or vanishing total electric dipole moments.

### 9.4.2  Ionic Processes

Ionic contributions to the dielectric response may result from ion migration, as reflected by the frequency-dependent electrical conductivity $\kappa^*(\nu)$. As expanded upon in detail in the theoretical section (Section 9.2), it is convenient to separate $\kappa^*(\nu)$ into the Ohmic loss due to the DC conductivity and residual effects due to conductance dispersion. The latter phenomenon is little understood. Its relevance depends on the time/frequency range under consideration:

- In the terahertz regime, translational processes due to intermolecular vibrations of ions in the cage of their neighbours are well established.

- In the microwave (megahertz/gigahertz) region, experiments and simulations do not decisively indicate conductance dispersion. If present, the measured dielectric loss will capture these ionic contributions.
- In the kilohertz regime and below, electrode polarisation has so far prevented any decisive experiments.

The relevance of intermolecular vibrations for dielectric spectra is well established by terahertz-TDS/FIR spectroscopy. An illustrative example is the mode at 1.7 THz in Figure 9.2, which was assigned to an intermolecular bending vibration in the hydrogen-bonded network of [EtNH$_3$][NO$_3$] [35, 37]. More complex spectra, for example, of [C$_4$mim][BF$_4$], obviously reflect a superposition of many vibrational modes [5, 33]. It seems that in the terahertz region, dielectric spectra are highly specific to the ionic liquids, presumably due to the sensitivity of the intermolecular vibrations to local ionic liquid structures. The presence of these translational modes in the terahertz regime is confirmed by computation [15, 17, 18, 36, 38]. In VOCs, intermolecular vibrations are dielectrically inactive, except for interaction-induced contributions in the terahertz regime, which can be significant [7].

As noted earlier, relaxation in the microwave regime should mainly reflect dipolar reorientation of the ions. It may, however, be speculated that the broad modes also cover conductance dispersion. Nevertheless, it seems that to a first approximation, the conductivity in the megahertz/gigahertz regime can be treated as a frequency-independent constant equal to the DC conductivity. The same conclusion can be drawn from simulated dielectric spectra of ionic liquids [16]. For molten NaCl [17], where the dielectric response is purely ionic and conductance dispersion is not superimposed by processes due to dipolar ions, MD simulations by Song [17] also do not signal conductance dispersion in the microwave regime [17].

Conductance dispersion in the megahertz/gigahertz or sub-megahertz region may be expected on grounds of the retarded response of the ionic environment to the motion of a central ion. This effect, occasionally denoted as "space charge polarisation" [12], resembles the *ion cloud relaxation* predicted in the case of dilute electrolyte solutions by Debye and Falkenhagen (DF), as early as in 1928 [49]. Meanwhile, DF theory has been extended to concentrated solutions [50, 51]. Debye and Falkenhagen have conjectured that, in an electric field, the Debye–Hückel (DH) ion cloud cannot follow immediately a moving ion and becomes asymmetric, which retards the motion of the ion. In an oscillating electric field, this effect depends on frequency.

Ion cloud relaxation is experimentally well established for solutions of multiply charged macro-ions, such as polyelectrolytes like DNA, where it gives rise to a strong mode in the kilohertz region [52, 53]. Ion cloud relaxation of much lower magnitude—and located in the megahertz regime ($\sim$25 MHz)—has been observed in the case of ionic micelles [54, 55]. In electrolyte solutions, such as NaCl–H$_2$O, ion cloud relaxation is theoretically expected in the megahertz/gigahertz regime, but it is fair to say that the reported evidence is very weak [56–58], presumably due to very low amplitudes in systems of low nett charge.

The predictive power of these results for neat ionic liquids is very limited. The concept of a DH-type ion cloud implies a monotonously decreasing charge density of the counterions around a central ion, as opposed to the oscillating charge density of charge-ordered layers of cations and anions in ionic liquids [3]. There is a DF-type theory for dielectric processes in solid and glassy ion conductors [59] that successfully rationalises conductance dispersion in terms of the structural relaxation of the environment after the jump of an ion to a different position [60]. It does not seem possible to infer from this theory the magnitude and timescale of conductance processes in the fluid regime of ionic liquids of low viscosity.

## 9.5   RELATION TO SOLVATION DYNAMICS

The characterisation and understanding of the dielectric response are crucial for an understanding of the solvation dynamics of ionic liquids, which plays a key role in reactions involving charge rearrangements. Solvation dynamics can be probed by solvation spectroscopy (time-resolved fluorescence spectroscopy), which uses solvatochromic dyes for observing the solvent reorganisation after excitation of the probe's dipole moment by photons [18,61]. Solvation dynamics should be closely related to the dielectric response.

Maroncelli's group has used dielectric continuum theory for computing solvation spectra of ionic liquids from dielectric spectra. By contrast to successful calculations for VOCs, such models have failed for ionic liquids [62]. This shortcoming may be due to several reasons, such as the lack of knowledge on the high-frequency (terahertz) portion of the spectra, limitations of continuum dielectric theory, or a different relevance of translational versus rotational dynamics for the dielectric and solvation responses. Song [17] has recently proposed a DH-type continuum model for ion concentrations up to neat ionic liquids, which offers the prospect of a better understanding of the role of the dielectric response in charge-controlled chemical reactions.

## 9.6   THE STATIC DIELECTRIC CONSTANT OF IONIC LIQUIDS

Dielectric relaxation spectroscopy is especially useful for determining static dielectric constants of conducting liquids [10, 11]. For VOCs, the prominent role of $\varepsilon$ for assessing and modelling solvent properties is well recognised. The definition of $\varepsilon$ in terms of the zero-frequency limit of the frequency-dependent dielectric dispersion curve yields a recipe for determining $\varepsilon$ of conducting liquids, where conventional capacitance methods fail. Because $\varepsilon$ is the sum of

**TABLE 9.1   Reported Experimental Static Permittivities of [C₄mim][BF₄]**

| Method | Frequency | $\varepsilon$ | Reference |
|--------|-----------|---------------|-----------|
| DRS | 200 MHz–20 GHz | 14.0 (11.7) | [10] |
| DRS | 200 MHz–89 GHz | 14.6 (12.2) | [31][a] |
| DRS | 1 MHz–20 GHz | 14.1 | [28] |
| Terahertz-TDS | 100–1200 GHz | 6.7 | [37] |
| Waveguide | 2.45 GHz | 8.7 | [62] |
| Voltammetry | 50 kHz | 68.89 | [63] |
| Capacitance | 1–3 MHz | 1600 | [64] |

[a] Recalculated from Reference 16.

the amplitudes of all contributing dielectric modes, in contrast to most other polarity probes, one can trace the mechanisms of the underlying processes. Such procedures have been exploited for more than 50 years for characterising static dielectric constants of electrolyte solutions, as, for example, documented in the 1973 monograph by Hasted [63].

For obvious reasons, static dielectric constants of ionic liquids measured by DRS are at least by an order of magnitude less accurate than data for VOCs obtained by conventional methods. Table 9.1 compares $\varepsilon$ values for [C₄mim] [BF₄] at, or near, 25 °C. To obtain a common basis, DRS results in earlier studies have been re-extrapolated in terms of the CC model. On this common basis, the results of three independent DRS experiments [10, 28, 32] agree to within 4%. By comparison, extrapolation of $\varepsilon$ from terahertz-TDS data [34] does not capture the complete dielectric response. A single-point experiment at 2.45 GHz [64] does not reflect static conditions. Voltammetry [65] and capacitance measurements [66] have also yielded unreliable results.

As a matter of concern, in DRS experiments, "static" refers to processes on the nanosecond timescale. Even in the most favourable case, where the dispersion curve was not perturbed by electrode polarisation down to 3 MHz [9], static conditions only refer to processes on a timescale shorter than 100 ns. While the absence of slower dynamics seems plausible for rotational motions, low-frequency translational contributions due to conductance dispersion cannot be excluded *a priori*.

The challenge is to discriminate electrode polarisation effects from sample-specific low-frequency conductance dispersion. Electrode polarisation is apparatus specific and can be identified (and sometimes reduced) by choosing a suitable geometry and material of the sample cell. While in some experiments electrode polarisation was observed at 100 MHz or even beyond (see also Figure 9.1) [4, 5], experiments by Nakamura and Shikata imply a lower bound for meaningful spectra of 20 MHz [28]. Input impedance measurements with a special sample cell enabled us to record the dielectric spectrum of [EtNH₃] [NO₃] to 3 MHz without interference from electrode polarisation [9]. These experiments do not provide evidence that processes in low-viscosity ionic liquids at the lower edge of the microwave regime are overlooked in experiments above 100 MHz.

The situation is different in the kilohertz region and below, where electrode polarisation rapidly increases with decreasing frequency. For example, in impedance measurements extending below the kilohertz regime, electrode polarisation has exceeded the sample-specific modes by up to five orders of magnitude [8]. The resulting dramatic increase in the dispersion curve should not be confused [12] with generic dielectric processes in ionic liquids. Perhaps temperature variation, as performed by Kremer's group [22, 23], can create conditions where conductance effects can be distinguished from systematic and spurious effects due to electrode polarisation.

Disregarding the speculations about low-frequency ionic processes, Table 9.2, Table 9.3, and Table 9.4 summarise static dielectric constants extrapolated by the CC model [67], superseding values published in earlier work [11]. Typically, the corrected values exceed earlier results extracted by the CD model [4, 10, 11, 26] by 10%, which does not alter the basic physical conclusions.

Table 9.2 and Table 9.3 show that, for some widely used aprotic ionic liquids with weakly polar or non-polar anions such as $[NTf_2]^-$ or $[BF_4]^-$, the static dielectric constants at 25 °C exhibit moderate values of $\varepsilon = 12-18$. In these cases, the main contribution to $\varepsilon$ results from the dipolar cations. Ionic liquids with higher dielectric constants can be designed by using highly polar anions, such as $[C_2H_5SO_4]^-$ (Table 9.3) [11, 46].

For protic ionic liquids based on monoalkylammonium, $[RNH_3]^+$, ions, $\varepsilon$ shows moderate increases relative to aprotic ionic liquids. On the other hand, hydroxy-functionalisation of alkyl chains can lead to dramatic increases in $\varepsilon$, as seen by comparison of results for ethylammonium ionic liquids with results for 2-hydroxyethylammonium ionic liquids in Table 9.4. The high static dielec-

**TABLE 9.2    Cation Dependence of the Static Dielectric Constants of Aprotic Ionic Liquids at 25 °C, Exemplified by Salts with Bis{(trifluoromethyl)sulfonyl}amide ($[NTf_2]^-$) as a Common Anion**

| Cation | $\varepsilon$ |
|---|---|
| $[C_2mim]^+$ | 12.0,[a] 12.3[b] |
| $[C_3mim]^+$ | 13.3[a] |
| $[C_4mim]^+$ | 14.0,[a] 13.7[b] |
| $[C_5mim]^+$ | 15.0[a] |
| $[C_4py]^+$ | 15.2[a] |
| $[C_4mpyr]^+$ | 14.7[a] |

[a] Reference 30, data recalculated from spectra in Reference 4.
[b] Reference 28.

**TABLE 9.3   Anion Dependence of the Static Dielectric Constants of Aprotic Ionic Liquids at 25 °C, Exemplified by Salts With [C$_2$mim]$^+$ as a Common Cation**

| Anion | $\varepsilon$ |
|---|---|
| [N(CN)$_2$]$^-$ | 11.7[a] |
| [NTf$_2$]$^-$ | 12.0,[b] 12.3[c] |
| [BF$_4$]$^-$ | 14.5,[a] 13.6[c] |
| [SCN]$^-$ | 13.7[b] |
| [OTf]$^-$ | 16.5[b] |
| [C$_4$H$_9$SO$_3$]$^-$ | 30.0[b] |
| [C$_2$H$_5$SO$_3$]$^-$ | 35.0,[b] 35.5[d] |

[a] Reference 27.
[b] Reference 30.
[c] Reference 28.
[d] Reference 46.

**TABLE 9.4   Static Dielectric Constants of Some Protic Ionic Liquids at 25 °C [30]**

| Ionic liquid | $\varepsilon$ |
|---|---|
| [C$_2$H$_5$NH$_3$][NO$_3$] | 26.2 |
| [C$_2$H$_5$NH$_3$][HCO$_2$] | 31.5 |
| [C$_4$H$_9$NH$_3$][HCO$_2$] | 23.0 |
| [HOC$_2$H$_4$NH$_3$][NO$_3$] | 60.9 |
| [HOC$_2$H$_4$NH$_3$][HCO$_2$] | 61.0 |
| [HOC$_2$H$_4$NH$_3$][lac] | 85.6 |

Reference 30; part of the data supersede values quoted in Reference 68.

tric constants of 2-hydroxyethylammonium ionic liquids resemble the high values observed for water or strongly hydrogen-bonded VOCs, such as amides, where dipole correlations cause a pronounced increase in $\varepsilon$ [7].

The dielectric constant is an important measure of the solvation capability, which may be compared with results deduced from other polarity probes. Well-known polarity probes are, for example, UV-Vis and fluorescence spectra of solvatochromic dyes, liquid–liquid distribution coefficients, inverse gas chromatography, and solvent effects on certain chemical reactions [69]. These probes reflect different facets of the solute–solvent interactions. The resulting empirical or semi-empirical polarity scales range from simple one-parameter approaches to multi-parameter representations, for example, based on linear free energy relationships [3, 11, 69, 70].

Many polarity parameters for ionic liquids have been transcribed into effective dielectric constants using correlations between $\varepsilon$ and other measures of the polarity, which were established from data for VOCs. When the first experimental values of the static dielectric constant of aprotic ionic liquids became available [10], we immediately noted that these were lower than estimated from most other polarity probes. The experimental values of the order of $\varepsilon = 12-18$ correspond to dielectric constants of alcohols of intermediate chain length such as pentanol. Most other probes suggest values for ionic liquids of the order of $\varepsilon = 30-40$, resembling those of polar VOCs such as methanol, ethanenitrile, and dimethyl sulfoxide [11]. The difference results from the fact that the static dielectric constant is a bulk property, while other polarity parameters reflect local probes, often purposely designed to map specific interactions such as the hydrogen-bonding ability of the cation or anion. Obviously, the dielectric constants do not correlate with these parameters.

One of the challenges is to relate the static dielectric constant to the molecular dipole moments of the ions. The static dielectric constant, or more accurately the sum of the amplitudes of the dipolar contributions, is related to the mean square fluctuation $\langle M^2 \rangle$ of the total electric dipole moment $M(t)$ of the sample. $\langle M^2 \rangle$ can be broken down into an effective molecular dipole moment. Assuming, for simplicity, that only the cation contributes to the static dielectric constant, as encountered in many cases, $\langle M^2 \rangle$ can be transcribed into an effective molecular dipole moment $\mu_{eff}$ by the Kirkwood formula, Equation (9.14) [7],

$$\langle M^2 \rangle = N\mu_{eff}^2 = Ng_K\mu_0^2, \tag{9.14}$$

where $\mu_0$ is the dipole moment of the isolated molecule and the Kirkwood factor $g_K$ reflects orientational correlations of the dipoles. For VOCs, this type of analysis has been performed in numerous cases [7].

Unfortunately, such procedures cannot be applied to ionic liquids in a straightforward manner because the dipole moment of a charged particle is ill defined [71]. It can be shown that the static dielectric constant and the mean squared total electric dipole moment $\langle M^2 \rangle$ are invariant against the origin of the atomic coordinates, but the molecular dipole moment $\mu_0$ and the Kirkwood factor $g_K$ depend on the origin chosen [15]. It is therefore only possible to calculate from $\varepsilon$ an unambiguous value of the effective dipole moment. If one applies standard procedures devised for VOCs, then one finds for aprotic ionic liquids values of the order of $\mu_{eff} = 3-6\,D$ ($1\,D = 3.3 \times 10^{-30}\,Cm$) [27], which corresponds to highly dipolar VOCs. At a first glance, this seems to contradict the observed low dielectric constants, but the low $\varepsilon$ values just reflect the low dipole densities caused by the large molar volumes of the bulky ions.

Again, there is no clear correlation of $\mu_{eff}$ with reported polarity parameters. This finding is not surprising for those polarity probes that map specific interactions such as the hydrogen-bonding (donating or accepting) ability of the ions. However, there is also no clear correlation of $\mu_{eff}$ with values

[70] for the semi-empirical Kamlet–Taft $\pi^*$ parameter [72], which is said to reflect the combined effect of electrostatic interactions and electronic polarisability.

Finally, we mention yet another difference between ionic liquids and VOCs. While, for VOCs, dielectric continuum theory provides a surprisingly good representation of their solvation properties, it is often inadequate for assessing solvation properties of ionic liquids on a quantitative level. A typical example is the solvent effect on chemical reactions. Transition state theory yields an expression for the rate constant in terms of two barriers, namely the intrinsic reaction barrier, $\Delta G_{in}^{\ddagger}$, and the solvent reorganisation barrier, $\Delta G_{solv}^{\ddagger}$ [69]. The latter is defined as the difference in free energy of solvation between the transition state and the reactants.

In a dielectric continuum approach, $\Delta G_{solv}^{\ddagger}$ is given by the Born solvation free energy, which depends on the static dielectric constant of the surrounding medium through an expression of the general form $(\varepsilon - 1)/(\varepsilon + 2) - (n^2 - 1)/(n^2 + 2)$ [69]. Experimental static dielectric constants predict, for simple imidazolium salts, values of $\Delta G_{solv}^{\ddagger} \cong 5\text{–}7\,\text{kJ}\cdot\text{mol}^{-1}$, while experimental determinations of $\Delta G_{solv}^{\ddagger}$ from fluorescence Stokes shift data yield $\Delta G_{solv}^{\ddagger} = 10\text{–}15\,\text{kJ}\cdot\text{mol}^{-1}$ [73]; that is, electrostatics can account for only 30–50% of the solvent reorganisation barrier.

## 9.7   CONCLUSIONS

In the microwave regime, the spectra of low-viscosity ionic liquids at $25\,^{\circ}\text{C}$ show quite common features. By contrast, terahertz spectra indicate a much more specific behaviour, presumably because intermolecular vibrations are highly specific to local interactions. The goal in writing this account has been to spotlight these dielectric properties from the experimental perspective and to highlight differences to dielectric properties of VOCs.

Although at ambient conditions the microwave portion of the dielectric response is experimentally well characterised for many ionic liquids, fundamental questions remain. One pressing problem concerns the detailed origin of the extraordinarily fast reorientation of the cations, and of the still faster reorientation and low relaxation amplitudes exhibited by many anions. Another challenge concerns the description of the dielectric response in terms of molecular dipole moments because, for charged species, the electric dipole moment cannot be defined in an unambiguous manner.

To complete the dielectric spectra of ionic liquids, it would be necessary to extend the experiments at their low-frequency edge, say, below 1 MHz; however, in this range, electrode polarisation and the high Ohmic loss so far have prevented meaningful experiments. The question for processes at lower frequencies is key to the extrapolation of the static dielectric constant of ionic liquids. To the best of the present knowledge, there is no evidence for such contributions, so that the assumption of such processes is speculative.

Space limitations have prompted us to limit the discussion to neat ionic liquids of low viscosity near 25 °C. Exciting developments are expected from an extension of DRS to a wider temperature range [32], which should also provide a link to glassy dynamics [22–24]. For example, the temperature dependence of the static dielectric constant needs accurate and systematic characterisation. Another exciting perspective is founded in an extension of DRS to mixtures between ionic liquids and VOCs, as some applications, such as chemical reactions or separation processes, concern mixtures of ionic liquids with VOCs [1–3, 48].

It seems that the proper characterisation and thorough understanding of dielectric spectra of ionic liquids will keep experimentalists and theoreticians busy for the foreseeable future.

## ACKNOWLEDGEMENTS

Over the years, the author's work has benefited from the engagement of many coworkers quoted in the literature references. Mian-Miang Huang helped in the preparation of the manuscript, Dr. Matthias Krüger prepared figures. Professors R. Buchner, R. Ludwig, M. Maroncelli, and O. Steinhauser are thanked for helpful discussions. The Deutsche Forschungsgemeinschaft is thanked for financial support of the author's work within the priority programme SPP 1191 ("Ionic Liquids").

## REFERENCES

1   Wasserscheid, P. and Welton, T. (eds.), *Ionic Liquids in Synthesis*, 2nd ed. (Wiley-VCH, Weinheim, 2008).

2   Plechkova, N. V., Rogers, R. D. and Seddon, K. R. (eds.), *Ionic Liquids: From Knowledge to Application*, ACS Symp. Ser., Vol. 1030 (American Chemical Society, Washington D. C., 2009).

3   Weingärtner, H., Understanding ionic liquids at the molecular level: Facts, problems, and controversies, *Angew. Chem. Int. Ed.* **47**, 654–670 (2008).

4   Daguenet, C., Dyson, P. J., Krossing, I., Oleinikova, A., Slattery, J., Wakai, C., and Weingärtner, H., The dielectric response of ionic liquids, *J. Phys. Chem. B.* **110**, 12682–12688 (2006).

5   Buchner, R., and Hefter, G., Interactions and dynamics in electrolyte solutions by dielectric spectroscopy, *Phys. Chem. Chem. Phys.* **11**, 8984–8999 (2009).

6   Kremer, F., and Schönhals, A., *Broadband Dielectric Spectroscopy* (Springer, Berlin, 2003).

7   Böttcher, A. V. J., and Bordewijk, P., *Theory of Dielectric Polarization*, Vol. 2 (Elsevier, Amsterdam, 1978).

8   Leys, J., Wübbenhorst, M., Menon, C. P., Rajesh, R., Thoen, J., Glorieux, C., Nockemann, P., Thijs, B., Binnemans, K., and Longuemart, S. J., Temperature dependence

of the electrical conductivity of imidazolium ionic liquids, *J. Chem. Phys.*, **128**, paper No. 064509 (2008).

9   Weingärtner H., Knocks, A., Schrader, W., and Kaatze, U., Dielectric spectroscopy of the room temperature molten salt ethylammonium nitrate, *J. Phys. Chem. A.* **105**, 8646–8650 (2001).

10  Wakai, C., Oleinikova, A., Ott, M., and Weingärtner, H., How polar are ionic liquids? Determination of the static dielectric constant by microwave dielectric spectroscopy, *J. Phys. Chem. B.* **109**, 17028–17030 (2005).

11  Weingärtner, H., The static dielectric constant of ionic liquids, *Z. Phys. Chem.* **220**, 1395–1405 (2006).

12  Chiappe, C., Malvaldi, M., and Pornelli, C. S., Ionic liquids: Solvation ability and polarity, *Pure Appl. Chem.* **81**, 767–776 (2009).

13  R. H. Cole, Dielectrics in physical chemistry, *Annu. Rev. Phys. Chem.* **40**, 1–29 (1989).

14  Hansen, J.-P., and McDonald, I. R., Statistical mechanics of dense ionized matter. 4. Density and charge fluctuations in a simple molten salt, *Phys. Rev. A* **11**, 2111–2123 (1975).

15  Schröder, C., and Steinhauser, O., On the dielectric conductivity of molecular ionic liquids, *J. Chem. Phys.* **131**, paper No. 114504 (2009).

16  Schröder, C., Wakai, C., Weingärtner, H., and Steinhauser, O., Collective rotational dynamics of ionic liquids: A computational and experimental study of 1-butyl-3-methylimidazolium tetrafluoroborate, *J. Chem. Phys.* **126**, paper no. 084511 (2007).

17  Song, X., Solvation dynamics in ionic fluids. An extended Debye-Hückel dielectric continuum model, *J. Chem. Phys.* **131**, paper no. 044503 (2009).

18  Shim, Y., and Kim H. J., Dielectric relaxation, ion conductivity, solvent rotation and solvation dynamics in a room temperature ionic liquid, *J. Phys. Chem. B* **112**, 11028–11038 (2008).

19  Moynihan, C. T., Boesch, L. P.; Laberge, N. L., Decay function for the electric field relaxation in vitrous ionic conductors, *Phys. Chem. Glasses* **14**, 122–125 (1973).

20  Howell, F. S., Bose, R. A., Macedo, P. B., and Moynihan, T. C., Relaxation in a glass-forming molten salt, *J. Phys. Chem.* **78**, 639–648 (1974).

21  Ngai, K. L., and Rendell, R. W., Interpreting the real part of the dielectric conductivity contributed by mobile ions in ionically conducting media, *Phys. Rev. B* **61**, 9393–9398 (2000).

22  Sangoro, J., Sergej, A., Naumov, S., Galvosas, P., Kärger, J., Wespe, C, Bordusa, F., and Kremer, F., Charge transport and mass transport in imidazolium-based ionic liquids, *Phys. Rev. E*, **77** paper no. 051202 (2008).

23  Jacob, C., Sangoro, J. R., Serghei, A., Naumov, S., Korth, Y., Kärger, J., Friedrich, C., and Kremer, F., Charge transport and glass dynamics in imidazole based liquids, *J. Chem. Phys.* **129**, paper no. 234511 (2008).

24  Ito, N., and Richert, R., Solvation dynamics and electric field relaxation in an imidazolium-$PF_6$ ionic liquid: From room temperature to the glass transition, *J. Phys. Chem. B.* **111**, 5016–5022 (2007).

25  Schrödle, S., Annat, G., MacFarlane, D. R., Forsyth, M., Buchner, R., and Hefter, G., Broadband dielectric response of the ionic liquid N-methyl-N-ethylpyrrolidinium dicyanamide, *Chem. Comm.* 1748–1750 (2006).

26 Weingärtner, H., Sasisanker, P., Daguenet, C., Dyson, P. J., Krossing, I, Slattery, J., and Schubert, T., The dielectric response of room temperature ionic liquids: Effect of cation variation, *J. Phys. Chem. B.* **111**, 4775–4780 (2007).

27 Hunger, J., Stoppa, A., Schrödle, S., Hefter, G., and Buchner, R., Temperature dependence of the dielectric properties and dynamics of ionic liquids, *ChemPhysChem* **10**, 723–733 (2008).

28 Nakamura, K., and Shikata, T., Systematic dielectric and NMR study of the ionic liquid 1-alkyl-3-methylammonium, *ChemPhysChem.*, **11**, 285–294 (2010).

29 Mizoshiri, M., Nagao, T., Mizoguchi, Y., and Yao, M., Dielectric permittivity of room temperature ionic liquids: A relation to the polar and nonpolar domain structures, *J. Chem. Phys.* **132**, paper no. 164510 (2010).

30 Weingärtner, H., and Huang, M.-M., unpublished data.

31 Izgorodina, E. I., Forsyth, M., and MacFarlane, D. R., On the components of dielectric constants of ionic liquids: Ionic polarization? *Phys. Chem. Chem. Phys.* **11**, 2452–2458 (2009).

32 Stoppa, A., Hunger, J., Thoman, A., Helm, H., Hefter, G., and Buchner, R., Interactions and dynamics in ionic liquids, *J. Phys. Chem. B* **112**, 4854–4858 (2008).

33 Turton, D. A., Hunger, J., Stoppa, A., Hefter, G., Thoman, A., Walther, M., Buchner, R., and Wynne, K., Dynamics of imidazolium ionic liquids from a combined dielectric relaxation and optical Kerr effect study: Evidence for mesoscopic aggregation, *J. Amer. Chem. Soc.* **131**, 11140–11146 (2009).

34 Yamamoto, K., Tani, M., and Hangyo, M., Terahertz time domain spectroscopy of imidazolium ionic liquids, *J. Phys. Chem. B* **111**, 4854–4859 (2007).

35 Krüger, M., Funkner, S., Bründermann, E., Weingärtner, H., and Havenith, M., Polarity fluctuations of the protic ionic liquid ethylammonium nitrate in the terahertz regime, *J. Chem. Phys* **132**, paper No. 101101 (2010).

36 Fumino, H., Wulf, A., and Ludwig, R., The cation-anion interaction in ionic liquids probed by far-infrared spectroscopy, *Angew. Chem. Int. Ed.* **47**, 3830–3834 (2008).

37 Fumino, K., Wulf, A., and Ludwig, R., Hydrogen bonding in protic ionic liquids: Reminiscent of water, *Angew. Chem. Int. Ed.* **48**, 3184–3186 (2009).

38 Schröder, C., and Steinhauser, O., Using fit functions in computational dielectric spectroscopy, *J. Chem. Phys.* **132**, paper No. 244109 (2010).

39 Schmuttenmaer, C. A., Exploring dynamics in the far infrared with terahertz spectrocopy, *Chem. Rev.* **104**, 1759–1779 (2004).

40 Wulf, A., Ludwig, R., Sasisanker, P., and Weingärtner, H., Molecular reorientation in ionic liquids: A comparative dielectric and magnetic relaxation study, *Chem. Phys. Lett.* **439**, 323–326 (2007).

41 Huang, M.-M., Bulut, S., Krossing, I., and Weingärtner, H., Are hydrodynamic models suitable for describing the reorientational dynamics of ions in ionic liquids? A case study of methylimidazolium tetra(hexafluoroisopropoxy) aluminates, *J. Chem. Phys.* **133**, 101101 (2010).

42 Tokuda, H., Hayamizu, K., Ishii, K., Susan, M. A. B. H., and Watanabe, M., Physical properties of room temperature ionic liquids. 1. Variation of anionic species, *J. Phys. Chem. B* **108**, 16593–16600 (2004).

43  Tokuda, H., Hayamizu, K., Ishii, K., Susan, M. A. B. H., and Watanabe, M., Physical properties of room temperature ionic liquids. 2. Variation of alkyl chain length in imidazolium cation, *J. Phys. Chem. B* **109**, 6103–6110 (2005).

44  Bulut, S., Klose, P., Huang, M.-M., Weingärtner, H., Dyson, P., Lawrency, G., Friedrich, C., Kümmerer, K., and Krossing, I., Synthesis of room temperature ionic liquids with the weakly coordinating $[Al(ORF)_4]^-$ anion ($RF = CH(CF_3)_2$) and the determination of their principal physical properties, *Chem. Eur. J.* **16**, 13139–13154 (2010).

45  Dote, J. C., Kivelson, D., and Schwartz, R. N., A molecular, quasi-hydrodynamic free-space model for molecular rotational relaxation in liquids, *J. Phys. Chem.* **85**, 2169–2180 (1981).

46  Hunger, J., Stoppa, A., Buchner, R., and Hefter, G., Dipole correlations in the ionic liquid 1-N-ethyl-3-N-methylimidazolium ethylsulfate and its binary mixtures with dichloromethane, *J. Phys. Chem. B* **113**, 9527–9537 (2009).

47  Weingärtner, H., Nadolny, H.G., and Käshammer, S., Dielectric properties of an electrolyte solution at low reduced temperature, *J. Phys. Chem. B* **103**, 4738–4743 (1999).

48  Huang, M.-M., Schneiders, K., Schulz P., Wasserscheid, P., and Weingärtner, H., Ion speciation driving chirality transfer in imidazolium-based camphorsulfonate ionic liquid solutions, *Phys. Chem. Chem. Phys.* **13**, 4126–4131 (2011).

49  Debye, P., and Falkenhagen, H., The dispersion of conductivity and dielectric constant of strong electrolytes, *Phys. Z.* **29**, 121–132 (1928).

50  Chandra, A., and Bagchi, B., Frequency dependence of ionic conductivity of electrolyte solutions, *J. Chem. Phys.* **112**, 1876–1886 (2000).

51  Chandra, A., Wei, D. Q., and Patey, G. N., The frequency-dependent conductivity of electrolyte solutions, *J. Chem. Phys.* **99**, 2083–2094 (1993).

52  Bordi, F., Cametti, C., and Colby, R. H., Dielectric spectroscopy and conductivity of polyelectrolyte solutions, *J. Phys.: Condens. Matter* **16**, R1423–R1463 (2004).

53  Madel, M., and Odijk, T. Dielectric properties of polyelectrolyte solutions, *Annu. Rev. Phys. Chem.* **35**, 75–108 (1984).

54  Buchner, R., Baar, C., Fernandez, P., Schrödle, S., and Kunz, W., Dielectric spectroscopy of micelle hydration and dynamics in aqueous ionic surfactant solutions, *J. Mol. Liquids* **118**, 179–187 (2005).

55  Grosse, C., Permittivity of a suspension of charged particles in electrolyte solution, *J. Phys. Chem.* **92**, 3905–3910 (1988).

56  J. E. Anderson, The Debye-Falkenhagen effect: Experimental fact or fiction?, *J. Non-Cryst. Solids* **172–174**, 1190–1194 (1994).

57  Van Beek, W. M., and Mandel, M., Static relative permittivity of some electrolyte solutions in water and methanol, *J. Chem. Soc., Faraday Trans. 1* **74**, 2339–2351 (1978).

58  Ghowsi, K., and Gale, R. J., Some aspects of the high frequency conductance of electrolytes, *J. Electrochem. Soc.* **136**, 2806–2811 (1989).

59  Funke, K., and Riess, I. Debye-Hückel-type relaxation processes in solid ionic conductors. 1. The model, *Z. Phys. Chem. Neue Folge* **140**, 217–232 (1984).

60  Funke, K. Debye-Hückel-type relaxation processes in solid ionic conductors. 2. Experimental evidence, *Z. Phys. Chem. Neue Folge* **154**, 251–295 (1987).

61 Samanta, A., Dynamic Stokes shift and excitation wavelength dependent fluorescence of dipolar molecules in ionic liquids, *J. Phys. Chem. B* **110**, 13704–13716 (2006).

62 Arzhantsev, S., Jin, H., Baker, G. A., and Maroncelli, M., Measurement of the complete solvation response of ionic liquids, *J. Phys. Chem. B* **111**, 4978–4989 (2007).

63 Hasted, J., *Aqueous Dielectrics* (Chapman & Hall, London, 1973).

64 Gollei, A., Vass, A., Pallai, E., Gerzson, M., Ludanyi, L., and Mink, J., Apparatus and method to measure dielectric properties of ionic liquids, *Rev. Sci. Instr.* **80**, 044703 (2009).

65 Wu, J., and Stark, J. P. W., Measurement of low-frequency relative permittivity of room temperature molten salts by triangular wave form voltage, *Meas. Sci. Techn.* **17**, 781–788 (2006).

66 Kosmulski, M., Marczewska-Boczkowska, K., Zukowski, P., Subocz, J., and Saneluta, C., Permittivities of 1-alkyl-3-methyl tetrafluoroborates and hexafluorophosphates, *Croat. Chem. Acta* **80**, 461–466 (2007).

67 Huang, M.-M., Jiang, Y., Sasisanker, P., Driver, G., and Weingärtner, H., Static dielectric constants of forty ionic liquids at 298.25 K, *J. Chem. Eng. Data* **56**, 1494–1499 (2011).

68 Huang, M.-M., and Weingärtner, H., Protic ionic liquids with unusually high dielectric permittivities, *Chem. Phys. Chem.* **9**, 2172–2173 (2008).

69 Reichardt, C., and Welton, T., *Solvents and Solvent Effects in Organic Chemistry*, 4th ed. (Wiley-VCH, Weinheim, 2011).

70 Hallet, J. P., and Welton, T., How polar are ionic liquids? *ECS Trans.* **16**, 33–38 (2009).

71 Wangsness, K., *Electromagnetic Fields*, 2nd ed. (John Wiley & Sons, New York, 1986).

72 Kamlet, M. J., Abboud, J. L., and Taft, R. W., The solvatochromic comparison method 6: the $\pi^*$ scale of solvent polarities, *J. Amer. Chem. Soc.* **99**, 8325–8327 (1977).

73 Jin, H., Baker, G. A., Arzhantsev, S., Dong, J., and Maroncelli, M., Solvation and rotational dynamics of coumarin 153 in ionic liquids: Comparisons to conventional solvents, *J. Phys. Chem B* **111**, 7291–7302 (2007).

# 10 Ionic Liquid Radiation Chemistry

JAMES F. WISHART

Chemistry Department, Brookhaven National Laboratory, Upton, New York, USA

## ABSTRACT

Many potential uses of ionic liquids, such as recycling of used nuclear fuel, aerospace applications, and radiation processing, involve exposure to ionising radiation. Radiation chemistry provides powerful tools for studying redox reaction mechanisms, charge transport processes, and the reactivity of highly energetic species, including the electrochemical processes that occur in ionic liquid-based energy storage devices. This chapter provides an introduction to radiation chemistry and its relevance to the expanding field of ionic liquids. It discusses what happens to ionic liquids when they are irradiated and evaluates whether some ionic liquids would be stable enough to use in nuclear fuel separations. It presents some practical aspects of using ionic liquids in radiolysis studies and nanoparticle synthesis, and concludes with prospects for future work.

## 10.1 INTRODUCTION: WHAT IS RADIATION CHEMISTRY?

Nowadays, the term "radiation chemistry" is unfamiliar to many chemists; however, it is the name applied to the chemistry induced by ionising radiation. Ionising radiation is all around us and it has important medical and industrial applications, including medical diagnostics, imaging and therapeutic treatment, the sterilisation of medical supplies and foodstuffs, the production of high-performance materials, industrial and construction inspection, and the removal of pollutants from waste water. Radiation chemistry is therefore very important to our daily lives, yet scientific interest in radiation chemistry has seen

*Ionic Liquids Further UnCOILed: Critical Expert Overviews*, First Edition.
Edited by Natalia V. Plechkova and Kenneth R. Seddon.
© 2014 John Wiley & Sons, Inc. Published 2014 by John Wiley & Sons, Inc.

several ups and downs since the field was inaugurated by Pierre and Marie Curie eleven decades ago [1].

From the Second World War up to the 1980s, international interest and effort in the study of radiation chemistry was very strong for two main reasons. First, the radiation chemistry of water and aqueous solutions is fundamentally important for comprehending radiation biology and medicine, including the effects of radiation on the human body and on the genetic code. Second, the effects of radiation as a product and a by-product of peaceful uses of the atom, including nuclear power, industrial applications, and medicine, needed to be understood. The radiation chemists of that era were an extremely talented and resourceful group who tackled many of the basic questions, such as the identification and reaction chemistry of the reactive transient species produced in water and other liquids by radiation, often using clever methods and chemistries that extended their mechanistic insights well beyond the range of the equipment of the time.

Consequently, by the 1980s and 1990s, many of the core issues had been addressed to some degree of satisfaction, and the emphasis of the field shifted to investigation of the radiation chemistry of organic solvents and to applying radiation chemistry techniques to other mechanistic investigations, most notably electron-transfer reactions, biochemistry, and radiation biology. Pulse radiolysis, the time-resolved kinetic measurement of reactions initiated by a pulse of radiation, is particularly useful for the mechanistic study of redox and redox-induced reactions. Financial support for the field also generally waned due to the global retreat away from nuclear power development, which had been a natural justification for stewardship of radiation chemistry expertise.

These trends have reversed during the last decade because of several factors. A new generation of ultra-fast accelerators for pulse radiolysis [2–4], triggered by and synchronised with versatile laser systems, have allowed direct observation of early-time processes that determine the distribution of the radiolytically produced primary transient species, whose chemistry determines all subsequent events. The development of advanced computational equipment and methods now permits detailed simulation of early events in radiation chemistry to support and extend the power of experimental observations. Societal concerns about the need to expand non-carbon-based energy sources have revived worldwide interest in nuclear power development, with an emphasis on responsible and sustainable management of the nuclear fuel cycle [5]. Recognition of the central role of radiation chemistry in the successful implementation of this programme has provided renewed support for the field.

In addition, new types of "neoteric" media, ionic liquids and supercritical solvents, became important targets of radiation chemistry studies in recent years. Supercritical water is of particular interest because of its use in advanced oxidation processes, and for its potential use as the primary coolant in one type of Generation IV nuclear reactor, scheduled for start-up around the year 2030. However, the gamut and operating ranges of known supercritical solvents are limited and experimentally restrictive, so from a certain perspective, there is only so much that can be learned from them. On the other hand, ionic liquids present an infinitely diverse array of materials to investigate for their radiolytic behaviour. Furthermore, the unusual properties of ionic liquids that distinguish them from conventional solvents, such as (1) their inherent binary nature, (2) their molecular-scale polar/non-polar heterogeneity combined with a degree of short-range order, and (3) their dynamical timescales extending over three to five orders of magnitude longer than normal liquids, make them particularly useful for studying fundamental radiation chemistry that is difficult to observe with regular solvents. These areas include the mechanisms of radiolytic energy deposition, the spatial distribution of primary radiolysis products, and the atypical reactivity and mobility of transient species in unrelaxed solvation states as compared with equilibrated ones.

Ionic liquids thus provide the opportunity to pick up and advance many of the fundamental mechanistic questions that had lain fallow since the early, "golden" years of radiation chemistry. This is not merely an issue of intellectual curiosity — due to the vast diversity of ionic liquids available to choose from — understanding the process of radiolytic product formation permits the control of product speciation through ionic liquid design. Such control by design can be extremely important for potential applications of ionic liquids in recycling spent nuclear fuel, for example [6, 7], where radiation exposure is unavoidable but radiation damage and loss of efficiency can be mitigated by the proper choice of materials.

## 10.2  THE RELEVANCE OF RADIATION CHEMISTRY TO IONIC LIQUID SCIENCE AND APPLICATIONS

The connections between ionic liquids and radiation chemistry rest on three pillars:

(a) Understanding the effects of cumulative radiation exposure on ionic liquid-based systems used in radiation fields, such as those experienced in processing nuclear materials, or in outer space applications (which could include propellants [8], instrumentation, or even lunar telescope mirrors [9]).

(b) The use of ionic liquids for the radiolytic preparation of advanced materials such as polymers and nanoparticles, often attempting to use the inherent local structural order of ionic liquids to create materials with specific morphologies.

(c) The application of pulse radiolysis techniques to the general study of reactivity in ionic liquids, as has been done successfully for many years in conventional solvents where chemistries have been devised to cleanly generate reducing or oxidising species to initiate desired reactions.

The fundamental knowledge needed to support these pillars is essentially the same: the identification of primary ionic liquid radiolysis products and the characterisation of their reactivities, as functions of ionic liquid composition.

This knowledge can be used to direct reaction pathways to produce well-defined intermediate species for the following:

1. The diversion of radiolytic damage into benign end points.
2. Polymerisation [10], polymer grafting [11], or nanoparticle growth [12].
3. The production of desired reducing, oxidising, or radical species for chemical kinetics studies.

The last point has far-reaching implications because many important future technologies using ionic liquids involve reactions that transfer charge [13], including dye-sensitised solar cells, batteries, electromechanical actuators, electrochromic displays, sensors, and catalysis. Pulse radiolysis provides a powerful technique to initiate charge-transfer reactions sometimes inaccessible by photochemical or electrochemical routes.

## 10.3   A BRIEF DESCRIPTION OF FUNDAMENTAL RADIATION CHEMISTRY AND IONIC LIQUIDS

A thorough description of radiation chemistry and its experimental techniques is well beyond the scope of this chapter. The interested reader is referred to several recent books on the subject [14–17], and to previous chapters and articles on its relationship to ionic liquids [7, 18–22].

Ionising radiations, in the form of photons (X- and gamma-rays) or energetic particles (electrons, alpha particles, heavy ions such as carbon $C^{6+}$, neutrons, etc.), deposit energy in materials (liquids, gases, or solids) through several types of physical interactions. The nett effect of these interactions is to excite the constituent molecules or atoms, often to the point of ejecting an electron into the surrounding medium, thus "ionising" the molecule and creating a "hole," or electron vacancy, on it.*

The incident radiation deposits its energy in increments of thousands to tens of thousands of $kJ \cdot mol^{-1}$ per interaction along its path, creating a track of ionised molecules. Secondary electrons generated by the ionisations often create their own ionisation track or spur, branched off the main track, creating a dendritic pattern of localised energy deposition. Away from the track, the medium is unaffected. This is a key point—radiolytic energy deposition is non-uniform and locally dense, depending on the type of incident radiation. Cross-reactions between initial radiolytic products depend on the density of species formed, and they determine the ultimate product yields. Consequently, product yields will be different for electrons compared with alpha particles since the latter deposit more energy per unit length of travel through a medium.

---

* In conventional radiation chemistry of neutral molecules, the hole is usually called a cation because it is positively charged. That usage will not be employed here in the context of ionic liquids to avoid confusion.

When electrons are ejected from their molecules, they have excess kinetic energy, which they lose through interactions with the surrounding medium until they come to rest. At that point, the molecules of the medium begin to reorient to solvate the introduced excess negative charge. In normal liquids, the solvation process is complete within a few picoseconds, but due to the slow dynamics of ionic liquids, the process can take a thousand times longer. During that time, the energetic, weakly localised pre-solvated electrons can react in ways that are energetically or dynamically inaccessible to the localised, solvated electrons they are evolving into. The relative importance of pre-solvated electron reactivity in ionic liquids is a major distinction with conventional solvents. The details and implications of this fact have been treated several times before, and for the sake of brevity here the reader is referred to the previous discussions [7, 18–24]. To summarise, pre-solvated electron reactivity dictates different reactivity profiles and product distributions than predicted from solvated electron reactivity. However, on the useful side, it can be exploited to control those distributions and to overcome the kinetic limitations of slow reactant diffusion in ionic liquids. In addition, the distributed dynamical properties of ionic liquids dilate the electron solvation timescale so that the mechanisms of the electron scavenging processes, which have been controversial subjects for decades, are exposed for detailed analysis with modern picosecond pulse radiolysis instrumentation [22, 25].

The fate of radiolytically induced holes is a very important aspect of radiation chemistry as well. In general, holes may recombine with their geminate electrons, extract electrons by oxidising solute molecules or other constituents of the ionic liquid, form radical dimers, or undergo bond scission due to loss of a valence electron. The chemistry of the holes formed by ionic liquid radiolysis is even more complex than that of normal solvents because of the enormous variety of cations and anions that can be used to make ionic liquids. Hole chemistry is often harder to follow as well since few hole species have conveniently accessible transient optical absorption spectra (most are in the UV). Some exceptions are bromide and thiocyanate ionic liquids, which form observable dimer radical anions $[Br_2]^-$ and $[(SCN)_2]^-$ [26, 27], and imidazolium cations [28, 29], which form ring-centred radical dications. To address this general deficiency, detailed electron spin resonance studies of a wide variety of irradiated ionic liquid glasses have been undertaken to determine the radical species produced and to pinpoint the sites where radiolytic damage occurs in different families of ionic liquids [7, 30–32]. The chemistry of a particular cation or anion can even be altered by the choice of counterion [33, 34]. Easily oxidisable anions, such as thiocyanate, can reductively quench holes that were initially formed on cations, effectively eliminating precursors to permanent damage fixation in favour of a relatively stable and reversible redox species.

Radiolysis product studies using analytical methods (such as electrospray mass spectral analysis) are extremely useful for connecting primary radiation chemistry studied by the time-resolved and cryogenic experiments described earlier to the quantitative accumulation of radiation damage in ionic liquid systems. The group of Moisy has pioneered this approach by examining gamma-irradiated salts of the $[C_4mim]^+$ cation with the $[NTf_2]^-$, $[OTf]^-$, $[PF_6]^-$, and $[BF_4]^-$ anions [35, 36], as well as $[N_{1\,4\,4\,4}][NTf_2]$ [37]. They have identified major fragmentation pathways, as well as the addition of the anion-derived fragments $-CF_3$ and $-F$ to the $[C_4mim]^+$ cation, another example of the counterion effects mentioned earlier. The product analysis results generally agree with the expectations based on the radicals observed in the irradiated ionic liquid glasses [7, 30–34], but some observed products have yet to be identified.

## 10.4    WOULD IONIC LIQUIDS BE STABLE ENOUGH FOR SPENT NUCLEAR FUEL RECYCLING?

The description of fundamental aspects of ionic liquid radiation chemistry given earlier underscores the complexity of the problem when one attempts to put all ionic liquids in one neat bundle. While it was commonplace during the early years of expansion in ionic liquid research to make broad generalisations about the properties, behaviour, and merits of ionic liquids, it is now abundantly clear to the field that exceptions are the rule [8, 38]. The term "ionic liquids" is almost as generic as "solvents." This is actually a sign of how robust the field has become. Despite this awareness, it is all too common to find publications on ionic liquid radiation chemistry that attempt to draw broad conclusions from investigations of one or two ionic liquids. Frequently, these liquids are known bad actors such as $[PF_6]^-$ and $[BF_4]^-$ salts, which are now understood to undergo hydrolysis to produce hydrofluoric acid (HF) [39, 40].

One such area where generalities should be avoided concerns the radiation stability of ionic liquids, which is a concern for spent nuclear fuel processing and space applications. However, it is important to remember that all materials will degrade to some extent under ionising radiation; the ultimate objective is to identify specific ionic liquids that can outperform the materials currently in use for a particular application, for example, making a more robust and efficient separation process. Various groups of researchers have applied different techniques to evaluate radiation stability, including nuclear magnetic resonance (NMR) spectroscopy, mass spectrometry (MS), and hydrogen gas evolution. The results present a reasonably coherent picture over the limited range of ionic liquids that have been studied so far, and have pointed to new families of ionic liquids to be next investigated.

The baseline study of ionic liquid radiation stability was published in 2002 by a large group of authors from the United Kingdom [41]. They studied the effect of alpha, beta (electron), and gamma radiation on $[C_4mim][NO_3]$, $[C_2mim]Cl$, and $[C_6mim]Cl$, and compared the ionic liquid results with those of the standard tributylphosphate/kerosene mixture used in the PUREX process to separate plutonium and uranium from spent nuclear fuel. Using

NMR spectroscopy to quantify the extent of radiation damage at a dose of 400 kGy (1 Gy = 1 J kg$^{-1}$), they were unable to observe degradation products above the estimated detection threshold of 1%, whereas the PUREX mixture shows 15% conversion to radiolysis products at that dose level. They concluded that the ionic liquids they studied were "relatively radiation resistant," particularly with respect to the PUREX mixture.

This resistance was attributed to the aromatic nature of the imidazolium cation since it is known from "classical" radiation chemistry that aromatic organic liquids are resistant to ionisation because a much larger portion of the radiolytic energy is absorbed as molecular excitations rather than ionisations. This difference can be quantified by measuring the radiolytic yield (called the $G$-value) of molecular hydrogen per unit of energy absorbed. For benzene, $G(H_2) = 3.8 \times 10^{-9}$ mol J$^{-1}$ [42], while $G(H_2)$ is much larger in aliphatic liquids, 5–6 $\times$ 10$^{-7}$ mol J$^{-1}$ [43], demonstrating that the effect of aromaticity has a strong stabilising effect.

Tarabek et al. [44] recently explored the influence of aromaticity on radiolytic stability over a representative sample of aromatic (imidazolium and pyridinium), aliphatic (pyrrolidinium and phosphonium), and protic ionic liquids, all with the same [NTf$_2$]$^-$ anion, by measuring H$_2$ yields under electron radiolysis. The results presented in Table 10.1 fit well with the aromaticity interpretation. The imidazolium and pyridinium ionic liquids have the lowest $G(H_2)$ values, substantially larger than benzene but comparable with butylbenzene, thus allowing for effect of the side chains of the cations. The aliphatic pyrrolidinium and protic triethylammonium ionic liquids have higher H$_2$ yields, indicating slightly higher radiation sensitivity, while the phosphonium ionic liquid with very long side chains exhibits a very large, hydrocarbon-like $G(H_2)$. It is therefore clear that incorporating aromaticity into ionic liquids for spent

**TABLE 10.1   Radiolytic Yields of Molecular Hydrogen in Ionic and Molecular Liquids**

| Liquid | $G(H_2)/10^{-8}$ mol·J$^{-1}$ | Reference |
|---|---|---|
| C$_6$mim][NTf$_2$] | 2.6 | [44] |
| [C$_6$(dma)$_n$py][NTf$_2$]$^a$ | 2.6 | [44] |
| [C$_4$mpyr][NTf$_2$] | 6.5 | [44] |
| [HN$_{222}$][NTf$_2$] | 7.2 | [44] |
| [P$_{88814}$][NTf$_2$] | 25 | [44] |
| [C$_4$mim][NO$_3$] | 6.5$^b$ | [41] |
| [C$_6$mim]Cl | 7.2$^b$ | [41] |
| Benzene | 0.38 | [42] |
| Benzene (alpha radiolysis) | 2.0$^b$ | [42] |
| Butylbenzene | 2.6 | [43] |
| Aliphatic hydrocarbons | 50–60 | [43] |

Yields are for electron or gamma radiolysis unless otherwise noted.
$^a$ [C$_6$(dma)$_n$py]$^+$ stands for 1-hexyl-4-dimethylaminopyridinium cation.
$^b$ Radiolysis using ~6 MeV alpha particles.

fuel recycling would help stabilise them towards ionising radiation. In addition to raising the cross section for excitation at the expense of ionisation, aromatic cations can also reduce the effects of ionisation by serving as relatively stable electron and hole traps, raising the probability of electron–hole recombination to avoid radiolytic damage.

As shown in Table 10.1, Allen et al. [41] measured $G(H_2)$ values for [$C_4$mim] [$NO_3$] and [$C_6$mim]Cl, but those yields are significantly higher than that for [$C_6$mim][$NTf_2$] because alpha particles were used instead of electrons. Heavy ion radiations deposit more energy per unit length than electron and gamma radiations, and the higher density of reactive intermediates thus generated results in more cross-reactions to form molecular hydrogen, among other things. The $G(H_2)$ in benzene is also significantly elevated under those conditions (Table 10.1). Alpha radiolysis of ionic liquids is an important subject to study because many of the isotopes relevant to spent fuel reprocessing are alpha emitters; however, the appropriate instrumentation for steady-state and time-resolved alpha radiolysis is relatively scarce.

The group of Moisy has done extensive work using high-performance liquid chromatography (HPLC), NMR and electrospray MS to quantify gamma radiation damage by measuring the loss of cations and anions from [$C_4$mim]$^+$ ionic liquids with [$NTf_2$]$^-$, [$OTf$]$^-$, [$PF_6$]$^-$, and [$BF_4$]$^-$ anions [36]. They obtained $G(-[C_4\text{mim}]^+)$ yields in the range of 2.8–3.7 × 10$^{-7}$ mol J$^{-1}$ and $G(-\text{anion}) =$ 1.0–2.2 × 10$^{-7}$ mol J$^{-1}$, using doses of up to 2 MGy. In an earlier report [37], they obtained $G(-[N_{1\,4\,4\,4}]^+) = 3.8 \times 10^{-7}$ mol J$^{-1}$ and $G(-[NTf_2]^-) =$ 2.5 × 10$^{-7}$ mol J$^{-1}$ for [$N_{1\,4\,4\,4}$][$NTf_2$]. (Parenthetically, it is also possible to calculate a yield of $G([NO_2]^-) = 1.5 \times 10^{-7}$ mol J$^{-1}$ for the formation of nitrite ion during gamma radiolysis of [$C_4$mim][$NO_3$] from the data in Allen et al. [41].) The magnitudes of these radiation damage yields are similar to those observed in polar liquids, which would seem to indicate that the stabilities of these particular ionic liquids are comparable with other solvents and not significantly superior. It is notable, however, that Moisy's group reported only very weak radiolysis product NMR signals (<0.5%) at doses up to 1.2 MGy [35] and needed 2 MGy doses to obtain quantifiable electrospray ionisation (ESI)-MS data for yield estimation [36, 37]. Their overall conclusion from this work, taking into account that actual spent fuel processing systems are exposed to doses of 0.1–1.0 MGy, is that the ionic liquids they studied are suitable for nuclear applications and may have benefits in terms of their low production of hydrogen gas [36].

## 10.5 SUITABILITY OF IONIC LIQUID PREPARATIONS FOR RADIATION CHEMISTRY STUDIES

Due to the nature of their formation, the primary species formed during the radiolysis of any material, not just ionic liquids, tend to be extremely energetic, unstable, and reactive. It has always been important for the study of chemical kinetics that the materials used should be free of any contaminants that could interfere with or catalyse a reaction, but radiolysis studies can be particularly sensitive due to the reactivity of the intermediates with contaminants. This presents a particular challenge for studying kinetics in ionic liquids because some of the standard solvent purification techniques, such as distillation and chromatography, are difficult or impossible.

Unlike conventional solvents, where it is normal to commercially obtain very pure materials with complete analytical data, the situation with commercial ionic liquids is uneven, with no broadly accepted purity standards set by industry or some governing body. Although the policies of some ionic liquid suppliers may be more thorough, stated ionic liquid purity criteria have largely focussed exclusively on water and halide content. Nonetheless, other serious contaminants could include unreacted starting materials, side products, fine particles or compounds extracted out of columns, and solvents used in processing, including for the removal of water. Stories of experimental troubles caused in ostensibly pure ionic liquids by lithium ion [45], alumina [45,46], silica [46], methylimidazole [47], and dichloromethane (used to remove water but inadequately pumped off) circulate among researchers in the field.

The situation is also complicated for those who choose to, or of necessity must, prepare their own ionic liquids. If one is up on the literature, it is possible to use the best recommended synthesis and purification techniques, and one has the advantage of knowing all the reagents and solvents used in the preparation. However, adequate analytical instrumentation or the time to use it may not be available for the individual principal investigator. One approach that is being taken by many people to address this is to collaborate in larger groups that share materials and the expertise to properly characterise them. These large collaborations are definitely a boon to the field and should be encouraged. From my observations, however, widespread distribution of common research materials does not guarantee that the quality of those materials is adequate.

For non-aromatic ionic liquids, one useful empirical criterion of ionic liquid purity for radiation chemistry purposes is the lifetime of the solvated electron, which is a very reactive species that can be scavenged by acids, aromatics, haloalkanes, and protonated amines. Figure 10.1 shows the decay kinetics of the solvated electron in 1-butyl-1-methylpyrrolidinium bis{(trifluoromethyl)sulfonyl}amide, followed by its optical absorption at 900 nm measured at the BNL Laser-Electron Accelerator Facility (LEAF) pulse radiolysis facility [2]. The four lots of "high purity" ionic liquid came from two suppliers, as listed in the legend. The samples from Supplier 1, lot "A" and Supplier 2 show simple pseudo-first-order electron decays, indicating the presence of a significant amount of adventitious scavenger, in the case of "A," a very large amount. After these data were shown to a representative of Supplier 1 at a conference, the supplier sent lot "B," which contained much less scavenger, as indicated in the figure. The next batch ordered, lot "C," showed even more ideal behaviour. The true form of the electron decay will look much like a second order reaction, but it is actually inhomogeneous electron–hole recombination within the radiation track. Preparations of [C$_4$mpyr][NTf$_2$] made in our laboratories

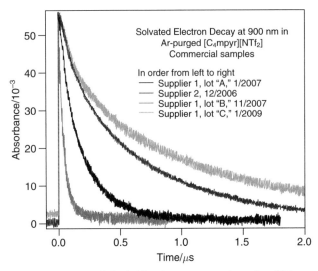

**Figure 10.1**  Electron pulse radiolysis kinetic traces monitored at 900 nm, showing the decay of solvated electrons in four commercial lots of 1-butyl-1-methylpyrrolidinium bis{(trifluoromethyl)sulfonyl}amide. The peak absorbances have been normalised for easier comparison of the kinetics. Samples synthesised in our own laboratory fall between lots "B" and "C."

show behaviours between lots "B" and "C." To my colleagues in ionic liquid radiation chemistry, I would suggest that if your aliphatic ionic liquid electron decays look like "A" and not "B" or "C," you should be concerned.

This story shows that ionic liquids suppliers will work with customers to provide the quality of materials they need, but because of the range of science being done, they cannot necessarily anticipate those needs. Nevertheless, it is important for the quality of research in the field—which ultimately will determine the outcome between success and failure—to establish baseline performance criteria for ionic liquid products intended for certain types of research and applications, just as with conventional solvents. The various research communities, physical chemists, electrochemists, catalysis researchers, and so on, should make their group needs known to industry, and explain these needs in terms of a business case that will satisfy all parties. On the bottom line, better research will lead to more business.

## 10.6  PRACTICAL IMPORTANCE: APPLYING FUNDAMENTAL IONIC LIQUID RADIATION CHEMISTRY TO NANOPARTICLE SYNTHESIS

Thanks to their unusual combinations of properties, ionic liquids are enabling exciting new areas of physical chemistry to be explored, including aspects of fundamental radiation chemistry that were previously inaccessible. Such basic knowledge is important for understanding how our world works, but it is especially gratifying if it can be directly applied to accomplishing a purpose such as the production of advanced materials.

Radiation chemistry has been applied to metal nanoparticle synthesis in conventional solvents for many decades. It has helped to explain and ulti- mately to perfect the mechanisms of silver photographic image formation [48], and it allows an exquisite level of control over nanoparticle synthesis, including bimetallic core/shell versus alloy composition and the growth of specific mor- phologies and complex three-dimensional structures [49]. Nanoparticles in ionic liquids are currently a topic of intense interest, so it comes as no surprise that radiolytic nanoparticle synthesis has been tried in ionic liquids. However, results can vary widely from one ionic liquid to another (changing cations or anions) due to basic differences in the radiation chemistries of different fami- lies of ionic liquids, which must be understood for proper application.

An interesting example of the impacts of different radiation chemistry mechanisms on radiolytic nanoparticle synthesis in ionic liquids was recently published by Kuwabata and coworkers [12], who examined gold nanoparticle synthesis in three ionic liquids, $[C_4mim][NTf_2]$, $[N_{1\,4\,4\,4}][NTf_2]$, and $[C_1C_3pip]$ $[NTf_2]$. Using gamma and electron beam radiolysis of 0.5 mmol $l^{-1}$ $NaAuCl_4$ $\cdot 2H_2O$ in each of the three ionic liquids and radiation doses of 6 or 20 kGy, gold nanoparticles were only observed in the imidazolium salt and not in the ammonium ones (except for a very small amount in $[C_1C_3pip][NTf_2]$ at the highest dose and dose rate).

The results are not adequately explained in the original report, but from our understanding of primary ionic liquid radiation chemistry, we know that radiolytically produced excess electrons would be rapidly captured by imid- azolium cations to form neutral radicals [24, 28, 29], which may form dimers by reacting with an additional cation [7, 31, 32], whereas the electron does not react with the cations in quaternary ammonium ionic liquids, and instead becomes a solvated species [23, 50, 51]. However, the diffusion-limited rate constant for solvated electron scavenging in $[N_{1\,4\,4\,4}][NTf_2]$ at room temperature is $\leq 2 \times 10^8$ $l$ $mol^{-1} \cdot s^{-1}$ [23], resulting in an effective lifetime for the electrons (in the absence of other reactions) of at least 10 microseconds at the gold concentration used. In practice, the electrons will not survive nearly that long because they will react either with inadvertent scavengers, or if the ionic liquid is very pure, by geminate recombination with transient species produced from the radiolytically formed holes on the timescale of a couple of microseconds.

Thus, it comes as no surprise that gold reduction will not be competitive in the ammonium ionic liquids under the conditions that were used. However, the remarkable feature of these experiments is that the reduction product of the imidazolium cation (whose exact form is unknown but the subject of active investigation [34]) is indeed capable of efficiently reducing tetrachlo- roaurate(III) anions to form gold nanoparticles. The final product of the imidazolium-derived species after gold reduction is not yet characterised, nor is the yield of metallic gold per unit of radiolytic energy absorbed. Supporting this analysis, another study showed that gold nanoparticles were successfully formed in a quaternary ammonium ionic liquid when the $Zn^{2+/+}$ couple was used as an electron shuttle, even though the gold concentration was five times lower [52].

The established knowledge base on ionic liquid radiation chemistry also suggests that radiolytic nanoparticle synthesis should proceed efficiently in pyridinium ionic liquids, since the pyridinium cation captures electrons to form a relatively stable pyridinyl radical that undergoes facile redox processes [53, 54]. As of this writing, no published information could be found on such studies. Ultimately, advanced radiolytic nanoparticle synthesis in ionic liquids may involve mixtures of two or more cations or anions, with certain ions selected for their radiolysis and redox properties, and others chosen to control nanoparticle growth and morphology through interactions with the particle surface [55].

## 10.7   FUTURE PROSPECTS

There are vast areas of radiation chemistry research to be explored using ionic liquids and vice versa. Ionic liquids provide a new and adaptable medium to test theories concerning the early events in radiolytic energy deposition and the effects of solvent relaxation on transient species energetics and reactivity. The potential for transfer of radiolytic damage (holes and electrons) between cations and anions makes each combination of ions a new chapter in the story. Ultimately, we wish to elucidate chemistries for particular families of ionic liquids that will convert the chaos of the initial radiolysis events into known yields of specific products, so that pulse radiolysis in ionic liquids can be used to study electron transfer, catalysis, charge transport through advanced devices, nanomaterial synthesis, durability in extreme environments, carbon fixation, clean fuel generation, nuclear separations, and many other technologies that are important for creating a sustainable world infrastructure.

## ACKNOWLEDGEMENTS

The author thanks Ilya A. Shkrob, Robert Crowell, Jasmine Hatcher, R. Dale Rimmer, Masao Gohdo, Tomasz Szreder, and Jay LaVerne for helpful discussions. This work was supported by the US Department of Energy, Office of Basic Energy Sciences, Division of Chemical Sciences, Geosciences, and Biosciences under contract No. DE-AC02-98CH10886.

## REFERENCES

1   Curie, P., and Curie, M., Effets chimiques produits par les rayons de Becquerel, *C. R. Hebd. Seances Acad. Sci.* **129**, 823–825 (1899).

2   Wishart, J.F., Cook, A.R., and Miller, J.R., The LEAF Picosecond Pulse Radiolysis Facility at Brookhaven National Laboratory, *Rev. Sci. Instrum.* **75**, 4359–4366 (2004).

3   Wishart, J.F., Tools for radiolysis studies, in *Radiation Chemistry: From Basics to Applications in Material and Life Sciences*, eds. J. Belloni, T. Douki, M. Mostafavi and M. Spotheim-Maurizot, L'Actualité Chimique Livres (EDP Sciences, Paris, 2008), pp. 17–33.

4   Belloni, J., Crowell, R.A., Katsumura, Y., Lin, M., Marignier, J.-L., Mostafavi, M., Muroya, Y., Saeki, A., Tagawa, S., Yoshida, Y., De Waele, V., and Wishart, J.F., Ultrafast pulse radiolysis methods, in *Recent Trends in Radiation Chemistry*, eds. J.F. Wishart and B.S.M. Rao (World Scientific Publishing Co., Singapore, 2010), pp. 121–160.

5   Grimes, R.W., and Nuttall, W.J., Generating the Option of a Two-Stage Nuclear Renaissance, *Science* **329**, 799–803 (2010).

6   Ha, S.H., Menchavez, R.N., and Koo, Y.M., Reprocessing of spent nuclear waste using ionic liquids, *Korean J. Chem. Eng.* **27**, 1360–1365 (2010).

7   Wishart, J.F., and Shkrob, I.A., The radiation chemistry of ionic liquids and its implications for their use in nuclear fuel processing, in *Ionic Liquids: From Knowledge to Application*, eds. R.D. Rogers, N.V. Plechkova, and K.R. Seddon, ACS Symposium Series (American Chemical Society, Washington, 2009), pp. 119–134.

8   Metlen, M.S.A., and Rogers, R.D., The second evolution of ionic liquids: From solvents and separations to advanced materials: Energetic examples from the ionic liquid cookbook, *Acc. Chem. Res.* **40**, 1182–1192 (2007).

9   Angel, R., Worden, S.P., Borra, E.F., Eisenstein, D.J., Foing, B., Hickson, P., Josset, J.L., Ma, K.B., Seddiki, O., Sivanandam, S., Thibault, S., and van Susante, P., A cryogenic liquid-mirror telescope on the moon to study the early universe, *Astrophys. J.* **680**, 1582–1594 (2008).

10  Qi, M., Wu, G., Sha, M., and Liu, Y., Radiation induced polymerization of MMA in imidazolium ionic liquids and their mixed solutions with organic solvents, *Radiat. Phys. Chem.* **77**, 1248–1252 (2008).

11  Hao, Y., Peng, J., Li, J.Q., Zhai, M.L., and Wei, G.S., An ionic liquid as reaction media for radiation-induced grafting of thermosensitive poly (N-isopropylacrylamide) onto microcrystalline cellulose, *Carbohydr. Polym.* **77**, 779–784 (2009).

12  Tsuda, T., Seino, S., and Kuwabata, S., Gold nanoparticles prepared with a room-temperature ionic liquid-radiation irradiation method, *Chem. Commun.* 6792–6794 (2009).

13  Wishart, J.F., Energy applications of ionic liquids, *Energy Environ. Sci.* **2**, 956–961 (2009).

14  Wishart, J.F., and Nocera, D.G. (eds.), *Photochemistry and Radiation Chemistry*, Advances in Chemistry, Vol. 254 (American Chemical Society, Washington, DC, 1998).

15  Belloni, J., Douki, T., Mostafavi, M., and Spotheim-Maurizot, M., eds., *Radiation Chemistry: From Basics to Applications in Material and Life Sciences*, L'Actualité Chimique Livres (EDP Sciences, Paris, 2008).

16  Wishart, J.F., and Rao, B.S.M. (eds.), *Recent Trends in Radiation Chemistry* (World Scientific Publishing Co., Singapore, 2010).

17 Hatano, Y., Katsumura, Y., and Mozumder, A. (eds.), *Charged Particle and Photon Interactions with Matter: Recent Advances, Applications, and Interfaces* (CRC Press, Boca Raton, 2010).

18 Wishart, J.F., Radiation chemistry of ionic liquids: Reactivity of primary species, in *Ionic Liquids as Green Solvents: Progress and Prospects*, eds. R.D. Rogers and K.R. Seddon, ACS Symposium Series, Vol. 856 (American Chemical Society, Washington, 2003), pp. 381–396.

19 Funston, A.M., and Wishart, J.F., Dynamics of fast reactions in ionic liquids, in *Ionic Liquids IIIA: Fundamentals, Progress, Challenges, and Opportunities, Properties and Structure*, eds. R.D. Rogers and K.R. Seddon, ACS Symposium Series, Vol. 901 (American Chemical Society, Washington, 2005), pp. 102–116.

20 Wishart, J.F., Funston, A.M., and Szreder, T., Radiation chemistry of ionic liquids, in *Molten Salts XIV*, eds. R.A. Mantz, P.C. Trulove, H.C. De Long, G.R. Stafford, R. Hagiwara and D.A. Costa (The Electrochemical Society, Pennington, NJ, 2006), pp. 802–813.

21 Takahashi, K., and Wishart, J.F., Radiation Chemistry and Photochemistry of Ionic Liquids, in *Charged Particle and Photon Interactions with Matter*, eds. Y. Hatano, Y. Katsumura and A. Mozumder (Taylor & Francis, 2010), pp. 265–287.

22 Wishart, J.F., Ionic liquids and ionizing radiation: Reactivity of highly energetic species, *J. Phys. Chem. Lett.* **1**, 3225–3231 (2010).

23 Wishart, J.F., and Neta, P., Spectrum and reactivity of the solvated electron in the ionic liquid methyltributylammonium bis(trifluoromethylsulfonyl)imide, *J. Phys. Chem. B* **107**, 7261–7267 (2003).

24 Takahashi, K., Sato, T., Katsumura, Y., Yang, J., Kondoh, T., Yoshida, Y., and Katoh, R., Reactions of solvated electrons with imidazolium cations in ionic liquids, *Radiat. Phys. Chem.* **77**, 1239–1243 (2008).

25 Cook, A.R., and Shen, Y.Z., Optical fiber-based single-shot picosecond transient absorption spectroscopy, *Rev. Sci. Instrum.* **80**, 073106 (2009).

26 Grodkowski, J., and Neta, P., Formation and reaction of $Br_2^{\bullet-}$ radicals in the ionic liquid methyltributylammonium bis(trifluoromethylsulfonyl)imide and in other solvents, *J. Phys. Chem. A* **106**, 11130–11134 (2002).

27 Grodkowski, J., Nyga, M., and Mirkowski, J., Formation of $Br_2^{\bullet-}$, $BrSCN^{\bullet-}$ and $(SCN)_2^{\bullet-}$ intermediates in the ionic liquid methyltributylammonium bis[(trifluoromethyl)sulfonyl]imide. Pulse radiolysis study, *Nukleonika* **50**, S35–S38 (2005).

28 Behar, D., Gonzalez, C., and Neta, P., Reaction kinetics in ionic liquids: Pulse radiolysis studies of 1-butyl-3-methylimidazolium salts, *J. Phys. Chem. A* **105**, 7607–7614 (2001).

29 Marcinek, A., Zielonka, J., Gębicki, J., Gordon, C.M., and Dunkin, I.R., Ionic liquids: Novel media for characterization of radical ions, *J. Phys. Chem. A* **105**, 9305–9309 (2001).

30 Shkrob, I.A., Chemerisov, S.D., and Wishart, J.F., The initial stages of radiation damage in ionic liquids and ionic liquid-based extraction systems, *J. Phys. Chem. B* **111**, 11786–11793 (2007).

31 Shkrob, I.A., and Wishart, J.F., Charge trapping in imidazolium ionic liquids, *J. Phys. Chem. B* **113**, 5582–5592 (2009).

32    Shkrob, I.A., Deprotonation and oligomerization in photo-, radiolytically, and electrochemically induced redox reactions in hydrophobic alkylalkylimidazolium ionic liquids, *J. Phys. Chem. B* **114**, 368–375 (2010).

33    Shkrob, I.A., Marin, T.W., Chemerisov, S.D., and Wishart, J.F., Radiation induced redox reactions and fragmentation of constituent ions in ionic liquids. 1. Anions., *J. Phys. Chem. B* **115**, 3872–3888 (2011).

34    Shkrob, I.A., Marin, T.W., Chemerisov, S.D., Hatcher, J., and Wishart, J.F., Radiation induced redox reactions and fragmentation of constituent ions in ionic liquids. 2. Imidazolium cations., *J. Phys. Chem. B* **115**, 3889–3902 (2011).

35    Berthon, L., Nikitenko, S.I., Bisel, I., Berthon, C., Faucon, M., Saucerotte, B., Zorz, N., and Moisy, P., Influence of gamma irradiation on hydrophobic room-temperature ionic liquids [BuMeIm]$PF_6$ and [BuMeIm]$(CF_3SO_2)_2N$, *Dalton Trans.* 2526–2534 (2006).

36    Le Rouzo, G., Lamouroux, C., Dauvois, V., Dannoux, A., Legand, S., Durand, D., Moisy, P., and Moutiers, G., Anion effect on radiochemical stability of room-temperature ionic liquids under gamma irradiation, *Dalton Trans.* 6175–6184 (2009).

37    Bossé, E., Berthon, L., Zorz, N., Monget, J., Berthon, C., Bisel, I., Legand, S., and Moisy, P., Stability of [MeBu$_3$N][Tf$_2$N] under gamma irradiation, *Dalton Trans.* 924–931 (2008).

38    Plechkova, N.V., and Seddon, K.R., Applications of ionic liquids in the chemical industry, *Chem. Soc. Rev.* **37**, 123–150 (2008).

39    Swatloski, R.P., Holbrey, J.D., and Rogers, R.D., Ionic liquids are not always green: Hydrolysis of 1-butyl-3-methylimidazolium hexafluorophosphate, *Green Chem.* **5**, 361–363 (2003).

40    Villagran, C., Deetlefs, M., Pitner, W.R., and Hardacre, C., Quantification of halide in ionic liquids using ion chromatography, *Anal. Chem.* **76**, 2118–2123 (2004).

41    Allen, D., Baston, G., Bradley, A.E., Gorman, T., Haile, A., Hamblett, I., Hatter, J.E., Healey, M.J.F., Hodgson, B., Lewin, R., Lovell, K.V., Newton, B., Pitner, W.R., Rooney, D.W., Sanders, D., Seddon, K.R., Sims, H.E., and Thied, R.C., An investigation of the radiochemical stability of ionic liquids, *Green Chem.* **4**, 152–158 (2002).

42    LaVerne, J.A., and Schuler, R.H., Track effects in radiation chemistry: Core processes in heavy-particle tracks as manifest by the hydrogen yield in benzene radiolysis, *J. Phys. Chem.* **88**, 1200–1205 (1984).

43    Foldiak, G., Radiolysis of liquid hydrocarbons, *Radiat. Phys. Chem. (1977)*, **16**, 451–463 (1980).

44    Tarabek, P., Liu, S.Y., Haygarth, K., and Bartels, D.M., Hydrogen gas yields in irradiated room-temperature ionic liquids, *Radiat. Phys. Chem.* **78**, 168–172 (2009).

45    Endres, F., El Abedin, S.Z., and Borissenko, N., Probing lithium and alumina impurities in air- and water stable ionic liquids by cyclic voltammetry and in situ scanning tunneling microscopy, *Z. Phys. Chem.* **220**, 1377–1394 (2006).

46    Clare, B.R., Bayley, P.M., Best, A.S., Forsyth, M., and MacFarlane, D.R., Purification or contamination? The effect of sorbents on ionic liquids, *Chem. Commun.* 2689–2691 (2008).

47 Schmeisser, M., and van Eldik, R., Thermodynamic and kinetic studies on reactions of Fe-III(meso-[tetra(3-sulfonatomesityl)porphin]) with NO in an ionic liquid. Trace impurities can change the mechanism!, *Inorg. Chem.* **48**, 7466–7475 (2009).

48 Belloni, J., and Remita, H., Metal clusters and nanomaterials, in *Radiation Chemistry: From Basics to Applications in Material and Life Sciences*, eds. J. Belloni, T. Douki, M. Mostafavi and M. Spotheim-Maurizot, L'Actualité Chimique Livres (EDP Sciences, Paris, 2008), pp. 97–116.

49 Remita, H., and Remita, S., Metal clusters and nanomaterials: Contribution of radiation chemistry, in *Recent Trends in Radiation Chemistry*, eds. J.F. Wishart and B.S.M. Rao (World Scientific Publishing Co., Singapore, 2010), pp. 347–383.

50 Wishart, J.F., Lall-Ramnarine, S.I., Raju, R., Scumpia, A., Bellevue, S., Ragbir, R., and Engel, R., Effects of functional group substitution on electron spectra and solvation dynamics in a family of ionic liquids, *Radiat. Phys. Chem.* **72**, 99–104 (2005).

51 Asano, A., Yang, J., Kondoh, T., Norizawa, K., Nagaishi, R., Takahashi, K., and Yoshida, Y., Molar absorption coefficient and radiolytic yield of solvated electrons in diethylmethyl(2-methoxy)ammonium bis(trifluoromethanesulfonyl) imide ionic liquid, *Radiat. Phys. Chem.* **77**, 1244–1247 (2008).

52 Chen, S.M., Liu, Y.D., and Wu, G.Z., Stabilized and size-tunable gold nanoparticles formed in a quaternary ammonium-based room-temperature ionic liquid under gamma-irradiation, *Nanotechnology* **16**, 2360–2364 (2005).

53 Behar, D., Neta, P., and Schultheisz, C., Reaction kinetics in ionic liquids as studied by pulse radiolysis: Redox reactions in the solvents methyltributylammonium bis(trifluoromethylsulfonyl)imide and *N*-butylpyridinium tetrafluoroborate, *J. Phys. Chem. A* **106**, 3139–3147 (2002).

54 Skrzypczak, A., and Neta, P., Diffusion-controlled electron-transfer reactions in ionic liquids, *J. Phys. Chem. A* **107**, 7800–7803 (2003).

55 Endres, F., Hofft, O., Borisenko, N., Gasparotto, L.H., Prowald, A., Al-Salman, R., Carstens, T., Atkin, R., Bund, A., and El Abedin, S.Z., Do solvation layers of ionic liquids influence electrochemical reactions? *Phys. Chem. Chem. Phys.* **12**, 1724–1732 (2010).

# 11 Physicochemical Properties of Ionic Liquids

QING ZHOU, XINGMEI LU, SUOJIANG ZHANG, and LIANGLIANG GUO

Beijing Key Laboratory of Ionic Liquids Clean Process, Key Laboratory of Green Process and Engineering, State Key Laboratory of Multiphase Complex System, Institute of Process Engineering, Chinese Academy of Sciences, Beijing, People's Republic of China

## ABSTRACT

This chapter discusses the key relationships of the physicochemical properties of ionic liquids, particularly melting point, density, viscosity, and surface tension, including an assessment of the current situation and identifying trends for ionic liquids in the future.

## 11.1 INTRODUCTION

Today, scientific developments occur at the intersection, penetration, and integration of multiple disciplines: from here new research areas are opened up and new growth points for these disciplines are born. It is exactly this background that has led to the emergence of ionic liquids, which have provided new opportunities for the development of chemical technology. A significant amount of research on ionic liquids has been performed, from fundamental studies to application, due to their unique characteristics, such as non-volatility, excellent catalytic performance, and good solubility for many materials. These unique physical and chemical properties depend on the special microstructure and the complex interactions within the ionic liquid systems. However, there are thought to be at least a million types of possible ionic liquids and a trillion ($10^{18}$) types of mixed-liquid systems [1]. Therefore, it is very challenging to

*Ionic Liquids Further UnCOILed: Critical Expert Overviews*, First Edition.
Edited by Natalia V. Plechkova and Kenneth R. Seddon.
© 2014 John Wiley & Sons, Inc. Published 2014 by John Wiley & Sons, Inc.

screen and design functional ionic liquids for specific applications from the countless possible combinations of cations and anions. Anyone developing ionic liquids for technological applications must face this challenge, and there is always the danger that a competitor will chance upon a better choice. Hope lies in the major efforts now being made to model and predict the properties of ionic liquids [2], although such predictive methods will take much time to develop, and sometimes they are applicable only to several ionic liquids. So, currently, research on the physicochemical properties of ionic liquids has been closely combined with development of their applications.

Along with the continuing emergence of new research on ionic liquids, producing huge amounts of data describing their properties, easy access to the exponentially increasing number of publications on ionic liquids during the past decade necessitates a comprehensive data collection of as many properties as possible. However, a large amount of data is dispersed in various journals, reports, books, patents, and so on. Some scientific research associations or companies have established databases of ionic liquid physicochemical properties in order to aid systematic evaluation and analysis, such as ILThermo [3], Institute of Process Engineering (IPE) (Chinese Academy of Sciences, Beijing) [4], Merck (Germany) [5], DDBST [6], and DelphIL [7]. According to incomplete statistics, there are more than 2000 different ionic liquids that have been synthesised during the last 20 years [4]. From Figure 11.1, imidazolium ionic liquids are the most popular ionic liquids extensively investigated by both academia and industry: the structures of other common cations are shown in Table 11.1, which account for 800, or about one-third of total number of ionic liquids. These ionic liquid databases provide access to the scientific data needed to refine universal property relationships, reveal the inner relationship between microscopic structure and macroscopic properties, design new ionic liquids effectively, and develop applications for ionic liquids.

In the following sections, we will discuss the key relationships of the physicochemical properties of ionic liquids, especially melting point, density,

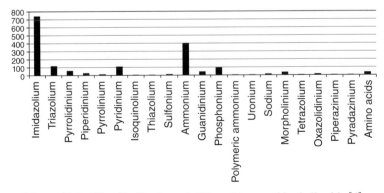

**Figure 11.1** The distribution of different types of ionic liquids [4].

**TABLE 11.1  The Structures of Common Cations found in Ionic Liquids**

| Entry | Name of Cations | Structure | Entry | Name of Cations | Structure |
|-------|-----------------|-----------|-------|-----------------|-----------|
| 1 | 1-Alkylimidazolium | | 14 | Phosphonium | |
| 2 | 1,3-Dialkylimidazolium | | 15 | Sodium[a] | $Na^+$ |
| 3 | Trialkylimidazolium | | 16 | Morpholinium | |
| 4 | Tetraalkylimidazolium | | 17 | Tetrazolium | |
| 5 | Pentaalkylimidazolium | | 18 | Oxazolidinium | |
| 6 | Double imidazolium | | 19 | Amino acids | |
| 7 | Triazolium | | 20 | Isoquinolinium | |

*(Continued)*

**TABLE 11.1** (*Continued*)

| Entry | Name of Cations | Structure | Entry | Name of Cations | Structure |
|-------|-----------------|-----------|-------|-----------------|-----------|
| 8 | Pyrrolidinium | | 21 | Sulfonium | |
| 9 | Piperidinium | | 22 | Uronium | |
| 10 | Pyrrolinium | | 23 | Piperazinium | |
| 11 | Pyridinium | | 24 | Pyridazinium | |
| 12 | Ammonium | | 25 | Polymeric ammonium | |
| 13 | Guanidinium | | | | |

[a] The anions Na$^+$ salts are [M(TiW$_{11}$O$_{39}$)$_2$]$_{13}$$^-$ (M = La, Ce, Pr, Sm, Gd, Dy, Er, Tm, Yb) and [MTiW$_{11}$O$_{39}$]$_5$$^-$ (M = Cr, Mn, Fe, Zn).

**TABLE 11.2 The Properties and Data Points for Ionic Liquids**

| Entry | Physicochemical Properties | Data Points |
|---|---|---|
| 1 | Melting point | 1473 |
| 2 | Glass transition temperature | 755 |
| 3 | Crystallisation temperature | 27 |
| 4 | Decomposition temperature | 910 |
| 5 | Solid–solid transition temperature | 55 |
| 6 | Freezing point | 85 |

viscosity, and surface tension, including an assessment of the current situation and identifying trends for ionic liquids in the future.

## 11.2  MELTING POINT

Melting point is one of most important properties of ionic liquids, and also one of the most reported phenomena. However, up to now, the collected data of melting points are only around 1400 (see Table 11.2); but many ionic liquids have no melting points, instead exhibiting a glass transition temperature. To complicate matters further, many room-temperature ionic liquids also exist as supercooled liquids, occasionally for extended periods of time (days or even weeks). For example, $[C_2mim][BF_4]$ exists as a supercooled liquid below the melting point [8]. As a result, melting point determinations are not always straightforward and some compounds may be incorrectly considered to be liquids if care is not taken during the course of the investigation [9]. Accurate values for melting points for ionic liquids are scarce as, like in the case of inorganic salts, melting point and glass transition temperatures can be strongly affected by the presence of impurities [10]. The presence of water, organic solvents (hexane, ethanenitrile, benzene, etc.), and halide impurities, even in small concentrations, has been long understood to alter the physical properties of ionic liquids [11–15]. Wilkes and coworkers investigated the freezing–melting behaviour of $[C_2mim][BF_4]$, $[C_4mim][BF_4]$, and $[C_3C_1mim][NTf_2]$. Contamination of ionic liquids by water and metal cations has variable impact on their freezing exotherms. But chloride impurities had little effect on the freezing–melting behaviour of the studied ionic liquids [16]. So, in our database, $[C_2mim][BF_4]$ has eight different melting points, ranging from 279.15 to 288.15 K, and $[C_2mim][NTf_2]$ has even 10 different melting points ranging from 252.15 to 270.15 K [4].

Figure 11.2 shows the melting points range of different series of ionic liquids. Up to now, it is known that the melting point of trialkylimidazolium ionic liquids is highest, with the melting point of 1,3-dimethyl-2-phenylimidazolium bromide at 556.15 K [17]. $[N_{1\ 8\ 8\ 8}][Me_3CC(O)CHC(O)(CF_2)_2CF_3]$ has the lowest melting point at 177.15 K [18].

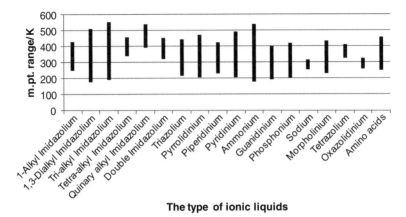

**The type of ionic liquids**

**Figure 11.2** The melting point ranges of typical series of ionic liquids.

The melting point ($T_m$) of an organic molecular compound is determined by the strength of its crystal lattice, which is controlled by three main factors: molecular symmetry, intermolecular forces, and conformational degrees of freedom of the molecule; this principle is also applicable to ionic liquids. Qualitatively, reducing ion symmetry, increasing the ion conformational degrees of freedom (e.g., utilising flexible substituents), and improving the charge distribution of the cation and/or anion are effective approaches for reducing the lattice energy of the salts, thus resulting in low melting materials [19]. This principle can be used in various types of ionic liquids. Detailed descriptions can be found in a large number of papers [20–25].

Up to now, the most popular family of ionic liquids are those ionic liquids based on the imidazolium cations. The modern resurrection of ionic liquid chemistry, in general, is largely due to the unusually low melting point of many imidazolium salts, and can be traced to the seminal report of Wilkes and Zaworotko [26]. As a result, an enormous number of different alkyl groups and anions have been explored. So, the melting point principle of imidazolium-based ionic liquids is more detailed than the other ionic liquids.

Several significant papers have discussed the melting point trend of alkyl substitution on methylimidazole [9, 24, 25], so the subject requires little detailed repetition. The smallest alkyl groups (methyl, ethyl, or propyl) result in salts having higher melting points, often being solids at room temperature. As can be seen in Figure 11.3, when the alkyl chain becomes longer (butyl to octyl), the melting point decreases and usually reaches a minimum somewhere in the range from butyl to octyl. Beyond this point, the melting point again increases until, with a much longer alkyl group (tetradecyl and higher), liquid crystalline compounds are often obtained. The most accepted explanation is that longer alkyl chains decrease the symmetry of the imidazolium cation and thereby interfere in efficient crystal packing: Coulombic forces are responsible for

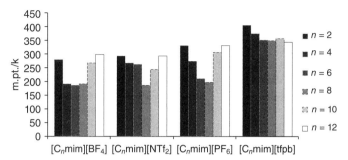

**Figure 11.3** The effect of alkyl chain length on the melting points of 1-alkyl-3-methy-limidazolium ionic liquids.

**TABLE 11.3    The Melting Points of Ionic Liquids, [A][X]**

| Entry | Cation (A) | Anion (X) | m.pt./K | Reference |
|-------|-----------|-----------|---------|-----------|
| 1 | Hmim | Cl | 345.15 | 27 |
| 2 | $C_1$mim | Cl | 398.15 | 28 |
| 3 | $C_1C_1$mim | Cl | 462.15 | 29 |
| 4 | Hmim | Br | 314.15 | 27 |
| 5 | $C_1$mim | Br | 382.65 | 30 |
| 6 | $C_2$im | Cl | 331.15 | 27 |
| 7 | $C_2$mim | Cl | 362.15 | 21 |
| 8 | $C_2C_1$mim | Cl | 461.15 | 21 |
| 9 | Hmim | $NTf_2$ | 282.15 | 27 |
| 10 | $C_1$mim | $NTf_2$ | 295.15 | 31 |
| 11 | $C_2$mim | $NTf_2$ | 270 | 31 |
| 12 | $C_2C_1$mim | $NTf_2$ | 293 | 31 |

much of the attractive forces in shorter alkyl group ionic liquids, and van der Waals forces are responsible for much of the attractive forces in the longer alkyl group ionic liquids. Anything that interferes in packing should decrease the melting point. When the alkyl chains are sufficiently long to become a significant component, the attractive van der Waals forces will lead to a steady increase in melting point.

Usually, an increase in molecular weight and the accumulation of charge causes an increase in the melting point. From Table 11.3, we can see that the addition of one methyl group may increase the melting point of an ionic liquid by 20–100 K. For example, the melting point of $[C_2C_1mim][NTf_2]$ is 23 K greater than that of $[C_2mim][NTf_2]$, and the melting points of [Hmim]Cl, $[C_1mim]$Cl, and $[C_1C_1mim]$Cl are 345.15, 398.15, and 462.15 K, respectively.

The increased branching of the alkyl group results in melting points higher than the linear analogues. An isopropyl group instead of an *n*-propyl group increases the melting point by 82 °C. Thus, [$^i$C$_4$mim][PF$_6$] (432.85 K) and [$^i$C$_3$mim][PF$_6$] (375.15 K) have higher melting points than [$^s$C$_4$mim][PF$_6$] (356.45 K) and [C$_3$mim][PF$_6$] (322.15 K), while [$^s$C$_4$mim][PF$_6$] (356.45 K) in turn has a higher melting point than simple [C$_4$mim][PF$_6$] (265.15 K) [3].

These principles are also applicable to other types of ionic liquids, such as pyridinium-based ionic liquids, ammonium-based ionic liquids, and phosphonium-based ionic liquids. Beyond simple alkyl groups, an increasing number of functionalised side chains have been reported: partially fluorinated alkyl groups have the tendency to increase the melting point slightly over their non-fluorinated counterparts. The ester, nitrile, amide, ether, or hydroxyl modified ionic liquids have melting points that are little different from those of the simple alkyl ionic liquids with similar molecular weight. As a result, there is not a clear correlation between the functionalisation of an ionic liquid and its melting point.

Explaining the anion effect is more difficult, owing to the presence of water and its interactions with the cation and anion. As is shown in Figure 11.4, imidazoles neutralised with HCl, HNO$_3$, or HBr gave relatively high values of $T_m$. In contrast, salts neutralised with bis{(trifluoromethyl)sulfonyl}amidic acid (HNTf$_2$) and bis{(pentafluoroethyl)sulfonyl}amidic acid (HNPf$_2$) were obtained as liquids [27]. For ionic liquids containing structurally similar anions

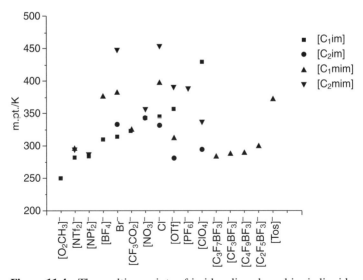

**Figure 11.4**   The melting points of imidazolium-based ionic liquids.

such as triflate ([OTf]$^-$) and [NTf$_2$]$^-$, the lower melting point of the [NTf$_2$]$^-$ salt could be attributed to electron delocalisation and the anion's inability to hydrogen-bond. In a similar manner, differences between [C$_2$mim][CF$_3$CO$_2$] and [C$_2$mim][C$_1$CO$_2$], besides the former having a melting point of 31 °C lower than the latter, can undoubtedly be attributed to the presence of fluorine atoms on that anion and their interaction with other anions and the cation. The melting point for [C$_2$mim][NO$_2$] is 17 °C higher than that for [C$_2$mim][NO$_3$], suggesting that anion structure also contributes to the thermal properties [26].

In order to investigate ionic liquid structural features that could lead to low melting point salts, several methods have been used in attempts to make quantitative predictions of their melting points, such as quantitative structure–property relationships (QSAR) [32–34], molecular mechanics (MM) simulations, as well as modifying "hole theory" or the "Parachor" [35–39]. However, these methods all have significant drawbacks, which limit their application for predicting the properties of unknown salts. These include the need for large experimental data sets to derive correlations, time-consuming computational methods, or the need for at least some experimental data from the ionic liquid under study [40]. Recently, Krossing and coworkers showed that the relatively low melting points of ionic liquids can be understood by a simple thermodynamic cycle based on lattice and solvation energies [41]. They assessed the Gibbs free energy of fusion as a predictor of the melting point using a Born–Fajans–Haber cycle, which was closed by the lattice and solvation Gibbs energies of the constituent ions in the molten salt [40, 42]. These were calculated using a combination of volume-based thermodynamics and quantum chemical calculations for the lattice free energies and the COSMO-RS (**CO**nductor-like**S**creening**MO**del for Real Solvents) solvation model. The two methods, the volume-based model (only ion volumes, $\sigma$, and $\tau$ as input) and the augmented method (using ion volumes, $\sigma$, $\tau$, and COSMO-RS output), were tested on several sets of ionic liquids, and a combination of all sets (67 ionic liquids) that span an experimental melting temperature range of 337 °C. The average error of the simpler, volume-based model is 36.4 °C and that of the augmented method is 24.5 °C. This method has no need for experimental input or tedious simulations, but relies on simple calculations feasible with standard quantum chemical program codes, and may further be augmented by COSMO-RS.

## 11.3   DENSITY

Density is an important physicochemical and crucial for the design of many technological processes, especially for the simulation of heat and mass transfer, and hydrodynamics calculations. Up to now, the total level of density data in ILThermo from IUPAC [3] and other databases [4] are 28.8% and 24.2%, respectively. In addition to some ionic liquids with cations of imidazolium,

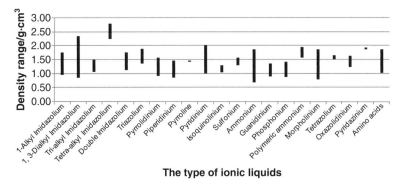

**Figure 11.5**   The density range of different types of ionic liquids [4].

pyrrolidinium, piperidinium, quaternary ammonium, guanidinium, quaternary phosphonium, and morpholinium, which densities lie in the range of 0.6–0.97 g cm$^{-3}$, almost all ionic liquids possess a density greater than 1 g cm$^{-3}$, some being close to, or more than, 2 g cm$^{-3}$ at 298.15 K (see Figure 11.5). This means that they can fully play a role as separation media in two-phase applications.

In terms of the designability of ionic liquids, the density is relatively easy to modulate and is influenced little by temperature shifts or impurities, such as halide, water, or solvent from the synthetic process. For pure ionic liquids, the values of density vary depending on the choice of cation and anion. Table 11.4 has listed several conventional ionic liquids, which are easy to analyse; these data reveal that the cationic structure influences the densities significantly. Generally, when connected with the same anion, the densities decrease progressively with increasing alkyl chain length of the cation, for example, for [C$_n$mim]X (X = [BF$_4$], [PF$_6$], or [NTf$_2$]). Figure 11.6 shows that the densities of the ionic liquids tend to approach those of linear alkanes, suggesting that the densities of those tail chains are promoted as $n$ increases. This may be because elongating the alkyl chain length increases the free volume within the ionic liquid, accordingly lowering its density. And this principle applies not only to imidazolium ionic liquids but also to pyridinium and quaternary ammonium ionic liquids.

The anion has a more remarkable influence to the density of ionic liquids. For the ionic liquids with cation [C$_4$mim]$^+$, the densities increase in the order

$$[C(CN)_3]^-, [O_2CMe]^-, [N(CN)_2]^-, [O_3SOC_8]^- < Cl^- < [NO_3]^-$$
$$< [BF_4]^-, [O_3SOC_2]^-, [CF_3CO_2]^- < [O_3SOC_1]^- < [AlCl_4]^-$$
$$< [OTf]^- < Br^- < [C_3F_7CO_2]^- < [PF_6]^- < [FeCl_4]^- < [NTf_2]^-,$$
$$[GaCl_4]^- < I^- < [ONf]^- < [NPf_2]^- < [FeBr_4]^- < [AuCl_4]^-.$$

**TABLE 11.4  The Densities of Conventional Ionic Liquids at 298.15 K**

| Ionic liquid | $\rho$/g·cm$^{-3}$ | Reference | Ionic liquid | $\rho$/g·cm$^{-3}$ | Reference |
|---|---|---|---|---|---|
| [C$_1$mim][BF$_4$] | 1.373 | 31 | [C$_1$mim][NTf$_2$] | 1.559 | 56 |
| [C$_2$mim][BF$_4$] | 1.28 | 8 | | 1.570 | 37 |
| | 1.279 | 43 | | 1.580 | 31 |
| | 1.280 | 37 | [C$_2$mim][NTf$_2$] | 1.51 | 80 |
| | 1.27 | 44 | | 1.518 | 43 |
| [C$_3$mim][BF$_4$] | 1.24 | 8, 37, 45 | | 1.519 | 37, 61 |
| [C$_4$mim][BF$_4$] | 1.19 | 46 | | 1.515 | 52 |
| | 1.21 | 8 | | 1.523 (293 K) | 43 |
| | 1.17 | 47 | [C$_3$mim][NTf$_2$] | 1.475 | 61 |
| | 1.21105 | 48 | | 1.473 | 81 |
| | 1.2012 | 49 | [C$_4$mim][NTf$_2$] | 1.43 | 20, 47 |
| | 1.208 | 50 | | 1.436 | 52, 61, 68, 73 |
| | 1.19735 | 51 | | 1.437 | 37, 82 |
| | 1.199 | 52 | | 1.433 | 31 |
| [C$_6$mim][BF$_4$] | 1.16 | 45 | | 1.429 | 56 |
| | 1.177 | 31 | | 1.44 (293 K) | 83 |
| | 1.1453 | 53 | | 1.439 (293 K) | 84 |
| | 1.101 | 54 | [C$_5$mim][NTf$_2$] | 1.403 | 61 |
| [C$_8$mim][BF$_4$] | 1.08 | 55 | | 1.412 | 85 |
| | 1.11 | 50, 56, 57 | [C$_6$mim][NTf$_2$] | 1.372 | 61 |
| | 1.092 | 54 | | 1.304 | 56 |
| | 1.10 | 58 | | 1.378 | 31 |
| | 1.1019 | 49 | | 1.364 | 54 |
| [C$_{10}$mim][BF$_4$] | 1.04 | 55, 59 | [C$_7$mim][NTf$_2$] | 1.357 | 73 |
| | 1.072 | 56 | [C$_8$mim][NTf$_2$] | 1.377 | 57 |
| [C$_4$mim][PF$_6$] | 1.35 | 20, 60 | | 1.344 | 61 |
| | 1.368 | 61, 62 | | 1.32 | 61 |
| | 1.3674 | 63 | | 1.317 | 68 |
| | 1.36657 | 64 | | 1.319 | 52 |
| | 1.35876 | 51 | | 1.321 | 37, 73, 86 |
| | 1.36 | 58, 65 | | 1.337 | 31 |

*(Continued)*

**TABLE 11.4** (*Continued*)

| Ionic liquid | $\rho/\text{g·cm}^{-3}$ | Reference |
|---|---|---|
| [C$_6$mim][PF$_6$] | 1.29 | 20 |
| | 1.292 | 52, 61 |
| | 1.2937 | 66 |
| | 1.2941 | 67 |
| [C$_7$mim][PF$_6$] | 1.302 | 68 |
| | 1.262 | 61 |
| | 1.274 | 69 |
| [C$_8$mim][PF$_6$] | 1.22 | 37 |
| | 1.234 | 52 |
| | 1.2357 | 66 |
| | 1.237 | 61, 69 |
| [C$_9$mim][PF$_6$] | 1.212 | 61 |
| [C$_{10}$mim][PF$_6$] | 1.14 | 59 |
| [C$_4$mim]Cl | 1.08 | 20, 60 |
| | 1.10 | 56, 58 |
| [C$_4$mim]Br | 1.32 | 58 |
| [C$_4$mim]I | 1.44 | 20, 60 |
| [C$_4$mim][O$_3$SOC$_1$] | 1.2 | 56 |
| | 1.2074 | 70 |
| | 1.2057 | 71 |
| | 1.211 | 52 |
| [C$_4$mim][O$_3$SOC$_2$] | 1.19893 | 72 |
| [C$_4$mim][O$_3$SOC$_8$] | 1.072 | 73 |
| | 1.0676 | 49 |
| [C$_4$mim][AlCl$_4$] | 1.2380 | 74, 75 |
| | 1.24 | 56 |
| [C$_4$mim][FeCl$_4$] | 1.3651 | 76 |
| | 1.38 | 37 |
| [C$_4$mim][GaCl$_4$] | 1.43 | 77 |
| [C$_4$mim][AuCl$_4$] | 2.146 | 78, 79 |
| | 2.350 | 78, 79 |
| [C$_4$mim][FeBr$_4$] | 1.98 | 37, 77 |

| Ionic liquid | $\rho/\text{g·cm}^{-3}$ | Reference |
|---|---|---|
| [C$_9$mim][NTf$_2$] | 1.299 | 61 |
| [C$_{10}$mim][NTf$_2$] | 1.271 | 61, 81 |
| | 1.279 | 73 |
| | 1.278 | 52 |
| [C$_{12}$mim][NTf$_2$] | 1.245 | 52 |
| [C$_{14}$mim][NTf$_2$] | 1.131 | 54 |
| | 1.201 | 52 |
| [$^i$C$_4$mim][NTf$_2$] | 1.428 (293 K) | 87 |
| [C$_2$F$_3$mim][NTf$_2$] | 1.656 (293 K) | 37, 87 |
| [C$_4$mim][C(CN)$_3$] | 1.0473 | 88 |
| [C$_4$mim][N(CN)$_2$] | 1.06 | 56 |
| [C$_4$mim][NO$_3$] | 1.058 | 54 |
| | 1.059 | 73 |
| | 1.159 | 54 |
| | 1.1565 | 89 |
| [C$_4$mim][NPf$_2$] | 1.514 | 37 |
| [C$_4$mim][OTf] | 1.29 | 56, 57 |
| | 1.296 | 68 |
| | 1.2976 | 53 |
| | 1.299 | 52 |
| | 1.3 | 54 |
| [C$_4$mim][ONf] | 1.473 | 56 |
| [C$_4$mim][O$_2$CMe] | 1.06 | 56 |
| | 1.053 | 52 |
| [C$_4$mim][CF$_3$CO$_2$] | 1.198 | 31 |
| | 1.209 | 57 |
| [C$_4$mim][C$_3$F$_7$CO$_2$] | 1.068 | 73 |
| | 1.217 | 90 |
| | 1.333 | 91, 92 |

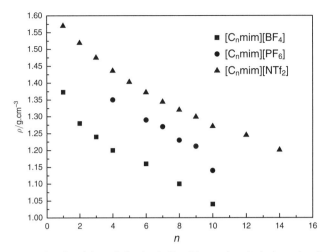

**Figure 11.6**  The densities of the ionic liquids vs. the chain length of cations.

The densities of ionic liquids with halides increase in the order of increasing volume, that is, $Cl^- < Br^- < I^-$, and $[FeCl_4]^- < [FeBr_4]^-$. However, for $[O_3SOC_n]^-$, which exhibits similar volume increases, the trend is the other way round, and the densities increase in the order $[O_3SOC_8]^- < [O_3SOC_2]^- < [O_3SOC_1]^-$. This result suggests that increasing the number of the carbon chain unit in anion decreases the densities of ionic liquids. This is the same trend observed for the cations. The results for the anions with similar structure, for example, $[AlCl_4]^-$, $[FeCl_4]^-$, $[GaCl_4]^-$, and $[AuCl_4]^-$, are caused by the difference in the mass of the central atom. It is generally believed that the densities of those ionic liquids with large volume, weak anions are relatively high, and that this trend has nothing to do with the cation. Therefore, when designing ionic liquids with different densities, the first selection should be an appropriate anion to determine the approximate density range, and then fine-tune this value through the choice cation.

Krossing and coworkers noticed a strong relationship between the molecular volumes $V_m$ of ionic liquids and their fundamental physical properties: density, viscosity, and conductivity [40]. The molar concentration of ionic liquids is related to $V_m$ by a power series, which leads to the relationship between $V_m$ and the density, Equation (11.1),

$$\rho = MgV_m^{-h}, \tag{11.1}$$

where $M$ is the molar mass, and $g$ and $h$ are empirical constants of best fit. This correlation is applicable to a wide variety of ionic liquids. However, the molar concentrations of the nitrile-functionalised ionic liquids do not fit with the data for the other salts. It is likely that functionalisation significantly changes the intermolecular interactions in the ionic liquid, and that this strongly affects the density changes that occur in the ionic liquids with temperature change, and during phase transitions [40].

Generally, it can be readily observed that an increase in temperature causes density to decrease, with a basically linear relationship, a trend almost universally observed. For example, Jacquemin et al. used this to calculate the density of ionic liquids at different temperatures, Equation (11.2) [93],

$$\rho = a + b(T - 273.15), \tag{11.2}$$

where $T$ is temperature (°C), $a$ is a coefficient, and $b$ is the density coefficient (in g cm$^{-3}$ K$^{-1}$). There are also some workers who used quadratic equations to fit the variation with temperature [94–97],

$$\rho = \sum_{i=0}^{2} a_i T^i. \tag{11.3}$$

Gardas and Coutinho [99] proposed an extension, Equation (11.4), to the Ye and Shreeve method [98] for the estimation of density over a wide range of temperatures and pressures,

$$\rho = \frac{M}{NV(a + bT + cP)}, \tag{11.4}$$

where $\rho$ is the density, $M$ is molecular weight, $N$ is the Avogadro constant, and $a$, $b$, and $c$ are coefficients. This estimation method for density was subsequently utilised elsewhere [88, 95, 100]. The Tait equation [101] has also been used to fit density at different temperatures and pressures [94, 96, 97, 102]:

$$\rho = \frac{\rho(T, P = 0.1\,\mathrm{MPa})}{\left\{ 1 - C \ln \dfrac{(B + P)}{(B + 0.1)} \right\}}, \tag{11.5}$$

where $\rho(T, P = 0.1\,\mathrm{MPa})$ is obtained from Equation (11.2) or Equation (11.3), and the coefficient $B$ is defined as $B = b_1 + (b_2/T)$. The coefficients $b_1$, $b_2$, and $C$ are obtained by fitting the Tait equation to experimental data.

## 11.4  VISCOSITY

Viscosity is another important property for ionic liquids. As with density, the fluid design of liquid–liquid extractors, distillation columns, reactors, process piping, and other units found in various chemical and pharmaceutical industries requires knowledge of the viscosities of fluids. At room temperature, ionic liquids display a broad range of viscosities, from 10 to several 1000 cP (even

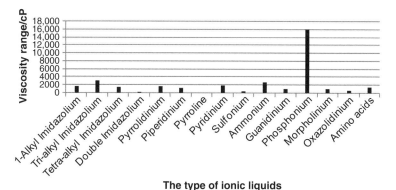

**Figure 11.7** The viscosity range of different types of ionic liquids. The data are from Reference 4.

several 10,000 cP) and significantly higher (see the Figure 11.7). Thus, the viscosities are usually higher by one to three orders of magnitude compared with traditional organic solvents. This brings many negative influences for the chemical industry operating processes and could become a limiting factor for ionic liquid scale-up applications. In fact, ionic liquids can be used with mixing solvents with lower viscosities in the actual process, lowering the high viscosity of ionic liquids, and often the reactants will also have this effect. In other applications, the higher viscosities of ionic liquids may be favourable, for example, lubrication or supported membrane separation [10].

The viscosity of ionic liquids is different from the density, in that it shows a remarkable influence of both temperature and impurity: a small increase in temperature or the presence of a trace of impurities can result in dramatic decreases in viscosity. It is now believed that the main impurity sources in room-temperature ionic liquids are the water content and the organic solvent content. So after measuring the viscosity, it is important to immediately measure the water content of the ionic liquid using Karl–Fischer titration or coulometry. In addition, investigations from Queen's University Ionic Liquid Laboratories (QUILL) suggested that the presence of even low concentrations of chloride in the ionic liquids substantially increases the viscosity, and in all cases, the viscosity increased dramatically with the concentration of chloride ions [11].

It is well known that hydrogen bonding can cause proton chemical shifts to move to lower field in nuclear magnetic resonance (NMR) spectra. The increase in viscosity is related to an increase in the cohesive forces via hydrogen bonding between the chloride and the protons of the imidazolium ring. This is the reason for different viscosities of the same ionic liquid in Table 11.5.

Viscosity is mainly determined by a combination of van der Waals forces, hydrogen bonding, and Coulombic interactions. The effect of cation head groups on viscosity can be seen from Table 11.5:

**TABLE 11.5 The Viscosities of Conventional Ionic Liquids at 298.15 K**

| Ionic Liquids | $\eta$/mPa·s | Reference | Ionic Liquids | $\eta$/mPa·s | Reference |
|---|---|---|---|---|---|
| [C₂mim][BF₄] | 78 | 58 | [C₂mim][NTf₂] | 36.5 | 73 |
| | 37 | 8, 37, 103 | | 45.9 | 111 |
| | 42 | 45 | | 33 | 112 |
| | 38 | 104 | | 34.2 | 113 |
| | 66 | 56 | | 34.7 | 114 |
| | 54.4 | 105 | | 37 | 57 |
| [C₃mim][BF₄] | 103 | 8, 37, 45 | [C₄mim][NTf₂] | 32.1 (293 K) | 115 |
| [C₄mim][BF₄] | 219 | 20, 46, 57, 60 | | 54.5 | 46 |
| | 180 | 8, 37, 45 | | 69 | 20 |
| | 233 | 50, 56 | | 80 | 56 |
| | 248 | 58 | | 35.9 | 116 |
| [C₅mim][BF₄] | 308 | 85 | | 50.5 | 73 |
| [C₆mim][BF₄] | 220 | 45 | | 45.6 | 113 |
| | 310 | 50, 56 | | 52 (293 K) | 47, 83, 87, 117 |
| | 380 | 58 | | 52 | 57 |
| [C₈mim][BF₄] | 440 | 50, 56 | [C₅mim][NTf₂] | 59 | 85 |
| | 492 | 58 | [C₆mim][NTf₂] | 80.1 | 73 |
| | 439 | 57 | | 71 | 57 |
| [C₁₀mim][BF₄] | 930 | 56 | [C₈mim][NTf₂] | 80.7 | 118 |
| [C₃mim][PF₆] | 312 | 106 | | 69.7 | 37 |
| | 450 | 107 | | 90.37 | 119 |
| | 371 | 108 | | 95 | 73 |

| Salt | Value | Ref. | Salt | Value | Ref. |
|---|---|---|---|---|---|
| $[C_4mim][PF_6]$ | 450 | 20 | $[C_{10}mim][NTf_2]$ | 87 | 57 |
|  | 393 | 109 |  | 92.51 | 86 |
|  | 397 | 60 |  | 90 | 37 |
|  | 493 | 58 |  | 120.2 | 73 |
| $[C_6mim][PF_6]$ | 560 | 109 |  | 88 | 37 |
| $[C_8mim][PF_6]$ | 585 | 20, 57 | $[^iC_4mim][NTf_2]$ | 83 (293 K) | 87 |
|  | 682 | 20, 57 | $[C_2F_3mim][NTf_2]$ | 248 (293 K) | 87 |
|  | 710 | 109 | $[C_4mim][NPf_2]$ | 111.6 | 37 |
|  | 810 | 56 | $[C_4mim][O_2CMe]$ | 139.7 | 73 |
| $[C_4mim]Cl$ | 3950 (303 K) | 110 | $[C_4mim][CF_3CO_2]$ | 70 | 120 |
|  | Solid | 58 |  | 76.9 | 90 |
| $[C_4mim]Br$ | Solid | 58 | $[C_4mim][OTf]$ | 90 | 56 |
| $[C_4mim]I$ | 1110 | 20, 56, 60 |  | 83.2 | 73 |
| $[C_4mim][O_3SOC_1]$ | 180 | 56 | $[C_4mim][ONf]$ | 373 | 56 |
|  | 188 | 73 | $[C_4mim][NO_3]$ | 165.27 | 89 |
|  | 213 | 72 | $[C_4mim][SCN]$ | 51.7 | 121 |
| $[C_4mim][O_3SOC_1]$ | 379.1 | 72 | $[C_4mim][N(CN)_2]$ | 37 | 56, 121 |
| $[C_4mim][O_3SOC_8]$ | 888.6 | 73 |  | 28.8 | 73 |
| $[C_4mim][AlCl_4]$ | 26 | 56 |  | 31.802 | 88 |
| $[C_4mim][FeCl_4]$ | 34 | 37 | $[C_4mim][C(CN)_3]$ | 27.318 | 88 |
| $[C_4mim][FeBr_4]$ | 62 | 37 |  |  |  |

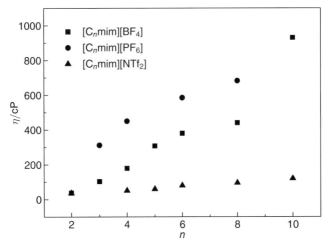

**Figure 11.8**    The viscosities of the ionic liquids vs. the chain length of cations.

1. It seems that the viscosities of the ionic liquids [$C_n$mim]X (X = [BF$_4$], [PF$_6$], or [NTf$_2$]) increase progressively with increasing alkyl chain length of the cation (see Figure 11.8). This result is contrary to the effect on densities because van der Waals interaction increases with increasing alkyl chain length.

2. It can be seen clearly by comparing the results for [$^i$C$_4$mim][NTf$_2$] and [C$_4$mim][NTf$_2$] that branching the alkyl chain reduces the rotational freedom and hence makes the ionic liquids more viscous.

3. As with density, the introduction of other elements or functionalities on to the cation may either increase or reduce the van der Waals interaction, thus increasing or reducing the viscosity of the ionic liquids. For example, introducing groups containing fluorine increases the viscosity significantly, cf., [C$_2$mim][NTf$_2$] and [C$_2$F$_3$mim][NTf$_2$] (see Table 11.5).

The effect of anion head groups on viscosity can also be seen from Table 11.5. For ionic liquids with the cation [C$_4$mim]$^+$, the viscosities increase in the order

$$[AlCl_4]^- < [C(CN)_3]^- < [N(CN)_2]^- < [FeCl_4]^- < [SCN]^- < [FeBr_4]^-,$$
$$[NTf_2]^- < [CF_3CO_2]^- < [OTf]^- < [NPf_2]^- < [O_2CMe]^- < [NO_3]^- < [BF_4]^-,$$
$$[O_3SOC_1]^- < [ONf]^-, [O_3SOC_2]^- < [PF_6]^- < [O_3SOC_8]^-.$$

For the ionic liquids containing the anion [O$_3$SOC$_n$]$^-$, the viscosities increase progressively with increasing alkyl chain length in the order (the same effect as noted with alkyl chains on the cation)

$$[O_3SOC_1]^- < [O_3SOC_2]^- < [O_3SOC_8]^-.$$

The viscosities of ionic liquids containing fluorinated alkyl chains became higher due to strong van der Waals interaction, such as in the series

$$[OTf]^- < [ONf]^-, [NTf_2]^- < [NPf_2]^-.$$

Although the ionic liquids with the $[NTf_2]^-$ anion, also containing a fluorinated alkyl chain, have strong van der Waals interaction, their viscosities are actually lower than other conventional ionic liquids. This is possibly because the reduced degree of the viscosity caused by the weak hydrogen bonds surpasses the increased degree of the viscosity caused by the van der Waals interactions. The results for the anions with similar structure, for example, $[AlCl_4]^-$ and $[FeCl_4]^-$, may be due to the difference in the mass of the central atom. This result was similar for densities.

The viscosities of ionic liquids may be reduced by methods other than structure modification. For example, the viscosities of ionic liquids can be decreased markedly with increasing temperature. Gardas and Coutinho developed a group contribution method for the viscosities of ionic liquids using an Orrick–Erbar approach, Equation (11.6) [99], and the viscosities of 29 ionic liquids were fitted:

$$\ln \frac{\eta}{\rho M} = A + \frac{B}{T}, \tag{11.6}$$

where $\eta$ and $\rho$ are the viscosity and density, respectively, $M$ and $T$ are the molecular weight and absolute temperature, respectively, and $A$ and $B$ are adjustable coefficients. However, the Orrick–Erbar method requires density data for the prediction of viscosity. To overcome this limitation and to attempt the development of an improved viscosity model with lower deviations in estimated viscosities, a new correlation, Equation (11.7), based on the Vogel–Tammann–Fulcher (VTF) equation [122–124] was proposed to predict viscosities of ionic liquids at atmospheric pressure as a function of temperature [88, 95, 125–128],

$$\ln \eta = A_\eta + \frac{B_\eta}{T - T_{0\eta}}, \tag{11.7}$$

$$A_\eta = \sum_{i=1}^{k} n_i a_{i,\eta}, \tag{11.8}$$

$$B_\eta = \sum_{i=1}^{k} n_i b_{i,\eta}, \tag{11.9}$$

where $T$ is the absolute temperature, and $A_\eta$, $B_\eta$, and $T_{0\eta}$ are adjustable coefficients. $A_\eta$ and $B_\eta$ can be obtained by a group contribution method according to Equation (11.8) and Equation (11.9), respectively, where $n_i$ is the number of groups of type $i$, $k$ is the total number of different groups in the molecule, and $a_{i,\eta}$ and $b_{i,\eta}$ are parameters. $B_\eta/T_{0\eta}$ is known as Angell strength parameter, so $T_{0\eta}$ is similar for all the ionic liquids studied. The calculated value of the viscosity is in good agreement with the corresponding experimental volume for the ionic liquids studied [88, 95, 125–128]. The experimental viscosity values for [C$_6$mim][PF$_6$] and [C$_8$mim][PF$_6$] at high pressure were fitted to a Tait-form equation by Tomida et al. [125],

$$\ln(\eta_p + \eta_0) = E\ln[(D+P)/(D+0.1)], \qquad (11.10)$$

where $\eta_p$ and $\eta_0$ are the viscosities at $P$ and 0.1 MPa, respectively, and $E$ and $D$ are adjustable parameters. If the value calculated with the VTF equation is substituted for $\eta_0$ in Equation (11.10), the viscosity at arbitrary temperature and pressure can be interpolated [126].

As mentioned previously, Krossing and coworkers proposed a strong relationship between the molecular volumes and the physical properties of ionic liquids [40]. Since the viscosity of a fluid is inversely proportional to the mobility of the ions within it, the viscosity is related to $V_m$ and $E_a$ according to Equation (11.11) and Equation (11.12), respectively [40],

$$\eta = ae^{bV_m}, \qquad (11.11)$$

$$\eta \propto e^{E_a/RT}, \qquad (11.12)$$

where $a$ is the empirical pre-exponential factor, $b$ is an empirical constant, and $V_m$ is the molecular volume. $E_a$ is the minimum energy that a particle must possess to move. The viscosity can be obtained using a VTF equation. Such a hypothesis is reasonable since the volume of the ions in a fluid is clearly related to the minimum energy required for them to move [40].

## 11.5  SURFACE TENSION

Surface tension is an important property in the study of physics and chemistry at free surfaces, as it affects, *inter alia*, the transfer rates of vapour absorption at the vapour–liquid interface [10]. It also is a measurement of the cohesive energy present at an interface and is usually quantified as a force/length measurement. This property is responsible for many of the dynamic behaviours of liquids, and is important to researchers and engineers in chemical process and reactor engineering fields. The values of surface tension are dependent on the liquid structure and orientation.

Experimental data for the surface tension of ionic liquids are very scarce and currently mostly limited to imidazolium-based ionic liquids, alkylammonium-based protic ionic liquids, and a dozen amino ionic liquids [74, 77, 129–133]. The values of surface tension lie in the range of 1.55–65.0 mN m$^{-1}$. Up to now,

**TABLE 11.6    The Values for Surface Tension of Ionic Liquids at 300.15 K**

| Entry | Cation, [A]$^+$ | Anion, [X]$^-$ | $\gamma$/mN m$^{-1}$ | Reference |
|-------|-----------------|----------------|----------------------|-----------|
| 1 | MeNH$_3$ | HCO$_2$ | 43.1 | 132 |
| 2 | EtNH$_3$ | HCO$_2$ | 38.5 | 132 |
| 3 | PrNH$_3$ | HCO$_2$ | 33.3 | 132 |
| 4 | BuNH$_3$ | HCO$_2$ | 31.9 | 132 |
| 5 | C$_2$mim | BF$_4$ | 54.4 | 136 |
| 6 | C$_2$mim | CH$_3$BF$_3$ | 45.2 | 136 |
| 7 | C$_2$mim | C$_2$H$_5$BF$_3$ | 42.5 | 136 |
| 8 | C$_2$mim | C$_3$H$_7$BF$_3$ | 38.0 | 136 |
| 9 | C$_2$mim | C$_4$H$_9$BF$_3$ | 34.2 | 136 |
| 10 | C$_2$mim | C$_5$H$_{11}$BF$_3$ | 33.8 | 136 |
| 11 | C$_2$mim | CH$_2$CHBF$_3$ | 44.3 | 136 |

the reported values of surface tension are all lower than for water (71.97 mN m$^{-1}$ at 298 K), except for the values reported for [C$_4$mim][Zn$_3$Cl$_7$] (78.3 mN m$^{-1}$ at 318 K) [134]. It should be noted, however, that the value for [C$_4$mim][ZnCl$_3$] is 57.49 mN m$^{-1}$ at 313.15 K [133], and the anions of both these materials are incorrectly formulated [135]. However, the values of surface tension for ionic liquids are generally higher than for many organic solvents.

The trends in surface tension changes for different types of ionic liquids have similar principles. As can be seen in Table 11.6, for ionic liquids with similar anions, the surface tension decreases with an increase in alkyl chain length of the imidazolium cation, while for the ionic liquids having the same cation, the surface tension also decreases with an increase in alkyl chain length in inions. As seen in Table 11.5, for the common cation [C$_2$mim]$^+$, the surface tension shows the order

$$[BF_4]^- >> [CH_3BF_3]^- > [CH_2CHBF_3]^- > [C_2H_5BF_3]^-$$
$$> [C_3H_7BF_3]^- > [C_4H_9BF_3]^- \sim [C_5H_{11}BF_3]^-.$$

An exception is that the value for surface tension of [C$_4$mim][Zn$_3$Cl$_7$] is higher than that of [C$_4$mim][ZnCl$_3$] (but again note that the anions of both these materials are incorrectly formulated [135]). Structure changes to either the cation or the anion had a similar effect on the surface tension, indicating that both ions are present at the surface and influence the surface tension. From these trends in surface tension with ion size, it may be inferred that the energy required to break the ionic liquid–air interface is related to the ion–ion interactions: the larger the ions, the lower their interactions with each other.

Drummond and coworkers studied a series of 25 protic ionic liquids [129] to determine the effect of structural changes on the surface tension, including the effect of hydroxyl groups, increasing alkyl chain lengths, branching, and the differences between inorganic and organic anions. It was found that for the branched protic ionic liquids, the branching decreased the surface tension, consistent with more hydrocarbon units per unit surface area. Thus, 2-methylpropylammonium methanoate and 2-methylbutylammonium methanoate have higher values of surface tension than those of butylammonium methanoate. The addition of a hydroxyl group has the same effect as the addition of an alkyl group, and will cause a significant increase in the surface tension, due to the combination of higher cohesive energy from the additional hydrogen bonding, and a reduction in the effective amount of hydrocarbon situated at the surface.

Hydroxyl groups at the end of the chain lead to higher surface tensions than those branched off the side of the chain. This was shown by the surface tension of ethanolammonium methanoate (65.0 mN m$^{-1}$ at 298.15 K) and ethylammonium glycolate (49.3 mN m$^{-1}$ at 298.15 K) being higher than those of 2-propanolammonium methanoate (46.2 mN m$^{-1}$ at 298.15 K) and ethylammonium lactate (39.3 mN m$^{-1}$ at 298.15 K), respectively. The reason is that the small anion causes little interruption to the surface packing of cations. The inorganic anions lead to relatively high surface tensions.

In general, the surface tension of many ionic liquids exhibits almost linear decreases while temperature increases. From Figure 11.9, we can see that the decreasing trend in surface tension is very clear for ionic liquids having $[C_2mim]^+$, $[C_3mim]^+$, or $[C_5mim]^+$, and the relationship is expressed by the Eötvös equation,

$$\gamma V_m^{2/3} = \kappa(T_c - T), \tag{11.13}$$

where $V_m$ is the molar volume of the liquid, according to the Eötvös equation, $T_c$ is critical temperature, and $k$ is an empirical constant.

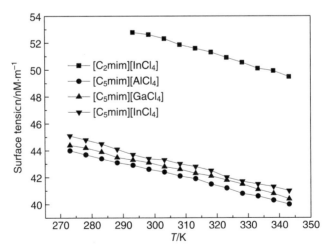

**Figure 11.9**  The values surface tension for several types of ionic liquids at different temperatures.

## 11.6    CONCLUSIONS

Ionic liquids have shown excellent performance in many areas, including catalysis, separations, organic synthesis, material preparation, and biomass conversion; this caused a rapid transition into an international scientific research hot spot. This resulted in a massive surge in the generation of physicochemical property data, which has provided a solid foundation for the screening and design of ionic liquids, equipment optimisation, and process simulation. In addition to the accumulation of data on ionic liquids, there has also been significant progress in modelling and predictive methodology, but there are still problems that cannot be ignored. Until now (1) a unified standard purity assessment method for analysing ionic liquids has not been established, which has created many inconsistencies in the measurement of physicochemical properties of ionic liquids; (2) standard determination methods of various physicochemical properties have not been established, resulting in small, and sometimes significant, differences in reported values for the same ionic liquid (although some recommendations do now exist) [118, 137]; (3) the current simulation work is based or developed on conventional models—theoretical models have not yet been established that are unique to ionic liquids, and there is no new theory breakthrough. Therefore, established accurate, standard analytical techniques and experimental methods, and a unified evaluation system, are problems to which we urgently need a solution in the research field of physicochemical properties of ionic liquids—only in this way will reliable data be accumulated to permit meaningful modelling. Then, by combining molecular dynamic simulations and quantitative calculations, comprehensive and systemic structure–property relationships will be revealed, new and simple prediction method will be established, and this will inevitably result in new breakthroughs in the fundamental understanding of ionic liquids.

### ACKNOWLEDGEMENTS

This work was supported by the National High Technology Research and Development Program of China (863 Program) (No. 2012AA063001), the National Natural Science Foundation of China (Grant Nos. 20776140, 20806083, and 20873152), and the Li Foundation (by a fellowship).

### REFERENCES

1  Seddon, K.R., Ionic liquids: Designer solvents?, in *The International George Papatheodorou Symposium: Proceedings*, eds. S. Boghosian, V. Dracopoulos, C.G. Kontoyannis, and G.A. Voyiatzis (Institute of Chemical Engineering and High Temperature Chemical Processes, Patras, 1999), pp. 131–135.

2  Izgorodina, E.I., "Theoretical approaches to ionic liquids: From past to future directions", in *Ionic Liquids UnCOILed: Critical Expert Overviews*, eds. N.V. Plechkova and K.R. Seddon (Wiley, Hoboken, New Jersey, 2013), pp. 181–230.

3  Ionic Liquids Database ILThermo. NIST Standard Online Reference Database #147, http://ilthermo.boulder.nist.gov/ILThermo/mainmenu.uix.

4  (a) Zhang, S., Sun, N., He, X., Lu, X., and Zhang, X., Physical properties of ionic liquids: database and evaluation, *J. Phys. Chem. Ref. Data* **35**, 1475–1517 (2006); Zhang, S., Lu, X., Zhou, Q., Li, X., Zhang, X., and Li, S., *Ionic Liquids: Physicochemical Properties* (Elsevier, Amsterdam, 2009).

5  The UFT/Merck Ionic Liquids Biological Effects Database, http://www.il-eco.uft.uni-bremen.de/.

6  Dortmund Data Base, DDBST: The Ionic Liquid Database, http://www.ddbst.com/ionic-liquids.html.

7  delph-IL, The Ionic Liquid Database, http://www.delphil.net/web/html/.

8  Nishida, T., Tashiro, Y., and Yamamoto, M., Physical and electrochemical properties of 1-alkyl-3- methylimidazolium tetrafluoroborate for electrolyte, *J. Fluorine Chem.* **120**, 135–141 (2003).

9  Handy, S.T., Room temperature ionic liquids: Different classes and physical properties, *Curr. Org. Chem.* **9**, 10, 959–988 (2005).

10  Rooney, D., Jacquemin, J., and Gardas, R., Thermophysical properties of ionic liquids, *Top Curr. Chem.* **290**, 185–212 (2009).

11  Seddon, K.R., Stark, A., and Torres, M-J., Influence of chloride, water, and organic solvents on the physical properties of ionic liquids, *Pure Appl. Chem.* **72**, 12, 2275–2287 (2000).

12  Baker, S.N., Baker, G.A., and Bright, F.V., Temperature-dependent microscopic solvent properties of "dry" and "wet" 1-butyl-3-methylimidazolium hexafluorophosphate: Correlation with ET(30) and Kamlet–Taft polarity scales, *Green Chem.* **4**, 165–169 (2002).

13  Widegren, J.A., Laesecke, A., and Magee, J.W., The effect of dissolved water on the viscosities of hydrophobic room-temperature ionic liquids, *Chem. Commun.* 1610–1612 (2005).

14  Saha, S., and Hamaguchi, H.O., Effect of water on the molecular structure and arrangement of nitrile-functionalized ionic liquids, *J. Phys. Chem. B* **110**, 2777–2781 (2006).

15  Silvester, D.S., and Compton, R.G., Electrochemistry in room temperature ionic liquids: A review and some possible applications, *Phys. Chem. Chem. Phys.* **220**, 1247–1274 (2006).

16  Van Valkenburg, M.E., Vaughn, R.L., Williams, M., and Wilkes, J.S., Thermochemistry of ionic liquid heat-transfer fluids, *Thermochim. Acta* **425**, 181–188 (2005).

17  Katritzky, A.R., Jain, R., Lomaka, A., Petrukhin, R., Karelson, M., Visser, A.E., and Rogers, R.D., Correlation of the melting points of potential ionic liquids (imidazolium bromides and benzimidazolium bromides) using the CODESSA program, *J. Chem. Inf. Comput. Sci.* **42**, 225–231 (2002).

18  Gupta, O.D., Twamleya, B., and Shreeve, J.M., Low melting and slightly viscous ionic liquids via protonation of trialkylamines by perfluoroalkyl β-diketones, *Tetrahedron Lett.* **45**, 1733–1736 (2004).

19  Zhou, Z.B., Matsumoto, H., and Tatsumi, K., Cyclic quaternary ammonium ionic liquids with perfluoroalkyltrifluoroborates: Synthesis, characterization, and properties, *Chem. Eur. J.* **12**, 2196–2212 (2006).

20  Huddleston, J.G., Visser, A.E., Reichert, W.M., Willauer, H.D., Broker, G.A., and Rogers, R.D., Characterization and comparison of hydrophilic and hydrophobic room temperature ionic liquids incorporating the imidazolium cation, *Green Chem.* **3**, 156–164 (2001).

21  Ngo, H.L., LeCompte, K., Hargens, L., and McEwen, A.B., Thermal properties of imidazolium ionic liquids, *Thermochim. Acta* **357**, 97–102 (2000).

22  Matsumoto, H., Matsuda, T., and Miyazaki, Y., Room temperature molten salts based on trialkylsulfonium cations and bis(trifluoromethylsulfonyl)imide, *Chem. Lett.* **29**, 12, 1430–1431 (2000).

23  Holbrey, J.D., Turner, M.B., Reichert, W.M., and Rogers, R.D., New ionic liquids containing an appended hydroxyl functionality from the atom-efficient, one-pot reaction of 1-methylimidazole and acid with propylene oxide, *Green Chem.* **5**, 731–736 (2003).

24  Mateus, N.M.M., Branco, L.C., Lourenco, N.M.T., and Afonso, C.A.M., Synthesis and properties of tetra-alkyl-dimethylguanidinium salts as a potential new generation of ionic liquids, *Green Chem.* **5**, 347–352 (2003).

25  Howarth, J., Hanlon, K., Fayne, D., and McCormac, P., Moisture stable dialkyl-imidazolium salts as heterogeneous and homogeneous Lewis acids in the Diels-Alder reaction, *Tetrahedron Lett.* **38**, 3097–3100 (1999).

26  Wilkes, J.S., and Zaworotko, M.J., Air and water stable 1-ethyl-3-methylimidazolium based ionic liquids, *Chem. Commun.* 965–966 (1992).

27  Ohno, H., and Yoshizawa, M., Ion conductive characteristics of ionic liquids prepared by neutralization of alkylimidazoles, *Solid State Ionics* **154–155**, 303–309 (2002).

28  Sheldon, R., Catalytic reactions in ionic liquids, *Chem. Commun.* 2399–2407 (2001).

29  Fannin, A.A., Floreani, D.A., King, L.A., Landers, J.S., Piersma, B.J., Stech, D.J., Vaughn, R.L., Wilkes, J.S., and Williams, J.L., Properties of 1,3-dialkylimldazollum chloride-aluminum chloride ionic liquids. 2. Phase transitions, densities, electrical conductivities, and viscosities, *J. Phys. Chem.* **88**, 2614–2621 (1984).

30  Katritzky, A.R., Jain, R., Lomaka, A., Petrukhin, R., Karelson, M., Visser, A.E., and Rogers, R.D., Correlation of the melting points of potential ionic liquids (imidazolium bromides and benzimidazolium bromides) using the CODESSA program, *J. Chem. Inf. Comput. Sci.* **42**, 225–231 (2002).

31  Ohlin, C.A., Dyson, P.J., and Laurenczy, G., Carbon monoxide solubility in ionic liquids: Determination, prediction and relevance to hydroformylation, *Chem. Commun.* 1070–1071 (2004).

32  Katritzky, A.R., Lomaka, A., Petrukhin, R., Jain, R., Karelson, M., Visser, A.E., and Rogers, R.D., QSPR correlation of the melting point for pyridinium bromides, potential ionic liquids, *J. Chem. Inf. Comp. Sci.* **42**(1), 71–74 (2002).

33  Sun, N., He, X., Dong, K., Zhang, X., Lu, X., He, H., and Zhang, S., Prediction of the melting points for two kinds of room temperature ionic liquids, *Fluid Phase Equilibr.* **247**(1–2), 137–142 (2006).

34 López-Martin, I., Burello, E., Davey, P.N., Seddon, K.R., and Rothenberg, G., Anion and cation effects on imidazolium salt melting points: A descriptor modelling study, *Chem. Phys. Chem.* **8**(5), 690-695 (2007).

35 Dlubek, G., Yu, Y., Krause-Rehberg, R., Beichel, W., Bulut, S., Pogodina, N., Krossing, I., and Friedrich, CH., Free volume in imidazolium triflimide ([C₃MIM][NTf₂] ionic liquid from positron lifetime: Amorphous, crystalline, and liquid states, *J. Chem. Phys.* **133**(12), No. 124502 (2010).

36 Woodward, C.E., and Harris, K.R., A lattice-hole theory for conductivity in ionic liquid mixtures: Application to ionic liquid plus water mixtures, *Phys. Chem, Chem. Phys.* **12**(5), 1172–1176 (2010).

37 Zhao, H., Liang, Z.C., and Li, F., An improved model for the conductivity of room-temperature ionic liquids based on hole theory, *J. Mol. Liq.* **149**, 55–59 (2009).

38 Domańska, U., Physico-chemical properties and phase behaviour of pyrrolidinium-based ionic liquids, *Int. J. Mol. Sci.* **11**, 1825–1841 (2010).

39 Domańska, U., and Bogel-Łukasik, R., Physicochemical properties and solubility of alkyl-(2-hydroxyethyl)-dimethylammonium bromide, *J. Phys. Chem. B* **109**, 12124–12132 (2005).

40 Slattery, M.J., Daguenet, C., Dyson, J.P., Schubert, J.S.T., and Krossing, I., How to predict the physical properties of ionic liquids: A volume-based approach, *Angew. Chem. Int. Ed.* **46**, 5384–5388 (2007).

41 Krossing, I., Slattery, M.J., Daguenet, C., Dyson, J.P., Oleinikova, A., and Weingärtner, H., Why are ionic liquids liquid? A simple explanation based on lattice and solvation energies, *J. Am. Chem. Soc.* **128**, 13427–13434 (2006).

42 Preiss, U., Bulut, S., and Krossing, I., In silico prediction of the melting points of ionic liquids from thermodynamic considerations: A case study on 67 salts with a melting point range of 337 °C, *J. Phys. Chem. B* **114**, 11133–11140 (2010).

43 Noda, A., Hayamizu, K., and Watanabe, M., Pulsed-gradient spin-echo ¹H and ¹⁹F NMR ionic diffusion coefficient, viscosity, and ionic conductivity of non-chloroaluminate room-temperature ionic liquids, *J. Phys. Chem. B* **105**, 4603–4610 (2001).

44 Matsumoto, K., and Hagiwara, R., Electrochemical properties of the ionic liquid 1-ethyl-3-methylimidazolium difluorophosphate as an electrolyte for electric double-layer capacitors, *J. Electrochem. Soc.* **157**, 5, A578–A581 (2010).

45 Zhou, Z.B., Matsumoto, H., and Tatsumi, K., Low-melting, low-viscous, hydrophobic ionic liquids: 1-Alkyl(alkyl ether)-3-methylimidazolium perfluoroalkyltrifluoroborate, *Chem. Eur. J.* **10**, 6581–6591 (2004).

46 Wu, B.Q., Reddy, R. G., and Rogers, R.D., Novel ionic liquid thermal storage for solar thermal electric power systems, in *Proceedings of Solar Forum 2001*, eds. S.J. Kleis and C.E. Bingham (American Society of Mechanical Engineers, Washington, DC, 2001), pp. 445–451.

47 Carda-Broch, S., Berthod, A., and Armstrong, D.W., Solvent properties of the 1-butyl-3-methylimidazolium hexafluorophosphate ionic liquid, *Anal. Bioanal. Chem.* **375**, 191–199 (2003).

48 Gao, H.Y., Qi, F., and Wang, H.J., Densities and volumetric properties of binary mixtures of the ionic liquid 1-butyl-3-methylimidazolium tetrafluoroborate with

benzaldehyde at $T = (298.15$ to $313.15)$ K, *J. Chem. Thermodyn.* **41**, 888–892 (2009).

49 Singh, T., Kumar, A., Kaur, M., Kaur, G., and Kumar, H., Non-ideal behaviour of imidazolium based room temperature ionic liquids in ethylene glycol at $T = (298.15$ to $318.15)$ K, *J. Chem. Thermodyn.* **41**, 717–723 (2009).

50 Baghdadi, M., and Shemirani, F., In situ solvent formation microextraction based on ionic liquids: A novel sample preparation technique for determination of inorganic species in saline solutions, *Anal. Chim. Acta* **634**, 186–191 (2009).

51 Qi, F., and Wang, H.J., Application of prigogine–flory–patterson theory to excess molar volume of mixtures of 1-butyl-3-methylimidazolium ionic liquids with n-methyl-2-pyrrolidinone, *J. Chem. Thermodyn.* **41**, 265–272 (2009).

52 Shimizu, K., Tariq, M., Gomes, M.F.C., Rebelo, L.P.N., and Lopes, J.N.C., Assessing the dispersive and electrostatic components of the cohesive energy of ionic liquids using molecular dynamics simulations and molar refraction data, *J. Phys. Chem. B* **114**, 5831–5834 (2010).

53 García-Miaja, G., Troncoso, J., and Romaní, L., Excess properties for binary systems ionic liquid + ethanol: Experimental results and theoretical description using the ERAS model, *Fluid Phase Equilibr.* **274**, 59–67 (2008).

54 Strechan, A.A., Kabo, A.G., Paulechka, Y.U., Blokhin, A.V., Kabo, G.J., Shaplov, A.S., and Lozinskaya, E.I., Thermochemical properties of 1-butyl-3-methylimidazolium nitrate, *Thermochim. Acta* **474**, 25–31 (2008).

55 Branco, L.C., Rosa, J.N., Ramos, J.J.M., and Afonso, C.A.M., Preparation and characterization of new room temperature ionic liquids, *Chem. Eur. J.* **8**, 16, 3671–3677 (2002).

56 Berthod, A., Ruiz-Ángel, M.J., and Carda-Broch, S., Ionic liquids in separation techniques, *J. Chromatogr. A* **1184**, 6–18 (2008).

57 Poolea, C.F., and Poole, S.K., Extraction of organic compounds with room temperature ionic liquids, *J. Chromatogr. A* **1217**, 2268–2286 (2010).

58 Lu, Y.B., Ma, W.Y., Hu, R.L., Dai, X.J., and Pan, Y.J., Ionic liquid-based microwave-assisted extraction of phenolic alkaloids from the medicinal plant Nelumbo nucifera Gaertn, *J. Chromatogr. A* **1208**, 42–46 (2008).

59 Fortunato, R., Afonso, C.A.M., Reis, M.A.M., and Crespo, J.G., Supported liquid membranes using ionic liquids: Study of stability and transport mechanisms, *J. Membrane Sci.* **242**, 197–209 (2004).

60 Visser, A.E., Reichert, W.M., Swatloski, R.P., Willauer, H.D., Huddleston, J.G., and Rogers, R.D., Characterization of hydrophilic and hydrophobic ionic liquids: Alternatives to volatile organic compounds for liquid-liquid separations, in *Symposium on Green Industrial Applications of Ionic Liquid*, eds. R.D. Rogers and K.R. Seddon, ACS Symp. Ser., Vol. 818 (American Chemical Society, San Diego, California, 2002), pp. 289–308.

61 Dzyuba, S.V., and Bartsch, R.A., Influence of structural variations in 1-alkyl(aralkyl)-3-methylimidazolium hexafluorophosphates and bis(trifluoromethylsulfonyl) imides on physical properties of the ionic liquids, *Chem. Phys. Chem.* **3**, 161–166 (2002).

62 Ishida, T., Nishikawa, K., and Shirota, H., Atom substitution effects of $[XF_6]^-$ in ionic liquids. 2. Theoretical study, *J. Phys. Chem. B* **113**, 9840–9851 (2009).

63   Harris, K.R., and Woolf, L.A., Temperature and pressure dependence of the viscosity of the ionic liquid 1-butyl-3-methylimidazolium hexafluorophosphate, *J. Chem. Eng. Data* **50**, 1777–1782 (2005).

64   Huo, Y., Xia, S.Q., and Ma, P.S., Solubility of alcohols and aromatic compounds in imidazolium-based ionic liquids, *J. Chem. Eng. Data* **53**, 2535–2539 (2008).

65   Sharma, A., Julcour, C., Kelkar, A.A., Deshpande, R.M., and Delmas, H., Mass transfer and solubility of CO and $H_2$ in ionic liquid. case of [Bmim][$PF_6$] with gas-Inducing stirrer reactor, *Ind. Eng. Chem. Res.* **48**, 4075–4082 (2009).

66   Pereiro, A.B., and Rodríguez, A., A study on the liquid–liquid equilibria of 1-alkyl-3-methylimidazolium hexafluorophosphate with ethanol and alkanes, *Fluid Phase Equilibr.* **270**, 23–29 (2008).

67   Valderrama, J.O., Reátegui, A., and Rojas, R.E., Density of ionic liquids using group contribution and artificial neural networks, *Ind. Eng. Chem. Res.* **48**, 3254–3259 (2009).

68   Lee, S.H., and Lee, S.B., The Hildebrand solubility parameters, cohesive energy densities and internal energies of 1-alkyl-3-methylimidazolium-based room temperature ionic liquids, *Chem. Commun.* 3469–3471 (2005).

69   Chun, S.K., Dzyuba, S.V., and Bartsch, R.A., Influence of structural variation in room-temperature ionic liquids on the selectivity and efficiency of competitive alkali metal salt extraction by a crown ether, *Anal. Chem.* **73**, 3737–3741 (2001).

70   García-Miaja, G., Troncoso, J., and Romaní, L., Excess enthalpy, density, and heat capacity for binary systems of alkylimidazolium-based ionic liquids + water, *J. Chem. Thermodyn.* **41**, 161–166 (2009).

71   Soriano, A.N., Doma, Jr. B.T., and Li, M.H., Measurements of the density and refractive index for 1-n-butyl-3-methylimidazolium-based ionic liquids, *J. Chem. Thermodyn.* **41**, 301–307 (2009).

72   Torrecilla, J.S., Palomar, J., García, J., and Rodríguez, F., Effect of cationic and anionic chain lengths on volumetric, transport, and surface properties of 1-alkyl-3-methylimidazolium alkylsulfate ionic liquids at (298.15 and 313.15) K, *J. Chem. Eng. Data* **54**, 1297–1301 (2009).

73   McHale, G., Hardacre, C., Ge, R., Doy, N., Allen, R.W. K., MacInnes, J.M., Bown, M.R., and Newton, M.I., Density-viscosity product of small-volume ionic liquid samples using quartz crystal impedance analysis, *Anal. Chem.* **80**, 5806–5811 (2008).

74   Zang, S.L., Zhang, Q.G., Huang, M., Wang, B., and Yang, J.Z., Studies on the properties of ionic liquid EMIInCl$_4$, *Fluid Phase Equilibr.* **230**, 192–196 (2005).

75   Jones de, A., Böes, E.S., and Stassen, H., Alkyl chain size effects on liquid phase properties of 1-alkyl-3-methylimidazolium tetrachloroaluminate ionic liquids-a microscopic point of view from computational chemistry, *J. Phys. Chem. B* **113**, 7541–7547 (2009).

76   Zhang, Q.G., Yang, J.Z., Lu, X.M., Gui, J.S., and Huang, M., Studies on an ionic liquid based on $FeCl_3$ and its properties, *Fluid Phase Equilibr.* **226**, 207–211 (2004).

77   Larriba, C., Yoshida, Y., and de Mora, J.F., Correlation between surface tension and void fraction in ionic liquids, *J. Phys. Chem. B* **112**, 12401–12407 (2008).

78  Matsumoto, K., Hagiwara, R., Yoshida, R., Ito, Y., Mazej, Z., Benkic, P., Zemva, B., Tamada, O., Yoshino, H., and Matsubara, S., Syntheses, structures and properties of 1-ethyl-3-methylimidazolium salts of fluorocomplex anions, *Dalton Trans.* 144–149 (2004).

79  Matsumoto, K., and Hagiwara, R., A new room temperature ionic liquid of oxy-fluorometallate anion: 1-Ethyl-3-methylimidazolium oxypentafluorotungstate (EMImWOF$_5$), *J. Fluorine Chem.* **126**, 1095–1100 (2005).

80  Matsumoto, H., Yanagida, M., Tanimoto, K., Nomura, M., Kitagawa, Y., and Miyazaki, Y., Highly conductive room temperature molten salts based on small trimethylalkylammonium cations and bis(trifluoromethylsulfonyl)imide, *Chem. Lett.* **29**, 8, 922–923 (2000).

81  Dzyuba, S.V., and Bartsch, R.A., Expanding the polarity range of ionic liquids, *Tetrahedron Lett.* **43**, 4657–4659 (2002).

82  Andreatta, A.E., Arce, A., Rodil, E., and Soto, A., Physico-chemical properties of binary and ternary mixtures of ethyl acetate + ethanol + 1-butyl-3-methyl-imidazolium bis(trifluoromethylsulfonyl)imide at 298.15 K and atmospheric pressure, *J. Solution Chem.* **39**, 371–383 (2010).

83  Gan, Q., Rooney, D., Xue, M.L., Thompson, G., and Zou, Y.R., An experimental study of gas transport and separation properties of ionic liquids supported on nanofiltration membranes, *J. Membrane Sci.* **280**, 948–956 (2006).

84  Carrera, G.V.S.M., Afonso, C.A.M., and Branco, L.C., Interfacial properties, densities, and contact angles of task specific ionic liquids, *J. Chem. Eng. Data* **55**, 609–615 (2010).

85  Xiao, D., Rajian, J.R., Hines, L.G., Jr., Li, S.F., Bartsch, R.A., and Quitevis, E.L., Nanostructural organization and anion effects in the optical kerr effect spectra of binary ionic liquid mixtures, *J. Phys. Chem. B* **112**, 13316–13325 (2008).

86  Andreatta, A.E., Arce, A., Rodil, E., and Soto, A., Physical properties of binary and ternary mixtures of ethyl acetate, ethanol, and 1-octyl-3-methyl-imidazolium bis(trifluoromethylsulfonyl)imide at 298.15 K, *J. Chem. Eng. Data* **54**, 1022–1028 (2009).

87  Bonhôte, P., Dias, A.P., Papageorgiou, N., Kalyanasundaram, K., and Grätzel, M., Hydrophobic, highly conductive ambient-temperature molten salts, *Inorg. Chem.* **35**, 1168–1178 (1996).

88  Carvalho, P.J., Regueira, T., Santos, L.M.N.B.F., Fernandez, J., and Coutinho, J.A.P., Effect of water on the viscosities and densities of 1-butyl-3-methylimidazolium dicyanamide and 1-butyl-3-methylimidazolium tricyanomethane at atmospheric pressure, *J. Chem. Eng. Data* **55**, 645–652 (2010).

89  Mokhtarani, B., Sharifi, A., Mortaheb, H.R., Mirzaei, M., Mafi, M., and Sadeghian, F., Density and viscosity of 1-butyl-3-methylimidazolium nitrate with ethanol, 1-propanol, or 1-butanol at several temperatures, *J. Chem. Thermodyn.* **41**, 1432–1438 (2009).

90  Tokuda, H., Tsuzuki, S., Susan, M.A.B.H., Hayamizu, K., and Watanabe, M., How ionic are room-temperature ionic liquids? An indicator of the physicochemical properties, *J. Phys. Chem. B* **110**, 39, 19593–19600 (2006).

91  Dupont, J., and Spencer, J., On the noninnocent nature of 1,3-dialkylimidazolium ionic liquids, *Angew. Chem., Int. Ed.* **43**, 5296–5297 (2004).

92  Singh, B., and Sekhon, S.S., Polymer electrolytes based on room temperature ionic liquid: 2,3-Dimethyl-1-octylimidazolium triflate, *J. Phys. Chem. B* **109**, 16539–16543 (2005).

93  Jacquemin, J., Husson, P., Padua, A.A.H., and Majer, V., Density and viscosity of several pure and water-saturated ionic liquids, *Green Chem.* **8**, 172–180 (2006).

94  Gardas, R.L., Freire, M.G., Carvalho, P.J., Marrucho, I.M., Fonseca, I.M.A., Ferreira, A.G.M., and Coutinho, J.A.P., $P\rho T$ measurements of imidazolium-based ionic liquids, *J. Chem. Eng. Data* **52**, 1881–1888 (2007).

95  Gardas, R.L., Ge, R., Goodrich, P., Hardacre, C., Hussain, A., and Rooney, D.W., Thermophysical properties of amino acid-based ionic liquids, *J. Chem. Eng. Data* **55**, 1505–1515 (2010).

96  Jacquemin, J., Husson, P., Mayer, V., and Cibulka, I., High-pressure volumetric properties of imidazolium-based ionic liquids: Effect of the anion, *J. Chem. Eng. Data* **52**, 2204–2211 (2007).

97  Jacquemin, J., Nancarrow, P., Rooney, D.W. Gomes, M.F.C., Husson, P., Majer, V., Pádua, A.A.H., and Hardacre, C., Prediction of ionic liquid properties. II. Volumetric properties as a function of temperature and pressure, *J. Chem. Eng. Data* **53**, 2133–2143 (2008).

98  Ye, C., and Shreeve, J.M., Rapid and accurate estimation of densities of room-temperature ionic liquids and salts, *J. Phys. Chem. A* **111**, 1456–1461 (2007).

99  Gardas, R.L., and Coutinho, J.A.P., A group contribution method for viscosity estimation of ionic liquids, *Fluid Phase Equilibr.* **266**, 195–201 (2008).

100  Soriano, A.N., Doma Jr., B.T., and Li, M.H., Density and refractive index measurements of 1-ethyl-3- methylimidazolium-based ionic liquids, *J. Taiwan Inst. Chem. Eng.* **41**, 115–121 (2010).

101  Dymond, J. H., and Malhotra, R., The Tait equation: 100 years on, *Int. J. Thermophys.* **9**, 941–951 (1988).

102  Gardas, R.L., Freire, M.G., Carvalho, P.J., Marrucho, I.M., Fonseca, I.M.A., Ferreira, A.G.M., and Coutinho, J.A.P., High-pressure densities and derived thermodynamic properties of imidazolium-based ionic liquids, *J. Chem. Eng. Data* **52**, 80–88 (2007).

103  Fletcher, S.I., Sillars, F.B., Hudson, N.E., and Hall, P.J., Physical properties of selected ionic liquids for use as electrolytes and other industrial applications, *J. Chem. Eng. Data* **55**, 778–782 (2010).

104  Zhou, Z.B., Matsumoto, H., and Tatsumi, K., Structure and properties of new ionic liquids based on alkyl- and alkenyltrifluoroborates, *Chem. Phys. Chem.* **6**, 1324–1332 (2005).

105  Hatakeyama, Y., Okamoto, M., Torimoto, T., Kuwabata, S., and Nishikawa, K., Small-angle x-ray scattering study of au nanoparticles dispersed in the ionic liquids 1-alkyl-3-methylimidazolium tetrafluoroborate, *J. Phys. Chem. C* **113**, 3917–3922 (2009).

106  Suarez, P.A.Z., Einloft, S., Dullius, J.E.L., de Souza, R.F., and Dupont, J., Synthesis and physical-chemical properties of ionic liquids based on 1-*n*-butyl-3-methylimidazolium cation, *J. Chim. Phys.* **95**, 1626–1639 (1998).

107 Baldelli, S., Influence of water on the orientation of cations at the surface of a room-temperature ionic liquid: A sum frequency generation vibrational spectroscopic study, *J. Phys. Chem. B* **107**, 6148–6152 (2003).

108 McEwen, A.B., Ngo, H.L., LeCompte, K., and Goldman, J.L., Electrochemical properties of imidazolium salt electrolytes for electrochemical capacitor applications, *J. Electrochem. Soc.* **146**, 5, 1687–1695 (1999).

109 Liu, J.F., Jiang, G.B., Chi, Y.G., Cai, Y.Q., Zhou, Q.X., and Hu, J.T., Use of ionic liquids for liquid-phase microextraction of polycyclic aromatic hydrocarbons, *Anal. Chem.* **75**, 5870–5876 (2003).

110 Kuang, Q.L., Zhang, J., and Wang, Z.G., Revealing long-range density fluctuations in dialkylimidazolium chloride ionic liquids by dynamic light scattering, *J. Phys. Chem. B* **111**, 9858–9863 (2007).

111 Ishikawa, M., Sugimoto, T., Kikuta, M., Ishiko, E., and Kono, M., Pure ionic liquid electrolytes compatible with a graphitized carbon negative electrode in rechargeable lithium-ion batteries, *J. Power Sources* **162**, 658–662 (2006).

112 Matsumoto, H., Sakaebe, H., Tatsumi, K., Kikuta, M., Ishiko, E., and Kono, M., Fast cycling of Li/LiCoO$_2$ cell with low-viscosity ionic liquids based on bis(fluorosulfonyl)imide [FSI]$^-$, *J. Power Sources* **160**, 1308–1313 (2006).

113 Stępniak, I., and Andrzejewska, E., Highly conductive ionic liquid based ternary polymer electrolytes obtained by in situ photopolymerisation, *Electrochim. Acta* **54**, 5660–5665 (2009).

114 Torriero, A.A.J., Siriwardana, A.I., Bond, A.M., Burgar, I.M., Dunlop, N.F., Deacon, G.B., and MacFarlane, D.R., Physical and electrochemical properties of thioether- functionalized ionic liquids, *J. Phys. Chem. B* **113**, 11222–11231 (2009).

115 Evans, R.G., Klymenko, O.V., Hardacre, C., Seddon, K.R., and Compton, R.G., Oxidation of N,N,N′,N′-tetraalkyl-para-phenylenediamines in a series of room temperature ionic liquids incorporating the bis(trifluoromethylsulfonyl)imide anion, *J. Electroanal. Chem.* **556**, 179–188 (2003).

116 Chen, H.S., He, Y.R., Zhu, J.W., Alias, H., Ding, Y.L., Nancarrow, P., Hardacre, C., Rooney, D., and Tan, C.Q., Rheological and heat transfer behaviour of the ionic liquid, [C$_4$mim][NTf$_2$], *Int. J. Heat Fluid Fl.* **29**, 149–155 (2008).

117 Deng, M.J., Chen, P.Y., Leong, T.I., Sun, I.W., Chang, J.K., and Tsai, W.T., Dicyanamide anion based ionic liquids for electrodeposition of metals, *Electrochem. Commun.* **10**, 213–216 (2008).

118 Marsh, K.N., Brennecke, J.F., Chirico, R.D., Frenkel, M., Heintz, A., Magee, J.W., Peters, C.J., Rebelo, L.P.N., and Seddon, K.R., Thermodynamic and thermophysical properties of the reference ionic liquid: 1-Hexyl-3-methylimidazolium bis[(trifluoromethyl)sulfony]amide (including mixtures) Part 1. Experimental methods and results, *Pure Appl. Chem.* **81**, 5, 781–790 (2009).

119 Alonso, L., Arce, A., Francisco, M., and Soto, A., Liquid–liquid equilibria for [C$_8$mim][NTf$_2$] + thiophene + 2,2,4-trimethylpentane or + toluene, *J. Chem. Eng. Data* **53**, 1750–1755 (2008).

120 Crosthwaite, J.M., Muldoon, M.J., Dixon, J.K., Anderson, J.L., and Brennecke, J.F., Phase transition and decomposition temperatures, heat capacities and viscosities of pyridinium ionic liquids, *J. Chem. Thermodyn.* **37**, 559–568 (2005).

121 Hansmeier, A.R., Ruiz, M.M., Meindersma, G.W., and De Haan, A.B., Liquid-liquid equilibria for the three ternary systems (3-methyl-n-butylpyridinium dicyanamide + toluene + heptane), (1-butyl-3-methylimidazolium dicyanamide + toluene + heptane) and (1-butyl-3-methylimidazolium thiocyanate + toluene + heptane) at $T = (313.15$ and $348.15)$ ;K and $p = 0.1$ MPa, *J. Chem. Eng. Data* **55**, 708–713 (2010).

122 Vogel, H., The low of the relation between the viscosity of liquids and the temperature, *Phys. Z.* **22**, 645–646 (1921).

123 Tamman, G., and Hesse, W., The dependence of viscosity upon the temperature of supercooled liquids, *Z. Anorg. Allg. Chem.* **156**, 245–257 (1926).

124 Fulcher, G..S., Analysis of recent measurements of the viscosity of glasses, *J. Am. Ceram. Soc.* **8**, 339–355 (1925).

125 Tomida, D., Kumagai, A., Kenmochi, S., Qiao, K., and Yokoyama, C., Viscosity of 1-hexyl-3-methylimidazolium hexafluorophosphate and 1-octyl-3-methylimidazolium hexafluorophosphate at high pressure, *J. Chem. Eng. Data* **52**, 577–579 (2007).

126 Gardas, R.L., and Coutinho, J.A.P., Group contribution methods for the prediction of thermophysical and transport properties of ionic liquids, *AIChE J.* **55**, 5, 1274–1290 (2009).

127 Sescousse, R., Le, K.A., Ries, M.E., and Budtova, T., Viscosity of cellulose-imidazolium-based ionic liquid solutions, *J. Phys. Chem. B* **114**, 7222–7228 (2010).

128 Cherif, E., and Bouanz, M., Density, viscosity and electrical conductivity of isobutyric acid-water with added ions in the critical regions, *Phys. Chem. Liq.* **47**, 626–637 (2009).

129 Greaves, T.L., Weerawardena, A., Fong, C., Krodkiewska, I., and Drummond, C.J., Protic ionic liquids-solvents with tunable phase behavior and physicochemical properties, *J. Phys. Chem. B* **110**, 22479–22487 (2006).

130 Greaves, T.L., and Drummond, C.J., Protic ionic liquids-properties and applications, *Chem. Rev.* **108**, 206–237 (2008).

131 Yang, J.Z., Tong, J., Li, J.B., Li, J.G., and Tong, J., Surface tension of pure and water-containing ionic liquid $C_5MIBF_4$ (1-methyl-3-pentylimidazolium tetrafluoroborate), *J. Colloid Interf. Sci.* **313**, 374–377 (2007).

132 Greaves, T.L., Weerawardena, A., Fong, C., and Drummond, C.J., Many protic ionic liquids mediate hydrocarbon-solvent interactions and promote amphiphile self-assembly, *Langmuir* **23**, 402–404 (2007).

133 Zhang, Q.G., and Wei, Y., Study on properties of ionic liquid based on $ZnCl_2$ with 1-butyl-3- methylimidazolium chloride, *J. Chem. Thermodyn.* **40**, 640–644 (2008).

134 Sun, S.G., Wei, Y., Fang, D.W., and Zhang, Q.G., Estimation of properties of the ionic liquid $BMIZn_3Cl_7$, *Fluid Phase Equilibr.* **273**, 27–30 (2008).

135 Estager, J., Nockemann, P., Seddon, K.R., Swadźba-Kwaśny, M., and Tyrrell, S., Validation of speciation techniques: A study of chlorozincate(II) ionic liquids, *Inorg. Chem.* **50**, 11, 5258–5271 (2011).

136  Tong, J., Hong, M., Guan, W., Li, J.B., and Yang, J.Z., Studies on the thermodynamic properties of new ionic liquids 1-methyl-3-pentylimidazolium salts containing metal of group III, *J. Chem. Thermodyn.* **38**, 1416–1421 (2006).

137  Chirico, R.D., Diky, V., Magee, J.W., Frenkel, M., and Marsh, K.N., Thermodynamic and thermophysical properties of the reference ionic liquid: 1-Hexyl-3-methylimidazolium bis[(trifluoromethyl)sulfonyl]amide (including mixtures). Part 2. Critical evaluation and recommended property values (IUPAC Technical Report), *Pure Appl. Chem.* **81**(5), 791–828 (2009).

# INDEX

Abbreviations, xv–xxiv
Ab Manan, Norfaizah, xiii, 117
Absorption, of vapour, 294
Absorption isotherms, 172–173
  of ammonia, 173
Absorption/stripping process, 91
Acid chloroaluminates, 17
Acidic catalysts, ionic liquids as, 14–19
Acidic ionic liquids, applications of,
  15–18
Acidic solvents, ionic liquids as, 14–19
Acidising techniques, 13
Acid removal, from organic mixtures,
  8–9
Acids. *See also* Amino acid entries;
  Lewis acid- entries
  conjugated, 231
  neutralization with protic, 232
Acylation, 25
Additive solvents, ionic liquids as, 12
Additives with specific properties, ionic
  liquids as, 12–14
Adhesion problem, 102, 103
Adsorption, cation, 168
Advanced biasing procedures, 153
Advanced materials, radiolytic
  preparation of, 261
Aggregates
  of ionic liquids, 207
  of ions, 218
Aggregation behaviour, 170
Air–ionic liquid interface, 295
Alcohol carbonylation, 21
Aldehydes, alkyl aromatic, 17–18
Aldol condensation, 25
Alkoxyamine preparation, 25–26

1-Alkyl-3-methylimidazolium cations,
  conformers of, 60–62
1-Alkyl-3-methylimidazolium ionic
  liquids
  phase behaviour of, 59–85
  thermal behaviours of, 59, 80
Alkyl aromatic aldehydes, 17–18
Alkylation, of aromatic rings, 17
Alkylation reactions, 19
  of paraffins, 16–17
Alkyl chain lengths, 280, 281, 284, 292,
  295
Alkyl chains
  fluorinated, 293
  ion-phobic, 102
Alkyl groups, melting points and,
  280–282
1-Alkylimidazolium cation, molecular
  structure of, 277
Alkylpyridinium dicyanamide, 4
Alkyl sulfate ionic liquids, 2–4
  long chain, 2–3
Alkyl sulfates, 246
  high purity, 3
  onium, 3–4
  phosphonium, 4
Alpha radiolysis, 266
Amines, isomeric, 106
Amine treatments, 10
Amino acid cations, molecular structure
  of, 277
Amino acid extraction, 205
Ammonia, absorption isotherms of, 173
Ammonium cation, molecular structure
  of, 278
Ammonium halides, 14

*Ionic Liquids Further UnCOILed: Critical Expert Overviews*, First Edition.
Edited by Natalia V. Plechkova and Kenneth R. Seddon.
© 2014 John Wiley & Sons, Inc. Published 2014 by John Wiley & Sons, Inc.

Analytical techniques, 297
Angell strength parameter, 294
Anion abbreviations, xvii–xviii
Anion–anion interactions, 171
Anion behaviour, 168
Anion–cation interactions, 171
Anion constituents, of polymerised ionic liquids, 94
Anion donor ability, 229–230
Anion effect, 282–283
Anion head groups, 290–291, 292
Anionic structures, perfluorinated, 232
Anionic surfactants, 13–14
Anions
    $CO_2$ solubility and, 42
    dipolar, 246
    fluorinated, 221, 222
    ionic liquids with cyanoborate, 4–5
    ionic liquids with phosphorus-containing, 4
    kosmotropic and chaotropic, 199
    oxidisable, 263
Anti-solvents, $CO_2$ as, 49
Antistatic agents, in polymers, 12–13
Aprotic ionic liquids, 235, 250, 251, 252
Aqueous solutions, molecular dynamics simulations of, 170
Aromatic aldehydes, alkyl, 17–18
Aromatic cations, 266
Aromatic extraction, 123
Aromatic hydrocarbons, carbonylation of, 17–18
Aromaticity, influence of, 265–266
Aromatic organic liquids, 265
Aromatic ring alkylation, 17
Aspartame synthesis, 194
**Asym** (asymmetric form) conformers, 61, 64, 78, 79, 80
Asymmetric framework method, 124, 134
Atomic configurations, 153
Atomistic simulations
    of cations, 156–157
    classical, 150

Barrier materials
    ionic liquid membranes as, 104–105
    ionic liquids as, 87

Baseline stability, of calorimetry trace, 67
BASF™ company, 8
BASIL™ technology, 8
Behaviour. *See also* Environmental behaviour; Materials behaviour; Phase behaviour entries; Phase transition behaviours; Properties; Swelling behaviour; Thermal behaviours
    aggregation, 170
    anion, 168
    dielectric, 237
    freezing–melting, 279
    of matter, 150
    premelting, 64–65
Bernard, Didier, xiii, 1
$\beta$ parameter, 160
Biasing procedures, advanced, 153
Binary gas solubilities, 173
Binary interaction parameters, 45
Binary ionic liquid + supercritical $CO_2$ systems, 52
Binary ionic liquid + supercritical $CO_2$ systems, 41–42
    phase behaviour predictions for, 45
Binary ionic liquid + supercritical $CHF_3$ systems, 42–43, 52
Binary (ionic liquid + supercritical fluid) systems
    classification of, 43–44
    liquid–vapour phase behaviour of, 43–44
    modelling of, 44–46
    phase behaviour of, 41–46
Binary solid–liquid equilibrium (SLE) systems, 123
Binary systems, phase behaviour of, 40
Biocatalysis
    single-phase and multiphase systems for, 197–198
    two-phase (biphasic) systems for, 195
    in whole cell systems, 203–206
Biocatalysts, history of, 193–194
Biocatalytic reactions
    future aspects of, 208–210
    in industrial processes, 193, 194
    influence on environment, 207
    in ionic liquids, 193–216

Biocatalytic syntheses, reported in ionic liquids, 196
Biodegradability, 206, 207
Biofuel production, from lignocellulosic biomass, 29
Biomass valorisation, patent applications concerning, 28–29
"Biomimetic" ionic liquids, 158
Biopolymers, 198
  additional, 28
  ionic liquids and, 26–28
Bioreactive systems, ionic liquid membranes for, 109
Biorefineries, 26
Biphasic processes, for alkoxyamine preparation, 25–26
Biphasic systems, 203–204
  for biocatalysis, 195
  reaction rates in, 50
Bondi method, 125, 127
Bonnet, Philippe, xiii, 1
Born–Fajans–Haber cycle, 283
Bromides, DSC measurements of, 62–65

[C₂mim]⁺-based ionic liquids
  physicochemical properties of, 221–226, 229–232, 232–233
  transference number and ionicity of, 226–229
[C₄mim][PF₆], phase behaviour of, 67–73
Calibration curves, 230, 232–233
Calorimetric curves, 68
Calorimetry trace, baseline stability of, 67
Carbon dioxide. *See* $CO_2$ entries; Supercritical $CO_2$
Carbon nanotubes (CNTs), 13, 168, 169
Carbon nanotube solvation, 169
Carbonylation
  of alcohol, 21
  of aromatic hydrocarbons and paraffins, 17–18
  continuous, 21
  of saturated hydrocarbons, 18
Catalyst inhibition, 204
Catalyst reuse, 107–108
Catalytic systems, ionic liquids as solvents for, 19–26
Cation abbreviations, xv–xvii

Cation adsorption, 168
Cation–anion interactions, 171
Cation changes, phase transitions linked with, 59–65
Cation constituents, of polymerised ionic liquids, 94
Cation families, force fields for, 157
Cation head groups, 289, 290–291
Cationic transference numbers, 226–227
Cation reorientation, 244, 245
Cations
  aromatic, 266
  atomistic simulations of, 156–157
  conformers of 1-alkyl-3-methylimidazolium, 60–62
  imidazolium-based, 163
  ionic liquid, 98, 111
  ionic liquids with lactam, 5
Cation structures
  in crystalline phases, 69–73
  found in ionic liquids, 277–278
CEES (2-chloroethyl ethyl sulfide) vapour transport, 104–105
Cell systems biocatalysis in, 203–206
Cell toxicity prediction, 207–208
Cellulase, hyperthermophilic, 209–210
Cellulose
  dissolution and processing of, 26–27
  sulfation/sulfonation of, 27
Chaotropic anions, 199
Charge–charge interactions, 168
Chemical potential, 132
  excess, 171
Chemical processing, in ionic liquid + supercritical fluid systems, 48–51
Chemical production, from lignocellulosic biomass, 29
Chemical transformations, microwave-assisted, 24
$CHF_3$ systems, binary ionic liquid + supercritical, 42–43, 52
Chiral ionic liquids
  applications of, 110
  with membranes, 110
Chloroaluminate(III) ionic liquids, 16, 17
  preparation of, 15
  regeneration of, 19
Chloroaluminates, acid, 17
Chlorolactams, 11–12

Classical atomistic simulations, 150
Classical potential functions, 150
Close-boiling mixtures treatment, 8
$CO_2$ (carbon dioxide). *See also*
 Supercritical $CO_2$
 as anti-solvent, 49
 as co-solvent, 49, 51
$CO_2$ absorption isotherms, 172–173
$CO_2$ capture
 in fossil fuel burning power plants,
 101
 ionic liquids for, 30
$CO_2$ gas permeabilities, 104
$CO_2$ mixtures, molecular dynamics
 simulation studies of, 170–171
$CO_2/N_2$ separations, 102
$CO_2$ removal, 10–11
 from process streams, 48
$CO_2$ separation, 99–104
 gelled membranes in, 103–104
 industrial applications of, 100–101
 MMMs in, 102–103
 poly(RTILs) in, 101–102
$CO_2$ solubility, 41–42, 46, 130
$CO_2$ systems
 binary ionic liquid + supercritical,
 41–42
 ternary ionic liquid + supercritical,
 46–48
Cofactor regeneration, 203
Cohesive energy density, 158–159
COIL conferences, viii
Cole–Cole (CC) model, 242, 243, 249, 250
Cole–Davidson (CD) model, 242, 243,
 250
Collective properties, of ionic liquids,
 154
Common ionic liquids, incremental
 changes to, 180
Composite liquid–polymer structures,
 95–97
Composite membranes, with three
 components, 95, 96
Compounds
 diblock, 130
 enantiomeric separation of, 110
 extraction of, 204–206
Computational equipment/methods,
 advanced, 260
Computations, engineering, 134–135

Computed viscosities, 162
Computer-aided design modelling
 (CADM), 121
"Conceptual" tools, 118
Conductance dispersion, 239, 247, 248
Conductivity. *See also* Direct current
 (DC) electrical conductivity;
 Electrical conductivity ($\sigma$;);
 Electrically conducting systems;
 Molar conductivity entries
 frequency-dependent, 238
 ionic, 163–164, 218
 thermal, 164
Conductivity maximum per ion, 170
Conductivity measurements, 220, 228.
 *See also* Molar conductivity
 measurements
Conductivity values, 222–224
COnductor-likeScreeningMOdel for
 Real Solvents (COSMO-RS), 88,
 132–133, 134, 283
Conformational cation changes, phase
 transitions linked with, 59–65
Conformational structures, 60
Conformers
 of 1-alkyl-3-methylimidazolium
 cations, 60–62
 multiple, 60, 81
Conjugated acids, 231
Conjunct polymers, 19
Contactors. *See* Membrane contactors
Continuous biphasic processes, with
 ionic liquids and supercritical $CO_2$,
 49–50
Continuous carbonylation, 21
Continuous fractional component (CFC)
 method, 172–173
Continuous processes. *See also*
 Continuous biphasic processes
 design of, 118
 with ionic liquids, supercritical $CO_2$,
 and miscibility switch phenomenon,
 50–51
Controlled density reduction, 50
Copolymers, hydrogenated, 12
Correlation functions, in molecular
 simulations, 152
Co-solvents, $CO_2$ as, 49, 51
Coulombic forces, 217
 melting points and, 280–281

Coulombic interactions, 81, 124, 230, 232
Crespo, Joăo G., xiii, 87
Critical expert overviews, ix–x
Critical points, 174
Cross-linkable/linked gemini RTILs
    (GRTILs), 94, 105. *See also* Room
    temperature ionic liquids (RTILs)
**Crystal** $\alpha$ phase, 68, 69–71
**Crystal** $\beta$ phase, 68, 69
Crystal–crystal phase transitions, 67,
    68–69, 81
**Crystal** $\gamma$ phase, 68, 69–71
Crystalline domains, 73–74, 76, 80
Crystalline grains, 79
Crystalline structure, 87–88
Crystallisation
    intermittent, 77–80, 81
    of ionic liquids, 62–65
    model for, 73–74
    rhythmic, 74–77, 80, 81
    smooth, 78
Crystallisation/melting, timescale of, 76
Crystallisation peak, 65
Crystallisation timescale, 76, 78
Crystals, simulations of, 178–179
Crystal structures, 72–73
    predicting, 178
Cubic equations of state, 44–45
Cumulative radiation exposure, 261
Cyanoborate anions, ionic liquids with,
    4–5
Cyclic experiments, 77

Databases, 119, 156
    force field, 179–180
    of ionic liquids, 276
Data production, 276
Debye–Falkenhagen (DF) theory, 247,
    248
Debye process, 240, 242
Degradation, of ionic liquids, 208
Demixing
    instantaneous, 51
    spinodal, 51
Denaturation, 199
Dense membranes, modified, 97
Density, 283–288. *See also* Energy
    density
    computing, 157–158
    at different temperatures, 288

Density functional theory (DFT), 60
Density measurements, 220
Density oscillations, 167
Depolymerisation, of starch, 28
Derivative quantity calculations, 158
Designer functionalities, for ionic liquids,
    118. *See also* Process design
Desulfurisation, of oil, 29
Desulfurisation technology, 10
1,3-Dialkylimidazolium cation, molecular
    structure of, 277
1,3-Dialkylimidazolium ionic liquids, 41,
    42, 52
    vapourisation enthalpy of, 158–159
Dialkyl sulfates, 4
Diblock compounds, 130
Dielectric behaviour, 237
Dielectric constants, 251–252
    static, 235, 236, 237, 238, 240, 241,
    248–253
Dielectric continuum approach, 253
Dielectric continuum theory, 248, 253
Dielectric dispersion, 238, 241
Dielectric dispersion curves, frequency-
    dependent, 235
Dielectric loss, 238, 241, 244
Dielectric modes, 236
Dielectric permittivity, frequency-
    dependent relative, 238
Dielectric properties, of ionic liquids,
    235–258
Dielectric relaxation, 236, 240. *See also*
    Ion cloud relaxation; Non-
    exponential relaxation
Dielectric relaxation spectroscopy
    (DRS), 236, 237, 248–249. *See also*
    DRS studies
    developments of, 254
Dielectric response, 235–236, 240, 241
    molecular processes affecting,
    243–248
Dielectric spectra, 236, 237, 253, 254
    of ionic liquids, 240–243
    phenomenological description of,
    240–243
Dielectric theory, 237–238
    of electrically conducting systems,
    237–240, 241
Differential scanning calorimetry (DSC),
    60, 220. *See also* DSC entries

Diffusivity, 99, 104, 227
  ionic, 224
Diffusivity measurements, 226
Dipolar anions, 246
"Dipolar" processes, 236
  affecting dielectric response, 243–246
Dipolar VOCs, 252. *See also* Volatile
  organic compounds (VOCs)
Dipole moments, 252
  electric, 253
  molecular, 253
Direct current (DC) electrical
  conductivity, 236
Dispersion
  conductance, 239, 247, 248
  dielectric, 238, 241
Dissociativity, 228
Dissolved cellulose, 26–27
  textile applications for, 27
Domains. *See also* Crystalline domains;
  Premelting domains
  hydrophobic and hydrophilic, 245
  polar and non-polar, 155
  rhythmic transition of, 76
Double imidazolium cation, molecular
  structure of, 277
Double layer, ionic liquid, 168
Double-layer stacking formation, 168
Downstream processes/processing, 197,
  203
DRS studies, 240. *See also* Dielectric
  relaxation spectroscopy (DRS)
DSC instruments, nano-Watt-stabilised,
  66. *See also* Differential scanning
  calorimetry (DSC)
DSC measurements, of bromides, 62–65
DSC/Raman spectroscopy
  measurements, 66–67
DSC traces, 75, 76, 77, 79
Dynamical relaxation processes, in
  molecular simulations, 151
Dynamic heterogeneity, 162
Dynamic hydrogen bonding, 167–168
Dynamic performance tests, 98
Dynamic water clusters, 105

Eco-efficiency analysis, 208
Effective ionic concentration, 233
Efficiency system design, 118

Electrical conductivity ($\sigma$), frequency-
  dependent, 238. *See also* Direct
  current (DC) electrical conductivity
Electrically conducting systems,
  dielectric theory of, 237–240, 241
Electric dipole moment, 253
Electrochemical applications, ionic liquid
  membranes for, 108–109
Electrochemical molar conductivity,
  227–228
Electrode polarisations, 236, 241,
  249–250
Electrolyte NRTL (e-NRTL) model, 124
Electron decay, 267–268
Electron reactivity, presolvated, 263
Electrons, solvated, 267, 269
Electron spin resonance studies, 263
Enantiomeric separation, of compounds,
  110
Energy density, cohesive, 158–159
Energy recycle design, 118
Engineering computations, 134–135
  for ionic liquids, 120
  for process design, 121–122
Engineering simulations, 117–148
Enhanced selectivity, ionic liquid systems
  with, 110
Enthalpy
  molar crystallisation, 78
  vapourisation, 155, 158–159, 174
Environmental behaviour, of ionic
  liquids, 210
Environmental impact, of ionic liquids,
  206–208
Environmentally benign solvents, 39
Environmentally friendly ionic liquids,
  207
Enzymatic catalysis, 106
Enzymatic fuel cells, development of,
  109
Enzymatic membrane reactors, ionic
  liquid membranes for, 109
Enzyme-catalysed reactions, water
  content/activity in, 200–201
Enzyme classes (ECs), 194, 195
Enzymes
  as biocatalysts, 193
  in biocatalytic syntheses, 196
  hydrolysing, 27

influence of ionic liquids on, 198–200
in ionic liquids, 194–197
patent applications concerning, 28–29
solubilisable, 109
Enzyme solubility, 198
Enzyme types, screening for, 209
EoS methods/models, 123, 128–132, 134.
*See also* Equations of state (EoS);
Group contribution EoS (GC-EoS)
model
Eötvös equation, 296
Equations of state (EoS). *See also* EoS
methods/models
cubic, 44–45
group contribution, 45
statistical-mechanics-based, 45
Equilibrating vapour phase, 201
Equilibrium MD (EMD) simulation,
161
Equimolar mixtures, MD studies of,
177–178
Equipment
for ionic liquid thermal analysis,
65–67
for mass transfer processes, 91
Estimation tools, 134–135
Ether glycols, 23–24
1-Ethyl-3-methylimidazolium ionic
liquids, 217, 219, 221, 226–227, 228,
229, 231–232, 233. *See also*
[$C_2$mim]$^+$-based ionic liquids
Excess chemical potential, 171
Expanded ensemble (EE) approach,
171–172
Experimental data, simulations *vs.*,
154–155
Experimental data collection, 122
Exploratory research, 181
"Extended Hamiltonian" methods, in
molecular simulations, 151
Extracting agents, ionic liquids as, 7–8
Extraction of compounds, 204–206
Extractive fermentation, 204
Extrapolation procedures, 162

Far infrared (FIR) methods, 236
Fermentation, extractive, 204
Fine particles recovery, 13
Fixed-charge models, 161, 166

Flowsheeting software packages, 122
Flowsheet programs, 118
Fluid catalytically cracked (FCC) off-gas,
17
Fluid multiphase systems, ternary, 43–44
Fluid theory models, statistical
association, 45–46
Fluorinated alkyl chains, 293
Fluorinated anions, 221, 222
Fluorination, 24–25
of halogenated compounds, 18
Fluorination catalysis, liquid phase, 18
Force field databases, 156, 179–180
Force field parameters, for ionic liquids,
157
Force fields, 178–180
for cation families, 157
developing, 179
for imidazolium-based ionic liquids,
178
in molecular simulations, 150–151
non-polarisable, 166
polarisable, 157, 161, 166
refined, 180
setting, 156–157
united-atom, 157
validation of, 178–179
Fossil fuel burning power plants, $CO_2$
capture in, 101
Free energy methods/schemes, 171, 172
multistage, 172
Freezing–melting behaviour, 279
Frequency analyses, 74
Frequency-dependent dielectric
dispersion curve, 235
Frequency-dependent electrical
conductivity, 238
Frequency-dependent relative dielectric
permittivity, 238
Frequency distribution curves, 76
Friedel–Crafts acylation, 25
Fuel cells
development of enzymatic, 109
ionic liquid membranes for, 108–109
Fugacity coefficients, 129
Functional group parameters, 126
Functionalised ionic liquids, 209
Function analysis, radial distribution,
168

Gamma radiation damage, 266
Gas absorption isotherms, 172–173.
    *See also* Vapour entries
Gases
    non-polar, 172
    separation of, 99–104
    solubility of, 129–130
Gas–liquid interfaces, 164–169
    ionic, 165–166
Gas permeabilities, $CO_2$, 104
Gas phase behaviour, 131–132
Gas phase quantum calculations, 159
Gas separation membrane performance,
    100
Gas solubility, 88–90, 131
Gas–vapour permeation applications, 97
Gatterman–Koch reaction, 17–18
GC non-random lattice-fluid EoS
    (GCNRLF-EoS) model, 131. *See
    also* Equations of state (EoS);
    Group contribution EoS (GC-EoS)
    model
Gelled ionic liquids, 95–97
Gelled membranes, in $CO_2$ separation,
    103–104
Gelled RTILs, 95–97. *See also* Room
    temperature ionic liquids (RTILs)
Gemini RTILs (GRTILs), 94. *See also*
    Room temperature ionic liquids
    (RTILs)
Gibbs energies ($G^E$), 129, 283
Gibbs excess ($G^E$) models, 123–128
Gibbs free energy models, 88
Glass transition temperatures, 279
Gold nanoparticles, 269–270
Granted patents, 28
Gravimetric balance, 41
Green-Kubo expression, 163–164
"Green" solvents, 40, 90
Group contribution EoS (GC-EoS)
    model, 45, 131–132. *See also*
    Equations of state (EoS)
**GT** (*gauche trans*) conformations, 62, 64,
    65, 69–70, 72, 73, 75–76
**G′T** (*gauche′ trans*) conformations, 62,
    69–73
Guanidinium cation, molecular structure
    of, 278
Guanidinium ionic liquids, 5

Guo, Liangliang, xiii, 275
*G*-values, 265–266

Halides, 14, 15
Haloalkanes, 16–17
Halogenated compound fluorination, 18
Halogen-free ionic liquid synthesis, 2
Hamiltonian methods, in molecular
    simulations, 151
"Hardware" tools, 120
Head groups
    anion, 290–291, 292
    cation, 289, 290–291
Heat capacity, 159. *See also* Temperature
    entries; Therm- entries
Heat transfer cycles, 77
Helmholtz energy, 129–130, 131
Henry's Law constants, 171, 172, 173
High-frequency processes, 242
High purity alkyl sulfates, 3
High-purity ionic liquids, 5–6, 267
Hofmeister series, 199–200
Holes, radiolytically induced, 263
Hollow fibre membrane contactor, 91
Homogeneous systems, 50
Homo-metathesis, 22
Hydrocarbons, carbonylation of
    saturated, 18
Hydrocarbon stream drying, 25
Hydrocarbon streams, sulfur compound
    removal from, 9–10
Hydrochloric acid recovery, 8–9
Hydrodesulfurisation process, 10
Hydrodynamic approaches, 244, 245
Hydroformylation, 20–21
Hydrogenated copolymers, 12
Hydrogenation, of conjunct polymers, 19
Hydrogen bonding, 289
    dynamic, 167–168
Hydrolysing enzymes, 27
Hydrolysis, of polyamides, 18
Hydrophilic domains, 245
Hydrophilic ionic liquids, 198–199
Hydrophilic poly(diol-RTIL)s, 104–105.
    *See also* Room temperature ionic
    liquids (RTILs)
Hydrophobic domains, 245
Hydrophobic ionic liquids, 105
Hydrophobicity, 199–200

Hydrosilation, 22–23
Hydroxyl groups, 296
Hyperthermophilic cellulase, 209–210
Hypothetical critical points, 174

Ideal heat capacity, 159
Imidazolium-based cations, 163
Imidazolium-based ionic liquids, 128,
    129, 130, 280. *See also* Imidazolium
    ionic liquids
    force field for, 178
    "refined" force fields for, 157
Imidazolium-based RTILs, 101, 104. *See
    also* Room temperature ionic liquids
    (RTILs)
Imidazolium halides, 15
Imidazolium ionic liquids, 3, 276. *See
    also* Imidazolium-based ionic liquids
Imidazolium ring, 165, 166, 167
Imidazolium ring plane, 169
Imidazolium salts, 22–23
Immobilisation procedures, quality of, 97
Immobilised ionic liquids, 6
Immobilised ions, 102
Impedance measurements, 226
Impurities
    in ionic liquids, 201–202
    toxicity effect of, 206–207
Incident radiation, 262
Industrial applications
    of $CO_2$ gas separation, 100–101
    of ionic liquid membranes, 90
Industrial processes, biocatalytic
    reactions in, 193, 194
Infinite-dilution activity coefficients, 132
"Infinite-dilution" solubility, 172
Instantaneous demixing, 51
Integrated absorption/stripping process,
    91
Integrated membrane–ionic liquid
    systems, 90
Interfacial simulations, 165–166
Intermittent crystallisation, 77–80, 81
Intermolecular interactions, 287–288
Intermolecular vibrations, 247
Inter-wall distance, 169
Ion aggregates, 218
Ion cloud relaxation, 247–248
Ionic compounds, polar, 7

Ionic concentration, effective, 233
Ionic conductivity, 163–164, 218
Ionic diffusivity, 224
Ionicity, 226, 228–229, 232
    correlation with ionic structures and
        physicochemical properties, 229–232
    in ionic liquids, 217–234
Ionicity values, 230
Ionic liquid + supercritical fluid systems
    chemical processing in, 48–51
    phase behaviour of, 41–46, 46–48
    reactions and separations in, 49–51
    separations in, 48–49
Ionic liquid abbreviations, xv
Ionic liquid additives, for oil drilling/oil
    wells, 13
Ionic liquid aggregates, 207
Ionic liquid–air interface, 295
Ionic liquid antistatic agents, in
    polymers, 12–13
Ionic liquid-based polymer membranes,
    108
Ionic liquid cations, 98
"Ionic liquid chemistry," 98
Ionic liquid complications, 134
Ionic liquid confinement, 168–169
Ionic liquid databases, 276
Ionic liquid degradation, 208
Ionic liquid deposition, 206
Ionic liquid design/designability, 218,
    276
Ionic liquid diffusion, 208
Ionic liquid diversity, 260–261, 276
Ionic liquid double layer, 168
Ionic liquid–gas interface, 165–166
Ionic liquid interfaces, with non-polar
    solvents, 166
Ionic liquid–liquid interface, 166
Ionic liquid lubricants, 12
Ionic liquid membrane technology,
    87–116
    future directions for, 109–111
Ionic liquid membranes
    applications of, 99–109
    as barrier materials, 104–105
    characterisation of, 97–99
    industrial applications of, 90
    structure and morphology of, 90–97
Ionic liquid mixtures, 170

Ionic liquid preparations, for radiation chemistry studies, 266–268
Ionic liquid properties
   origin of, 217–234
   temperature dependencies of, 222–224
   tuneability of, 111
Ionic liquid radiation chemistry, 259–274
   future prospects for, 270
   nanoparticle synthesis and, 268–270
Ionic liquid radiolysis products, 261–262
Ionic liquid recycling, 10
Ionic liquids. *See also* Room temperature ionic liquids (RTILs)
   as acidic catalysts and solvents, 14–19
   additional applications for, 11–12
   as additive solvents, 12
   as additives with specific properties, 12–14
   advantages of, 209
   alkyl sulfate, 2–4
   applications of, 28, 88
   applications of acidic, 15–18
   aprotic, 235, 250, 251, 252
   biocatalytic reactions in, 193–216
   biocatalytic syntheses reported in, 196
   "biomimetic," 158
   biopolymers and, 26–28
   cation–anion constituents of polymerised, 94
   cations in, 111
   cation structures found in, 277–278
   chloroaluminate, 16, 17
   for $CO_2$ capture, 30
   continuous biphasic processes with supercritical $CO_2$ and, 49–50
   continuous processes with, 50–51
   crystallisation of, 62–65
   with cyanoborate anions, 4–5
   definitions and properties of, 87–90
   densities of conventional, 285–286
   1,3-dialkylimidazolium, 41, 42
   dielectric properties of, 235–258
   dielectric spectra of, 240–243
   engineering computations for, 120
   environmental behaviour of, 210
   environmental impact of, 206–208
   environmentally friendly, 207
   enzymes in, 194–197
   as extracting agents, 7–8

force field parameters for, 157
functionalised, 209
gelled, 95–97
generalities concerning, 264
"green" character of, 90
guanidinium, 5
high-purity, 5–6, 267
hydrophilic, 198–199
hydrophobic, 105
imidazolium, 3, 276
imidazolium-based, 128, 129, 130, 280
immobilised, 6
improved use of, 2–6
impurities in, 201–202
incremental changes to common, 180
influence on enzymes and substrate, 198–200
interest in, 1–2
ionicity in, 217–234
with lactam cations, 5
Lewis acidic, 15
as ligands, 21
LLE simulations of, 176
low-temperature, 15
low-viscosity, 240, 253
as lubricants, 29–30
with magnetic properties, 110–111
as membranes, 87
microwave-assisted chemical transformations with, 24
molecular dynamics simulations of, 167
molecular simulation of, 149–192
molecular simulations of pure, 170
in multi-component systems, 170–171
neutral, 25
onium, 24
organic compound recovery from, 48–49
partial crystallisation of, 6
performance of, 297
in petrochemistry, 1–37
phase behaviour of 1-alkyl-3-methylimidazolium, 59–85
phase transition behaviours of, 73–80
with phosphorus-containing anions, 4
physicochemical properties of, 275–307

polymer membranes and, 93
pre-selection of, 132
preparation of chloraluminate(III), 15
preparing, 267
production under ultrasonication, 6
products and processes related to, 117, 118
properties and data points for, 279
properties of, 40, 153–155
protic, 250–251, 295–296
pyridinium, 3, 270
as reaction media, 50
"refined" force fields for imidazolium-based, 157
regeneration of chloraluminate(III), 19
room temperature, 59
self-diffusiveness/self-diffusivities of, 152, 154, 159–161, 161–162, 170
separation processes using, 6–12
solubility in, 171–173
as solvents for catalytic systems, 19–26
as solvents in SLMs, 106
with specific functionalities, 118
starch and, 27–28
static dielectric constant of, 248–253
sulfate, 2
supercritical fluids in, 39–57
task-specific, 92–93, 94
terahertz spectra of, 243
thermal analyses of, 60
thermal measurements of, 66
thermodynamic models for, 122–133
thermodynamic properties of, 134
unique properties of, 59–60
volatile organic compounds *vs.*, 253
water-immiscible, 197, 198
Ionic liquid science, radiation chemistry and, 261–262
Ionic liquid–solid interface, 166–169
Ionic liquid stability, nuclear fuel recycling and, 264–266
Ionic liquid structural features, 283
Ionic liquid/supercritical fluid systems, phase behaviour of, 40–48
Ionic liquid/supporting membrane systems, 106–107
Ionic liquid synthesis, 219
    halogen-free, 2

Ionic liquid systems
    with enhanced selectivity, 110
    stimuli-responsive, 110–111
Ionic liquid thermal analysis
Ionic liquid–vacuum interface, 165
"Ionic" processes, 236
    affecting dielectric response, 246–247
Ionic structures, ionicity correlation with, 229–232
Ionising radiation, 259, 262
Ion pairing, 246
Ion-pair modes, 246
Ion-phobic alkyl chains, 102
Irradiation, microwave, 24, 28. *See also* Radiation entries
Isomeric amines, 106
Isomerisation, 18
Isomers, rotational, 60–61, 62
Isopropyl groups, melting points and, 281–282
Isoquinolinium cation, molecular structure of, 277
Isothermic–isobaric Gibbs ensemble MC (GEMC) method, 172, 175. *See also* Monte Carlo (MC) simulations
Isotherms, absorption, 172–173

Kamlet–Taft parameter, 253
Karl Fischer titration, 201, 289
Kilohertz regime, 247, 250
Kirkwood formula, 252
Kohlrausch–Williams–Watts (KWW) function, 242
Kosmotropic anions, 199
Kragl, Udo, xiii, 193
Kroon, Maaike C., xiii, 39

Lactam cations, ionic liquids with, 5
Lactams, 11–12
*Lactobacillus* species, 204, 207
Lattice-fluid (LF-EoS) model, 131. *See also* Equations of state (EoS)
Lennard–Jones fluids, 166
Lewis acidic ionic liquids, 15
Lewis acids, 14
Lewis bases, 20–21
Lewis basicity, 217, 218–219, 222, 230
Librational processes, 244
Life cycle analysis, 208

Ligand liquids, 21

Light gas transport properties, of three-component composite membranes, 102, 103

Lignocellulosic biomass, biofuel and chemical production from, 29

Lipases, 194–195

Liquid clathrates, 14

Liquid crystalline solutions, 26

Liquid–gas interfaces, 164–169
  ionic, 165–166

Liquid–liquid equilibrium (LLE) systems, 123, 124, 125, 127, 131, 132, 149, 175–176

Liquid–liquid interfaces, 164–169

Liquid–liquid phase separation, 8

Liquid membranes, supported, 103–104

Liquid phase fluorination catalysis, 18

Liquid phase separations, 105–108

Liquid–polymer structures, composite, 95–97

Liquids, supercooled, 79, 279. *See also* Ionic liquid entries

Liquid–solid interfaces, 164–169

Liquid structure, 155

Liquid–vapour phase behaviour, of binary systems, 43–44

LLE calculations, 176. *See also* Liquid–liquid equilibrium (LLE) systems

LLE prediction, 133

Local composition models, 88

Local melting domains, 78

Log *P* value, 207–208

Long chain alkyl sulfate ionic liquids, 2–3

Lower critical endpoint (LCEP), 44

Low-temperature ionic liquids, 15

Low-viscosity ionic liquids, 240, 253

Low-viscosity VOCs, 236. *See also* Volatile organic compounds (VOCs)

Lu, Xingmei, xiii, 275

Lubricants, ionic liquid, 12, 29–30

Macroscopic electric dipole moment (**M[t]**), 237

Macroscopic viscosity, 225

Maginn, Edward J., xiii, 149

Magnetic properties, ionic liquids with, 110–111

Market needs, matching, 119

Mass transfer limitations, 50, 197–198

Mass transfer processes, equipment for, 91

Materials behaviour, process design and, 119

Mathias–Klotz–Prausnitz (MKP) mixing rule, 128

Matrix acidising technique, 13

Matter, behaviour of, 150

Maxwell equation predictions, 103

Maxwell's equations, 237–238

MC codes, in molecular simulations, 153. *See also* Monte Carlo (MC) simulations

MD codes, in molecular simulations, 151. *See also* Molecular dynamics (MD) simulations

Mean square displacement (MSD), 160, 169
  in molecular simulations, 152

Melting
  model for, 73–74
  rhythmic, 74–77, 80, 81

Melting/crystallisation, timescale of, 76

Melting domains, local, 78

Melting points, 279–283
  computing with MD simulations, 176–177
  predicting, 176
  thermodynamic, 176–177

Melting process, 63–65

Melting trace, 74

Membrane contactors, 90–93
  major drawback of, 92

Membrane disruption, 208

Membrane–ionic liquid systems, 90

Membrane performance, 100

Membrane reactors, enzymatic, 109

Membranes. *See also* Ionic liquid membranes; Polymer–electrolyte membrane (PEM); Polymer membranes; Three-component composite membranes (MMMs)
  chiral ionic liquids with, 110
  gelled, 103–104
  ionic liquids as, 87
  Nafion®, 108
  proton exchange, 108

separation selectivity of, 99
structural integrity of, 103
supported liquid, 103–104
swelling behaviour of, 98–99
two-component, 102
Membrane systems, 90
Membrane technology, related to ionic
 liquids, 87–116
Metal halides, 14
Metal nanoparticle synthesis, radiation
 chemistry in, 269
Metal salts, 7, 8
Metathesis, 22
Methane-to-methanol conversion, 24
Methanol, methane conversion to, 24
Methyl groups, melting points and,
 281
Microwave-assisted chemical
 transformations, with ionic liquids,
 24
Microwave experiments, 242
Microwave irradiation, 24, 28
Microwave regime, 240, 247, 250, 253
Microwave spectra, of ionic liquids,
 240–243
Miscibility switch phenomenon,
 continuous processes with, 50–51
Mixture properties, 179
Mixtures
 of ionic liquids, 170
 molecular dynamics simulation studies
  of $CO_2$, 170–171
 ternary, 198
Mixture thermodynamic properties,
 predicting, 154
Modelling. *See also* Cole–Cole (CC)
 model; Cole–Davidson (CD) model;
 Computer-aided design modelling
 (CADM); Electrolyte NRTL
 (e-NRTL) model; EoS methods/
 models; Fixed-charge models;
 Modified UNIFAC (Do) model;
 Multi-scale modelling approach;
 Non-random two-liquid (NRTL)
 activity coefficient model; NRTL
 segmented activity coefficients
 (NRTL-SAC) model; Polarisable
 continuum model (PCM); PR +
 Stryjek–Vera (PRSV) EoS model;

Quadrupolar models;
 Thermodynamic models;
 UNIQUAC method/model
 of binary (ionic liquid + supercritical
  fluid) systems, 44–46
 of phase behaviour, 88
Modelling groups, 181
Modified dense membranes, 97
Modified UNIFAC (Do) model, 127
Molar conductivity, 225, 226
 electrochemical, 227–228
Molar conductivity ratios, 217, 219
Molar crystallisation enthalpy, 78
Molecular dipole moments, 252, 253
Molecular dynamics (MD) simulations,
 150–153, 165
 of aqueous solutions, 170
 to compute melting points, 176–177
 drawbacks of, 151–152
 of ionic liquids, 167
 standard liquid, 180–181
 "statistical" properties and, 152
Molecular dynamics (MD) simulation
 studies
 of $CO_2$ mixtures, 170–171
 of equimolar mixtures, 177–178
Molecular mechanics (MM) simulations,
 283
Molecular processes, affecting dielectric
 response, 243–248
Molecular simulations
 advances in, 181
 defined, 150
 future perspectives on, 173–181
 goals of, 153–156
 of ionic liquids, 149–192
 of pure ionic liquids, 170
Molecular volumes, 287, 294
Molecule configurations, in molecular
 simulations, 151
Monte Carlo (MC) simulations, 150–153,
 172–173, 175, 176
 advantages of, 152–153
Morpholinium cation, molecular
 structure of, 277
Morphologies of ionic liquids as
 membranes, 87
Motion equations, in molecular
 simulations, 151

Multi-component systems, ionic liquids in, 170–171
Multiphase systems
  for biocatalysis, 197–198
  ternary fluid, 43–44
Multiple conformers, 60, 81
Multi-scale modelling approach, 120–121
Multi-scale simulations, 117
Multistage free energy methods, 172
Multi-walled carbon nanotubes (MWCNTs), 168, 169

Nafion® membranes, 108
Nanofiltration technology, 107
Nanomaterials, 29
Nanoparticles, 269
Nanoparticle synthesis
  ionic liquid radiation chemistry and, 268–270
  radiation chemistry in, 269
  radiolytic, 269, 270
Nanotubes (NTs), 168, 169
Nano-Watt-stabilised DSC instruments, 66. *See also* Differential scanning calorimetry (DSC)
Natural gas purification, 10–11
Natural gas sweetening, 101
"Neoteric" media, 260–261
Nernst–Einstein (NE) equation, 163, 225
Network analysers, vectorial, 240
Neutral functional groups, 126
Neutral ionic liquids, 25
Neutralization, with protic acids, 232
Newtonian viscosity, 162
Nickel-catalysed olefin oligomerisation, 20
Nishikawa, Keiko, xiii, 59
Nitrogen-containing polar compound separation, 11
Noble, Richard D., xiii, 87
Non-equilibrium MD (NEMD) simulation, 162
Non-exponential relaxation, 245
Non-Gaussian parameters, 160
**Non-planar** conformers, 61
Non-polar domains, 155
Non-polar gases, 172

Non-polarisable force fields, 166
Non-polar solute–ionic liquid systems, 127
Non-polar solvents, ionic liquid interface with, 166
Non-random two-liquid (NRTL) activity coefficient model, 123–128, 134
  UNIQUAC method *vs.*, 126
Novel products, developing, 119
NRTL segmented activity coefficients (NRTL-SAC) model, 125. *See also* Non-random two-liquid (NRTL) activity coefficient model
Nuclear fuel recycling, ionic liquid stability and, 264–266
Nuclear magnetic resonance (NMR) measurements, 228. *See also* Pulsed-gradient spin-echo (PGSE)–nuclear magnetic resonance (NMR) measurements
Nuclear magnetic resonance (NMR) spectroscopy, 160, 264, 265

Ohmic loss, 239
Oil desulfurisation, 29
Oil drilling/oil wells, ionic liquid additives for, 13
OKE spectra, 244. *See also* Optical Kerr effect (OKE)
Olefinic compounds, hydroformylation reaction of, 20–21
Olefinic oligomers, 17
Olefin oligomerisation, 15–16
  transition-metal catalysed, 20
Olefins, separation of, 7–8
Oligomerisation
  of olefins, 15–16, 17
  transition-metal catalysed olefin, 20
Oligomers, olefinic, 17
Olivier-Bourbigou, Hélène, xiv, 1
One-phase systems, 203. *See also* Single-phase systems
Onium alkyl sulfates, 3–4
Onium ionic liquids, 24
Onium salts, 6
Operation models, 122
Optical Kerr effect (OKE), 242. *See also* OKE spectra
Optimization techniques, 122

Organic compound recovery, from ionic liquids with supercritical $CO_2$, 48–49

Organic liquids, aromatic, 265

Organic mixtures, acid removal from, 8–9

Organic molecular compounds, melting points of, 280

Organic-rich phase, in ternary ionic liquid + supercritical $CO_2$ systems, 47

Organic solute extraction, 39

Organic solvent nanofiltration technology, 107

Organic solvents, 195
  advantages of, 193, 194

Organic substrates, solubility of, 197

Organomodified polysiloxanes, 22–23

Orrick–Erbar method, 293

Overviews, value of, ix–x

Oxazolidinium cation, molecular structure of, 277

Oxidisable anions, 263

Oxygen-containing polar compound separation, 11

Paraffin alkylation, 16–17

Paraffins carbonylation of, 17–18

Parameters
  assigning, 156
  non-Gaussian, 160

Partial crystallisation, of ionic liquids\, 6

Patent applications, 28–29

Patents, 2, 18, 19
  granted, 28

Peak-top temperature, 78, 79

Peng–Robinson (PR) EoS, 128–129. *See also* Equations of state (EoS)

Penicillin G extraction, 205–206

Pentaalkylimidazolium cation, molecular structure of, 277

Perfluorinated anionic structures, 232

Permeability ($P$), defined, 99. *See also* $CO_2$ gas permeabilities

Permittivities
  frequency-dependent relative dielectric, 238
  static, 249

Permittivity formalism, 239

Peters, Cor J., xiv, 39

Petrochemistry, ionic liquids in, 1–37

Phase behaviour
  of 1-alkyl-3-methylimidazolium ionic liquids, 59–85
  of binary (ionic liquid + supercritical fluid) systems, 41–46
  of binary or ternary systems, 40
  of $[C_4mim][PF_6]$, 67–73
  of gases, 131–132
  of ionic liquid/supercritical fluid systems, 40–48
  methods to determine, 40–41
  modelling, 88
  of ternary (ionic liquid + supercritical fluid) systems, 46–48

Phase behaviour data, 52

Phase behaviour predictions, for binary ionic liquid $CO_2$ systems, 45

Phase change properties, 179

Phase toxicity, 207

Phase transition behaviours, novel, 73–80

Phase transition peaks, 68

Phase transitions, 67–69
  linked with conformational cation changes, 59–65
  slow, 81

Phosphonium alkyl sulfates, 4

Phosphonium cation, molecular structure of, 277

Phosphorus-containing anions, ionic liquids with, 4

Photolysis, 25

Physical properties, predicting, 119, 15–164

Physicochemical properties, 232. *See also* Density entries; Melting points; Surface tension; Viscosities ($\eta$)
  of $[C_2mim]^+$-based ionic liquids, 221–226
  ionicity correlation with, 229–232
  of ionic liquids, 217, 219, 275–307
  methodology for measuring, 219–221

Pigamo, Anne, xiv, 1

Piperazinium cation, molecular structure of, 278

Piperidinium cation, molecular structure of, 278

**Planar** conformers, 61

Plechkova, Natalia V., x
Plutonium, processing spent, 264–265
Polar domains, 155
Polar ionic compounds, 7
Polarisable continuum model (PCM), 125
Polarisable force fields, 157, 161, 166
Polarisation
  electrode, 249–250
  space charge, 247
Polarity parameters, 252
Polarity probes, 251–252
Polar substances, 194
Polar VOCs, 252. *See also* Volatile organic compounds (VOCs)
Polyalphaolefins (PAOs), 15–16
Polyamide hydrolysis, 18
Poly(diol-RTIL)s, 104–105. *See also* Room temperature ionic liquids (RTILs)
Polymer–electrolyte membrane (PEM), 108–109
Polymer gas separation membrane performance, 100
Polymeric ammonium cation, molecular structure of, 278
Polymer ionic liquid membranes, 93–95
Polymerisation inhibitors, 25–26
Polymerised ionic liquids, cation–anion constituents of, 94
Polymerised membrane gemini RTILs (poly[GRTILs]), 94. *See also* Room temperature ionic liquids (RTILs)
Polymerised membrane RTILs (poly[RTILs]), 93–94, 95, 96, 97, 99. *See also* Poly(RTIL) composite membrane materials
  in CO$_2$ separation, 101–102
Polymer membranes
  ionic liquid-based, 108
  ionic liquids and, 93
Polymers
  conjunct, 19
  ionic liquid antistatic agents in, 12–13
Polymer synthesis, 23–24
Poly(RTIL) composite membrane materials, 104. *See also* Room temperature ionic liquids (RTILs)
Polysiloxanes, organomodified, 22–23

Polytetramethylene ether glycols, 23, 24
Polytrimethylene ether glycols, 23–24
Pore loadings, 169
Porous membrane contactors, 90–93
Post-predictive simulations, 154
Potential functions, classical, 150
Power plants, CO$_2$ capture in fossil fuel burning, 101
Predicted solubility, 132
Predicting cell toxicity, 207–208
Predicting crystal structures, 178
Predicting melting points, 176
Predicting physical properties, 119, 153–164
Predicting relative trends, 161
Predictions, qualitative trend, 158
Predictive models, 88
Premelting behaviour, 64–65
Premelting domains, 73–74
Presolvated electron reactivity, 263
Pressure differences, in gas–vapour permeation applications, 97
Pressure sensitive compositions, 14
Pressure tensor, 161
Primary ionic liquid radiolysis products, 261–262
Process design
  engineering computations for, 121–122
  materials behaviour and, 119
Process simulation tools, 117
Process simulators, 134
Process streams, CO$_2$ removal from, 48
"Process synthesis" tools, 118
Process systems engineering (PSE), 120
Process toxicity, 208
Product design, 119. *See also* Process design
Product design applications, 121
Properties. *See also* Additives with specific properties; Behaviour; Dielectric properties; Ionic liquid properties; Light gas transport properties; Magnetic properties; Mixture properties; *P-V-T* properties; Physical properties; Physicochemical properties; Phase change properties; Self-diffusiveness/self-diffusivities;

"Statistical" properties;
Thermodynamic properties;
Transport properties; Viscosities ($\eta$)
collective, 154
of ionic liquids, 40, 59–60, 87–90, 111,
153–155
optimising solvent, 235
origin of ionic liquid, 217–234
of room temperature ionic liquids, 59
Property models, 122
trends in, 155–156
Property predictions, 119, 153–164
Protective garment materials, 104–105
Protein renaturation, 199
Protic acids, neutralization with, 232
Protic ionic liquids, 250–251, 295–296
Proton exchange membranes, 108
Proton relay molecules, 108
PR + Stryjek–Vera (PRSV) EoS model,
128. *See also* Equations of state
(EoS); Peng–Robinson (PR) EoS
Pulsed-gradient spin-echo (PGSE)–
nuclear magnetic resonance (NMR)
measurements, 218, 220, 224. *See
also* Nuclear magnetic resonance
(NMR) entries
Pulse radiolysis, 260, 261, 262, 270
Pure-component parameters, 46
Pure group parameters, 45
Pure ionic liquids
molecular simulations of, 170
VLE of, 174–175
PUREX process, 264–265
Purification processes, 11–12
Purity criteria, 267, 268
*P-V-T* properties, 131
Pyridazinium cation, molecular structure
of, 278
Pyridinium cation, molecular structure
of, 278
Pyridinium ionic liquids, 3, 270
Pyridinium salts, 22–23
Pyrrolidinium cation, molecular structure
of, 278
Pyrrolinium cation, molecular structure
of, 278

Quadrupolar models, 172
Qualitative trend predictions, 158

Quantitative structure–property
relationships (QSAR), 283
Quantum calculations, gas phase, 159
Quantum chemical calculations, 123,
125, 132–133
Quantum chemistry methods, 150

Radial distribution function, 155
Radial distribution function analysis, 168
Radiation. *See also* Irradiation
incident, 262
ionising, 259, 262
Radiation biology/medicine, 260
Radiation chemistry, 259–261. *See also*
Ionic liquid radiation chemistry
fundamentals of, 262–264
of ionic liquids, 259–274
ionic liquid science and, 261–262
in metal nanoparticle synthesis, 269
Radiation chemistry studies, ionic liquid
preparations for, 266–268
Radiation chemistry techniques, 260
Radiation damage, gamma, 266
Radiation exposure, cumulative, 261
Radical polymerisation inhibitors,
25–26
Radiolysis, pulse, 260, 261, 262, 266
Radiolysis products, primary ionic liquid,
261–262
Radiolysis product studies, 263–264
Radiolysis studies, 267
Radiolytically induced holes, 263
Radiolytic damage, 270
Radiolytic energy deposition, 262
Radiolytic nanoparticle synthesis, 269,
270
Radiolytic preparation, of advanced
materials, 261
Radiolytic yield (*G*-value), 265–266
Raman active bands, 70
Raman bands, 69
Raman scattering intensities, 69
Raman spectra, 71–72
Raman spectroscopic studies, 60
Raman spectroscopy, 98
Raman spectroscopy/DSC
measurements, 66–67
Reaction media, ionic liquids as, 50
Reaction rates, in biphasic systems, 50

Reactions and separations, in ionic liquid + supercritical fluid systems, 49–51
Reaction/separation media, 52
Reactors, enzymatic membrane, 109
Redlich–Kwong (RK) EoS model, 129. *See also* Equations of state (EoS)
"Refined" force fields, 180
  for imidazolium-based ionic liquids, 157
Regeneration, of chloraluminate(III) ionic liquids, 19
Registration, Evaluation, Authorization and restriction of CHemical substance (REACH) regulations, 208
Regular solution theory (RST), 88–89
Relative dielectric permittivity, frequency-dependent, 238
Relative trends, predicting, 161
Renaturation, of proteins, 199
Reorientational dynamics, 165
Research, on biocatalytic reactions, 208–210
Returning temperature, 77
Reverse-NEMD (RNEMD) simulation, 162
Reviews, value of, ix–x
Rhythmic crystallisation.melting, 74–77, 80, 81
Robeson plots, 100, 101
Room temperature ionic liquids (RTILs), 88, 93. *See also* Gelled RTILs; Gemini RTILs (GRTILs); Polymerised membrane RTILs (poly[RTILs]); Poly(RTIL) composite membrane materials; RTIL entries
  imidazolium-based, 101, 104
  interest in, 218
  novel phase transition behaviours of, 73–80
  unique properties of, 59
Rooney, David, xiv, 117
Root mean square deviation (RMSD), 125, 126, 133
Rotational isomers, 60–61, 62
RTIL displacement problem, 103–104. *See also* Room temperature ionic liquids (RTILs)

RTIL membranes, supported, 104
RTIL "tuning," 99

*Saccharomyces cerevisiae*, 204, 205
Saturated hydrocarbons, carbonylation of, 18
Saturated salt solution equilibration, 201
Scanning electron microscopy (SEM), 98
Screening, for enzyme types, 209
Seddon, Kenneth R., x
Selectivity, 99
  improving and controlling, 194
  ionic liquid systems with enhanced, 110
Self-diffusion coefficients, 220, 224, 225, 227
Self-diffusiveness/self-diffusivities
  in calculating viscosity, 161–162
  of ionic liquids, 152, 154, 159–161, 170
Self-dissociativity, 218
Separation media, 52
Separation membrane performance, 100
Separation processes, using ionic liquids, 6–12
Separations
  $CO_2/N_2$, 102
  in ionic liquid + supercritical fluid systems, 48–49
  in the liquid phase, 105–108
  of nitrogen-containing polar compounds, 11
  of olefins, 7–8
  of oxygen-containing polar compounds, 11
Separations and reactions, in ionic liquid + supercritical fluid systems, 49–51
Separation selectivity, of membranes, 99
Shah, Jindal K., xiv, 149
Shear rates, 162
Simulations
  of crystals, 178–179
  engineering, 117–148
  experimental data *vs.*, 154–155
  interfacial, 165–166
  multi-scale, 117
  post-predictive, 154
Single particle time correlation functions, in molecular simulations, 152
Single-phase systems, for biocatalysis, 197–198, 203

SLMs containing ionic liquids (SILMs), 93, 97, 105, 106. *See also* Supported liquid membranes (SLMs)
in $CO_2$ separation, 100–101
Slow phase transitions, 81
Smooth crystallisation, 78
"Snap together" chemistry, 94
Sodium cation, molecular structure of, 277
Soft SAFT (statistical associating fluid theory) EoS model, 130. *See also* Equations of state (EoS); Truncated perturbed chain polar statistical associating fluid theory (tPC-PSAFT) EoS model
Software packages flowsheeting, 122
"Software" tools, 120
Solid–liquid equilibrium (SLE) systems, 123, 149, 176–178
Solid–liquid interfaces, 164–169
Solubilisable enzymes, 109
Solubility
   binary gas, 173
   of $CO_2$, 41–42, 46, 130
   enzyme, 198
   of gases, 129–130, 131
   "infinite-dilution," 172
   in ionic liquids, 171–173
   measurement of, 99
   of organic substrates, 197
   predicted, 132
   substrate, 194, 198
   surfactant, 198
Solubility parameters, 88–89
Solute–ionic liquid systems, non-polar, 127
Solutes, target, 98, 105, 106
Solute transport, 93
Solvated electrons, 267, 269
Solvation dynamics, related to dielectric response, 248
Solvation process, 263
Solvent conformations, 132
Solvent polarity parameter, 221
Solvent properties, optimising, 235
Solvent reuse, 107–108
Solvents, 91–92
   advantages of organic, 193, 194
   for cellulose, 26–27
   environmentally benign, 39

"green," 40, 90
organic, 195
tailored, 88
Solvents for catalytic systems, ionic liquids as, 19–26
Space charge polarisation, 247
Spectral parameterisation, 242–243
Spinodal demixing, 51
Square-well chain-fluid EoS (SWCF-EoS) model, 130–131. *See also* Equations of state (EoS)
Stability, of calorimetry trace, 67
Stacking formations, double-layer, 168
Standard liquid MD simulations, 180–181
Starch
   depolymerisation of, 28
   ionic liquids and, 27–28
Static conditions, 249
Static dielectric constants, 235, 236, 237, 238, 240, 241, 250–251, 252, 253
   of ionic liquids, 248–253
Static permittivities, 249
Statistical association fluid theory models, 45–46
Statistical-mechanics-based equations of state, 45
"Statistical" properties, molecular dynamics simulations and, 152
Stein, Florian, xiv, 193
Stejskal–Tanner equation, 220
Stimuli-responsive ionic liquid systems, 110–111
Stokes–Einstein–Debye (SED) equation, 245
Stokes–Einstein relationship, 162
Structural integrity, of membranes, 103
Sub-picosecond dynamics, 243
Substrate accumulation, 203
Substrates
   influence of ionic liquids on, 198–200
   organic, 197
Substrate solubility, 194, 198
Sulfate ionic liquids, 2
Sulfation/sulfonation, of cellulose, 27
Sulfonium cation, molecular structure of, 278
Sulfur compound removal, from hydrocarbon streams, 9–10
Sulfuric acid recovery, 9

Supercooled liquids, 79, 279
Supercritical CO$_2$, 39
  continuous biphasic processes with
    ionic liquids and, 49–50
  continuous processes with, 50–51
  organic compound recovery from ionic
    liquids with, 48–49
  properties of, 40
Supercritical fluid + ionic liquid systems,
    phase behaviour of, 41–46, 46–48
Supercritical fluids
  commonly used, 40
  in ionic liquids, 39–57
Supercritical fluid systems, phase
    behaviour of, 40–48
Supercritical solvents, 260–261
Superheating, 178
Supported ionic liquid-phase (SILP)
    technology, 21, 29
Supported liquid membranes (SLMs),
    90–93, 103–104. *See also* SLMs
    containing ionic liquids (SILMs)
  development of, 110
  major drawback of, 92
Supported RTIL membranes, 104.
    *See also* Room temperature ionic
    liquids (RTILs)
Surface charge densities, 168
Surface diffusion coefficients, 167
Surface negative charge, 168
Surface tension, 165, 294–296
Surfactants, anionic, 13–14
Surfactant solubility, 198
Suzuki reactions, 107–108
Sweetening, of natural gas, 101
Swelling behaviour, of membranes,
    98–99
**Sym** (symmetric form) conformers, 61,
    64, 78, 80
Syntheses. *See also* "Process synthesis"
    tools
  aspartame, 194
  of ionic liquids, 2, 219
  of polymers, 23–24
  selective improvement of, 209

Tailored solvents, 88
Tait equation, 288
Target solutes, 98, 105, 106

Task-specific ionic liquids, 92–93, 94
Techniques, abbreviations for, xviii–xx
Temperature dependencies, of ionic
    liquid properties, 222–224, 225
Temperatures. *See also* Heat entries;
    Therm- entries
  density at different, 288
  glass transition, 279
  viscosity and, 289
Terahertz regime, 246
Terahertz spectra, 253
  of ionic liquids, 243
Ternary fluid multiphase systems, 43–44
Ternary (ionic liquid + CO$_2$ + water)
    systems, 47–48
Ternary (ionic liquid + CO$_2$ + organics)
    systems, 46–47, 52
Ternary ionic liquid + supercritical CO$_2$
    systems, 46–48
Ternary (ionic liquid + supercritical
    fluid) systems, phase behaviour of,
    46–48
Ternary LLE systems, 127. *See also*
    Liquid–liquid equilibrium (LLE)
    systems
Ternary mixtures, 198
Ternary systems, 123–124
  phase behaviour of, 40
Tetraalkylimidazolium cation, molecular
    structure of, 277
Tetrazolium cation, molecular structure
    of, 277
Textile applications, for dissolved
    cellulose, 27
Thermal analyses, 220. *See also* Heat
    entries; Temperature entries
  equipment for, 65–67
  of ionic liquids, 60
Thermal behaviours, of 1-alkyl-3-
    methylimidazolium ionic liquids, 59,
    80
Thermal conductivity, 164
Thermal energy, 76
Thermal fluctuations, 80
Thermal histories, 66, 81
  equipment for, 65–67
Thermal phenomena, 80–81
Thermal properties, of [C$_2$mim]$^+$-based
    ionic liquids, 221–222

Thermal stability, 209
of [C₂mim]⁺-based ionic liquids, 221, 222
Thermodynamically rigorous approaches, 177
Thermodynamic information, in molecular simulations, 151
Thermodynamic integration methods, 171, 172
Thermodynamic melting points, 176–177
Thermodynamic models, for ionic liquids, 122–133
Thermodynamic pathways, traversing, 177
Thermodynamic properties, 157–159
of ionic liquids, 134
predicting, 153–155
Thermodynamic water activity, 200
Three-component composite membranes (MMMs), 95, 96
in $CO_2$ separation, 102–103
light gas transport properties of, 102, 103
Time-dependent information, in molecular simulations, 151
Timescales, in molecular simulations, 151–152
Tokuda, Hiroyuki, xiv, 217
Toxicity
phase, 207
process, 208
Toxicity effects, of impurities, 206–207
Toxicity evaluation, 206
Transference numbers, 226–229
cationic, 226–227
Transition-metal catalysed olefin oligomerisation, 20
Transition state theory, 253
Transport properties, 159–164, 179, 232
predicting, 153–155
Trialkylimidazolium cation molecular structure of, 277
Triazolium cation, molecular structure of, 277
Trimethylpentene to dimethylhexene (TMP/DMH) molar ratio, 16
Truncated perturbed chain polar statistical associating fluid theory

(tPC-PSAFT) EoS model, 129, 130. *See also* Equations of state (EoS)
**TT** (*trans trans*) conformations, 62, 64, 65, 69, 71, 72, 73, 75–76
Tuneability, of ionic liquid properties, 111
Two-component membranes, 102
Two-phase systems, 203. *See also* Biphasic systems
for biocatalysis, 195
"Two-phase"–"three-phase"–"two-phase"–"one-phase" transition, 47, 49

Ultrasonication, ionic liquid production under, 6
UNIFAC method/model, 123, 126–127, 133
UNIQUAC method/model, 123, 125–126, 127–128, 133
non-random two-liquid (NRTL) activity coefficient model *vs.*, 126
United-atom force field, 157
Upper critical endpoint (UCEP), 44
Uranium, processing spent, 264–265
Uronium cation, molecular structure of, 278

Vacuum–liquid interfaces, 165
Valorisation of biomass, patent applications concerning, 28–29
van der Waals forces, melting points and, 281
van der Waals function, 131
van der Waals interaction parameters, 156
van der Waals interactions, 167, 168, 218–219, 230, 231, 292, 293
"Vanilla" MD simulations, 180–181
Vapour absorption, 294. *See also* Gas entries
Vapourisation enthalpy, 155, 158–159, 174
Vapour–liquid equilibrium (VLE) systems, 124, 125, 126, 132, 149
of pure ionic liquids, 174–175
Vapour phase, in ternary ionic liquid + supercritical $CO_2$ systems, 47
Vapour phase studies, 174
Vapour pressures, 174
Vapours, separation of, 99–104

Vapour transport and capture, 93
Vectorial network analysers, 240
Vehicle mechanisms, 218
Vibrations, intermolecular, 247
Viscosities ($\eta$), 161–162, 202, 217, 233, 236, 288–294. *See also* Low-viscosity ionic liquids
   computed, 162
   macroscopic, 225
   Newtonian, 162
   temperature and, 289
Viscosity measurements, 220
Viscosity values, 222–224, 294
VLE modelling, 128, 130, 131, 133. *See also* Vapour–liquid equilibrium (VLE) systems
Vogel–Fulcher–Tammann (VFT) equations, 222, 224, 225, 226, 293, 294
Vogel–Tammann–Fulcher (VTF) equations. *See* Vogel–Fulcher–Tammann (VFT) equations
"Void-induced" melting method, 177
Volatile organic compounds (VOCs), 236, 237, 240, 242, 247, 248, 249
   ionic liquids *vs.*, 253
   polar and dipolar, 252
Volume fluctuations, 158

Watanabe, Masayoshi, xiv, 217
Water content/activity, in enzyme-catalysed reactions, 200–201
Water-immiscible ionic liquids, 197, 198
Water transport, 105
Water vapour, 106
Weingärtner, Hermann, xiv, 235
Whole cell systems, 193
   biocatalysis in, 203–206
Widom test particle insertion method, 171, 172
Wilson equation, 126
Wishart, James F., xiv, 259
Wong-Sandler mixing rules, 128

X-ray photoelectron spectroscopy (XPS) techniques, 97
X-ray powder diffraction (XRD), 27
Xylenes, 17

Ye and Shreeve method, 288

Zeolites, 102, 103
Zero-frequency limit, 238
Zero shear rate, 162
Zhang, Suojiang, xiv, 275
Zhou, Qing, xiv, 275